Evaporites and Hydrocarbons

Evaporites and Hydrocarbons

Edited by
B. Charlotte Schreiber

Columbia University Press
New York

This research was supported in part by a grant from The City University of New York PSC-CUNY Research Award Program.

Library of Congress Cataloging-in-Publication Data

Evaporites and hydrocarbons/edited by B. Charlotte Schreiber.
 p. cm.
 Bibliography: p.
 Includes index.
 ISBN 0-231-06530-2
 1. Evaporites. 2. Organic geochemistry. I. Schreiber, B. Charlotte.
QE471.'15.E8E925 1988
552'.5—dc19

88-13998
CIP

Columbia University Press
New York   Guildford, Surrey
Copyright © 1988 B. Charlotte Schreiber

Casebound editions of Columbia University Press books are Smyth-sewn and are printed on permanent and durable acid-free paper

Contents

Evaporites and Hydrocarbons

Introduction

B. CHARLOTTE SCHREIBER

The study of evaporites began over one hundred years ago with an experimental chemical emphasis: petrologic studies by the nonchemists seem to have been largely isolated from these experimental studies. The chemists, in turn, ignored much of the rock record, except for the chemical compositions of seawater and the associated chemical analysis of the evaporite sequences. Confounding the situation, because so few fossils are recognized within evaporites, little to no biological or paleontological assessments were made at all, and in fact evaporitic environments were considered abiotic or sterile. Without clear biological markers, the stratigraphy was necessarily chaotic. Consequently the accepted depositional model held that the biota which form open-marine carbonates can successfully live, produce reefs, and so forth, in the upper waters of a stratified hypersaline sea together with synchronous deposition of massive evaporites within the basin.

This was the situation until about twenty-five years ago, with the initiation of the actualistic studies by D. J. Shearman and his students at Imperial College (ca. 1963). From these studies, primarily in the coastal (supratidal) areas of the Persian (Arabian) Gulf, the sedimentologists have come to a very different view of evaporitic deposits. More recent studies by a group of French and Spanish geologists, led by F. Orszag-Sperber and J. M. Rouchy (1980), by G. Busson (1982), and by F. Orti-Cabo and Busson (1984), have demonstrated another depositional setting in which the physical, chemical, and biological conditions and the resulting types of evaporitic sediments forming in shallow hypersaline water bodies have been elucidated. Our present approach to understanding evaporitic sediments thus recognizes that one must take many aspects of study into account, not only the chemistry of such sediments. While geochemical considerations remain important, they now deal with the chemical descriptions of the processes in formative water bodies, which are the actual sites of precipitation. They also must now consider, first, the sediments themselves, then the direct understanding of the nature of the diagenesis of these sediments, and finally the recent considerations of climatologic, hydrologic, and biological observations (flora and fauna) that have been added to both sedimentologic and geochemical analysis. The biological studies have also become tied into the generation of characteristic biomarkers and the formation of evaporite-generated hydrocarbons.

The biota of evaporative waters, the least-known aspect of the formation of evaporites, are prodigious in numbers, although they are representative of a low faunal and floral diversity (see Kendall and Warren, article 2; Schreiber, article 4; and Evans and Kirkland, article 5). Evaporative waters certainly are not sterile, but simply carry life forms either without hard parts or with hard parts that are readily destroyed during diagenesis; hence preservation potential is extremely low. Organisms with preservable hard parts, which live in the surface layers of many saline and mesosaline water bodies (particularly some genera of foraminifera, gastropods, pelecypods, and diatoms), are easily destroyed or chemically altered when the waters become increasingly saline. This occurs because with a rise in salinity there is usually a marked rise in pH. This does not initially damage the carbonates, but silica then becomes relatively soluble and the diatom tests (opal A) are destroyed. Subsequently, anaerobic bacteria, living in the bottom muds of these ponds, generate acidic conditions, and consequently the dissolved silica again precipitates as an early cement in the mud. The increasing acidity also destroys most of those calcareous components that have accumulated within the bottom sediments. The final bottom deposit is commonly high in organic content but contains almost no recognizable biotic remains, except for the most resistant shell components and occasional pseudo-morphs. Only pollen, untouched by this chemistry, and carried in from surrounding areas, remains as testimony to the presence of life.

The major reasons for interest in evaporites by petroleum geologists is that they may serve in several important geological capacities, as a partial hydrocarbon source (Kirkland and Evans 1981 and article 5; Sonnenfeld 1985; Warren 1986), as a significant seal for many hydrocarbon accumulations, and as the cause for structural traps for many deposits (Kehle, article 7). Evaporites may also be removed from the record by solution, providing significant porosity or forming solutional structures in their stead (Parker 1967; see also Kendall, article 1).

These general relationships of evaporites and associated sediments may be studied best in the subsurface by the use of geophysical instruments, which are the eyes of the petroleum industry. Where the deposits are complex and have made interpretation difficult, a combination of depositional modeling of evaporites and the associated carbonates and/or clastics, together with stratigraphic and petrological studies in addition to the geophysical analysis, can unravel the history of these deposits (Kendall, article 1, and Nurmi, article 8).

EVAPORITE DEPOSITS

Evaporites form in many regions of the world, anywhere the rate of evaporation exceeds the rate of water input. The general pattern of evaporite deposition in the world today is illustrated in figure 1. It is readily seen that while many evaporites are marine marginal (Kendall and Warren, article 2; Handford, article 3), there are presently vast areas of the world, particularly within interior basins, that receive nonmarine evaporite deposition. This type of deposition, within continental settings, was very clearly marked during the early phases of the Triassic-Jurassic rifting of Pangea, when the lowermost por-

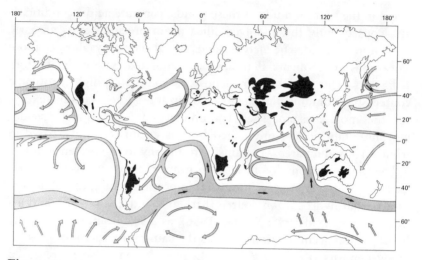

Figure 1
World map illustrating areas of Recent and Quaternary evaporite formation (marine and nonmarine). Patterns of surface circulation indicated on oceans (Drewry et al. 1974; adapted in Schreiber 1986).

tions of the deposits in the Atlantic and the Gulf Coast basins, as well as in other rift basins, were formed. The importance of continental climate cannot be overemphasized (Kinsman 1975). The formation of such rift-related deposits is a curious one, because during the period of evaporite deposition many of the rift basins appear to have lain in areas depressed below sea level. Within such low-lying basins the atmospheric pressure is elevated, and hence it seems that evaporation should be inhibited. Why the evaporite deposits? Evaporation in depressed basins actually remains fairly rapid due to strong adiabatic circulation, as clearly seen in the Dead Sea graben and as pointed out in a discussion of rift basins by Manspeizer (1981).

In rift areas most of the deposition is composed of materials carried in by surface runoff and groundwater drainage, but in some regions magmatic fluids apparently become significant contributors of ionic input, as in the Rift Basins of Africa, for example. Many early-rift deposits are composed of clastics, carbonates, and evaporites, formed with or altered by such additional influx (Hardie 1978; Eugster 1986; Spencer 1987). The later stages of rift development are commonly marked by deposition of marine-fed evaporites, and they pass into marine carbonate and/or clastic deposits, but hydrothermal alteration and/or input is to be expected to persist for some millions of years after drift has begun.

Rates of deposition and accumulation for evaporites (if dissolution does not occur) are generally very high, greater than for most other sediments. Moderate evaporation of seawater (200–250 cm per year in the Mediterranean region) results in deposition rates of 2 to 5 cm per year for gypsum, and 2 to 10+ cm per year for halite. This high rate of formation takes place when the evaporitic water body is shallow and the surface area is large, per volume of water. Wet years produce less evaporation, and naturally accumulation is lower. Evaporites, however, need not be restricted to warm climates. Arid portions of both the Arctic and Antarctic are also presently the sites of evaporite accumulation, albeit this is a slow process in these locales. Deposition rates for evaporites formed within most interior basins (even in arid regions) is slower than in marine-fed basins—not because evaporation is slow, but because for equal volumes of water the minerals dissolved in the feed water are initially very dilute. In most instances the rate of deposition of evaporites within those interior basins without marine influx is apparently at least one or two orders of magnitude lower than within marine-fed ones.

There are also deposits of *nonevaporite* calcium sulfate forming in the world today, which are generally similar in composition to evaporitic deposits. Primary anhydrite is known from hot brines present on the floor of the Red Sea (Bischoff 1969). The source for these sulfates is from hot springs which pass through older evaporites present on the Red Sea floor. Not unlike the Red Sea deposits are the newly discovered gypsum deposits presently forming within the Bannock Deep (eastern Mediterranean). In this deep, mineralized waters (at ambient temperatures) are emerging from Messinian evaporites in the Mediterranean Ridge into a brine-filled, stratified basin (Scientific Staff of the *Bannock* 1985). These brines (charged with both NaCl and $CaSO_4$) form an anoxic, stratified water body within the basin and then react with the overlying, normally oxygenated bottom waters, forming gypsum at the interface between the water bodies (Corselli and Agib 1987). Other nonhydrothermal deposits form at other sites in the world (Briskin and Schreiber 1978), and more will undoubtedly become known as we study new areas. Such sulfates (and possibly chlorides as well) are located in otherwise nonevaporative sequences, and are not true evaporites.

STRUCTURE

The physical behavior of evaporites during burial is rather different than that of most other sediments. For example, when halite forms within the evaporitic setting, the porosity of NaCl is approximately 30% to 50%. The porosity diminishes very rapidly, within a few meters of burial, as early epitaxial overgrowth fills the spaces between the initial crystals. At depths of only 30 to 40 m, most of the porosity and permeability is lost and only isolated intracrystalline voids, filled with fluid inclusions, remain. These inclusions, however, may comprise 10% of the whole by volume, and commonly persist within the halite throughout its burial history, or at least until deformation, elevation in temperature, or migrating fluids affect the beds.

When halite and potassic salt beds are stressed, by unequal loading or by tectonism, they deform very easily because of their low rigidity. When the salt strata are out of equilibrium because of the applied stresses, their relatively low density comes into play, and salt domes or rolls are formed—unless the overlying beds are thick and well indurated and are sufficiently competent to resist the movement. Initially these competent capping beds take up stress by internal compaction, recrystallization, and pressure solution and may begin to slide along on the weak, plastic salt layers without faulting while the salt layers develop intralayer isoclinal folds. But once the competent beds begin to yield, the density differences (buoyancy of salt) result in vertical motion. Buoyancy alone does not cause the salt to form domes and rolls, and many thick salt deposits have never deformed, as in the Michigan Basin (Kehle, article 7). When formed, the domes, rolls, and diapirs commonly generate structurally controlled sites for hydrocarbon traps. Examples of such traps abound in the Gulf Coast of the United States and Mexico, the North Sea, and throughout the Middle East. An excellent and graphic analysis of salt movement may be seen in a recent *Scientific American* article (Talbot and Jackson 1987).

Halite is not the only evaporite to generate deformational structures. Gypsum, the hydrated and most common form of sulfate, generally converts to anhydrite upon 400 to 600 m of burial, and loses up to 38% of its initial volume. During this conversion, the water of hydration gradually moves out into the surrounding sediments. Where halite lies above or updip of the gypsum, it suffers a measure of solution generated by the passage of these waters. Additionally, the sediments overlying the gypsum subside, and the area in which dehydration takes place is marked by local sedimentary thickening above the thinned sulfates. In situations where this water is trapped, the process of dehydration may create local overpressure which can facilitate sliding and deformation along the water-supported zone (Heard and Rubey 1966).

Strata that are already in the form of anhydrite, on the other hand, have a different set of physical characteristics. Anhydrite is quite dense (density of 2.963–2.98), is not buoyant, unlike halite (density of 2.135–2.164), and does not, in itself, form diapiric structures. But at depth, it also has a very low rigidity modulus. Once it has reached burial temperatures above 150° to 180° C, it yields to stress almost as easily as does halite (Müller and Briegel 1978, 1981). Thus sulfate, as well as halite, may serve as a ready décollement surface in areas of regional compression and thrusting. The role of either halite or calcium sulfate as a glide surface in thrusting is well known in the Jura, parts of the Alps, the Apennines, and the Himalayas, as well as in many other parts of the world.

NOW WHAT?

Controversies concerning the appropriate models for evaporite formation seem to be reaching a reasonable resolution. We now recognize a continuum of evaporite environments, ranging from nonmarine to marine, from continental desert to hypersaline seas. We find deposits containing evaporites, carbonates, and siliciclastics coming together into general sedimentary models rather than being viewed as segmented studies in different subjects. There seems to be a strong direction in current studies of evaporites toward an interdisciplinary mix of geochemistry, biology, and sedimentation, in which the work area is no longer a laboratory beaker, but within evaporative water settings as they exist in nature. What we have not solved, and what must be addressed, is just how to apply what we know and are now learning to stratigraphic interpretation.

STRATIGRAPHY: A PROBLEM FOR THE FUTURE

Because we have just come to the point where there is some under-standing of the many environments of evaporite deposition, and there is simply not enough data in most areas, the synthesis of evaporite stratigraphy cannot yet be accomplished for many basins. Particu-larly important are the many questions of the stratigraphic relations between the evaporites and presumed laterally equivalent facies. At a given locale these relations cannot readily be observed either in the field or in the subsurface, and so they remain ambiguous. We are forced to use sedimentary models which piece together the sedimen-tary record in order to infer the stratigraphy. But once we have models for deposition, we find that sedimentologic studies have stratigraphic implications and vice versa (see Kendall, article 1).

REFERENCES

Bischoff, J. L. 1969. Red Sea geothermal brine deposits: Their mineralogy, chemistry, and genesis. In E. T. Degans and D. A. Ross, eds., *Hot Brines and Recent Heavy Metal Deposits in the Red Sea*, pp. 368–401. Berlin: Springer-Verlag.

Briskin, M. and B. C. Schreiber. 1978. Authigenic gypsum in marine sedi-ments. *Marine Geology* 28:37–49.

Busson, G., ed. 1982. *Géologie Méditerranéenne. Annales de l'université de Provence* 9 (4):303–591.

Corselli, C. and F. S. Aghib. 1986. Brine formation and gypsum precipita-tion in the Bacino Bannock (eastern Mediterranean). *Marine Geology* 75:185–199.

Drewry, G. E., A. T. S. Ramsay, and A. G. Smith. 1974. Climatically con-trolled sediments, the geomagnetic field, and trade wind belts in Phanero-zoic time. *Jour. Geology* 82:531–553.

Eugster, H. 1986. Lake Magadi: A model for rift valley hydrochemistry and sedimentation. In L. E. Frostick, R. W. Renaut, I. Reid, and J. J. Tiercelin, eds., *Sedimentation in African Rifts* , pp. 177–190. Geol. Soc. Spec. Publication no. 25. Oxford: Blackwell.

Hardie, L. A. 1978. Evaporites, rifting, and the role of CaCl$_2$ hydrothermal brines (abs.). *Geol. Soc. Amer. 1978* Program, p. 416.

Heard, H. C. and W. W. Rubey. 1966. Tectonic implications of gypsum dehydration. *Bull. Geol. Soc. Amer.* 77:741–760.

Hite, R. J. 1970. Shelf carbonate sedimentation controlled by salinity in the Paradox Basin, southeast Utah. In J. L. Rau and L. F. Dellwig, eds., *Third Symposium on Salt* 1:48–66. Cleveland: Northern Ohio Geological Society.

Kinsman, D. J. J. 1975. Rift valley basins and sedimentary history of trailing continental margins. In A. G. Fischer and S. Judson, eds., *Petroleum and Plate Tectonics*, pp. 83–126. Princeton, N.J.: Princeton University Press.

Kirkland, D. and R. Evans. 1981. Source-rock potential of the evaporitic environment. *Bull. AAPG* 65:181–190.

Manspeizer, W. 1981. Early Mesozoic basins of the central Atlantic passive margins. In A. B. Bally, ed., *Geology of Passive Continental Margins: History, Structure, and Sedimentologic Record*, pp. 4/1–4/60. AAPG Education Course Notes no. 19. Tulsa, Okla.

Müller, W. H. and U. Briegel. 1978. The rheological behaviour of polycrystalline anhydrite. *Ecologae Geol. Helv.* 71:397–407.

Müller. W. H. and U. Briegel. 1981. Deformation experiments on anhydrite rocks of different grain sizes; rheology and microfabric. *Tectonphysics* 78:527–544.

Orszag-Sperber, F. and J. M. Rouchy, eds. 1980. *Géologie Méditerranéenne. Annales de l'université de Provence* 7:1–154.

Orti-Cabo, F. and G. Busson, eds. 1984. Introduction to the sedimentology of the coastal salinas of Santa Pola (Alicante, Spain). *Revista d'Investigacions Geologiques* 38/39:1–235.

Parker, J. M. 1967. Salt solution and subsidence structures, Wyoming, North Dakota, and Montana. *Bull. AAPG* 51:1929–1947.

Schreiber, B. C. 1986. Arid shorelines and evaporites. In H. G. Reading, ed., *Sedimentary Environments and Facies*, 2d ed., pp. 189–228. Oxford: Blackwell.

Scientific Staff of the Cruise Bannock 1984–12. 1985. Gypsum precipitation from cold brines in an anoxic basin in the eastern Mediterranean. *Nature* 314:152–154.

Shearman, D. J. 1963. Recent anhydrite, gypsum, dolomite, and halite from the coastal flats of the Arabian shore of the Persian Gulf. *Proc. Geol. Soc. London* no. 1607, pp. 63–65.

Sonnenfeld, P. 1985. Evaporites as oil and gas source rocks. *Journal Petroleum Geology* 8:253–271.

Spencer, R. 1987. The origin of major potash evaporites (abs.). Proceedings *SEPM Mid-Yearly Meeting* 4:79–80.

Talbot, C. J. and M. P. A. Jackson. 1987. Salt tectonics. *Sci. Amer.* 257:70–79.

Warren, J. K. 1986. Shallow-water evaporitic environments and their source rock potential. *Jour. Sed. Petrol.* 56:442–454.

1

Aspects of Evaporite Basin Stratigraphy

ALAN C. KENDALL

Interpretation of evaporite sequences must involve information about stratigraphic relationships between the evaporites and neighboring (and commonly laterally equivalent) facies. Sometimes these other facies are porous and hydrocarbon-charged or have been mineralized. There is thus economic justification for examining the facies relationships and the possible genetic links between the evaporites and adjacent strata. Unfortunately these relationships are only rarely observed directly and are invariably, to some extent, ambiguous. To a large extent lack of opportunity for direct observation has been counterbalanced by a trust in depositional or basinal models that explain (and therefore predict) these relationships. It is unfortunate that conclusions based upon sedimentologic studies of evaporites commonly

Alan C. Kendall is an Associate Professor in the Department of Geology, University of Toronto.

do not agree with stratigraphically derived conclusions and models. The subject of evaporite stratigraphy has commonly been over-looked. Whereas sedimentologists studying clastic or carbonate rocks have made great strides in genetic stratigraphy, those working with evaporites have hardly got their feet wet (no pun intended). In many respects we have hardly progressed beyond the point reached by Sloss (1953) and Krumbein and Sloss (1963). In these works the main attributes of genetic evaporite stratigraphy were identified. They recognized the following points:

1. Evaporites commonly occur in cycles, and each cycle component must represent a stage in the evaporative concentration of sea-water (figure 1.1).
2. The vertical succession is matched by the lateral relationships in

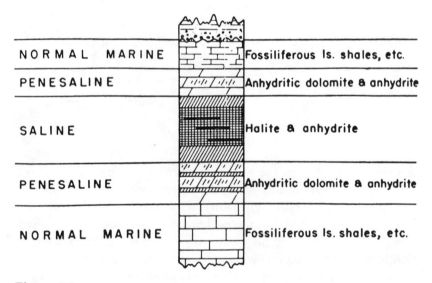

Figure 1.1
The ideal evaporite cycle (from Krumbein and Sloss 1963) showing a lower "advancing" ("regressive" of Hite 1970) hemicycle with rocks indicative of increasing salinity upwards, and an upper "relaxing" ("transgressive" of Hite) hemicycle with rocks suggesting decreasing salinity. Complete cycles of this type are most uncommon, the upper hemicycle usually being absent. Cycles in the Paradox Basin approach this completeness, yet the transgressive carbonates are not marine (see figure 1.6b).

evaporite deposits, with more saline components in the basin center surrounded more or less concentrically by less saline facies; or the less saline components are preferentially located near basin entrances, with more saline facies developed more distally (figure 1.2). (Although not so stated, it is clear that Krumbein and Sloss 1963 considered that Walther's Law applied to evaporite sequences, so that for them, items 1 and 2 were related.)

3. Evaporites form within silled basins that restrict the inflow and prevent total loss of the concentrated brines (figure 1.3). Krumbein and Sloss concluded that the dynamic equilibrium reflux model of King (1947), Scruton (1953), and Briggs (1957) explained the lateral (and synchronous) sequences observed in evaporite bodies.
4. Finally, Krumbein and Sloss (1963) recognized that evaporites occur in three main settings: basin centers, basin margins, and shelves. Interior cratonic basins were considered to hold more evaporites than marginal basins.

Some of Krumbein and Sloss' (1963) conclusions have subsequently undergone substantial revision. That evaporites exclusively represent marine precipitates is disputed by Hardie (1984), not only for such obvious examples as the trona deposits of the Eocene Green River Formation, but also for "classic" marine evaporites such as parts of the Permian Salado Formation of New Mexico and the Elk Point evaporites of western Canada. Specifically, Hardie has suggested that many potash deposits owe their main characteristics to evaporation of hydrothermal waters (calcium chloride brines) rather than seawater (Hardie 1978). Naturally such views are controversial, and since evaporites are the products of brine concentration, clearly there can be no more important a variable than the type of brine and its origin. Identification of this factor probably has the greatest implications for the stratigraphy of an evaporite, since it controls the types of minerals produced, their distribution, and sequence. In some instances the problem appears simple: the evaporites occur in cycles with marine carbonates, so that a marine origin for the brine seems most appropriate. In other instances the problem of evaporite origin has been long disputed. Evaporites associated with Permo-Triassic redbeds in both North America and Europe present an example. Taylor (1983) summarized the different explanations suggested for such evaporites in central England. They have been interpreted as playa lake deposits with brines derived from continental runoff; as deposits

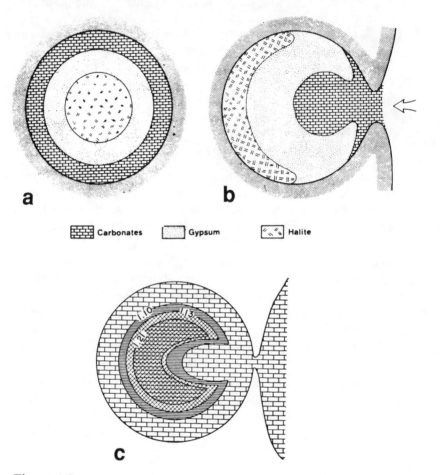

Figure 1.2
Patterns of evaporite distribution in evaporite basins: (a) Concentric
or "bulls-eye" pattern, with most-saline facies in basin center surrounded concentrically by facies tracts of less-saline mineralogy; (b)
"tear-drop" or excentric pattern, with most-saline facies present in
distal parts of basin, and a trend towards lower salinity facies toward
the basin inlet; and (c) Briggs' (1958) evaporite basin model, a combination of the two preceding patterns. (Diagrams (a) and (b) from
Hsü 1972; (c) from Krumbein and Sloss 1963, modified from Briggs
1958.)

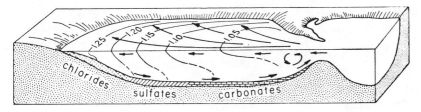

Figure 1.3
The silled evaporite-basin model (from Krumbein and Sloss 1963, after Briggs 1957) with increasing brine salinities away from the basin inlet, flow of surface waters into basin, and flow of denser basal brines toward the inlet. The lateral variation in surface brine density is mirrored in the lateral segregation of precipitates on the floor of the evaporite basin.

of a hypersaline sea cut off from the ocean by a sill; and as the products of marine flooding of a vast flat plain with evaporites concentrated in the deeper basinal areas. Utilizing stable isotope values of sulfates and carbonates, Taylor (1983) concluded that both marine and continental brine sources were important, but varied in significance at different stratigraphic horizons, and (for a given horizon) varied in different locations.

The silled-basin interpretation for all evaporites was refuted in the mid-1960s with the discovery of calcium sulfates associated with dolomite in Recent sabkhas (Curtis et al. 1963; Shearman 1966). These discoveries were quickly applied to the ancient; Shearman (1966), for example, interpreted Upper Jurassic anhydrite-bearing cycles from southern England as sabkha sequences. Evaporite sedimentology has suffered from large pendulum swings: the sabkha interpretation seemingly swept all before it, as one after another evaporite deposit was described or reinterpreted as having formed in this kind of environment. The fact that many of these evaporites had a basin-central distribution, whereas marine-marginal sabkha sections ought to be better developed at the basin periphery, seemed to pass without notice. Today the pendulum has shifted again—perhaps too far—and many evaporites, including many previously identified as deep-water and/or sabkha in origin, are now recognized as having formed in shallow subaqueous environments. Deep-water evaporites appear to be relatively rare, yet the silled-basin model for evaporite formation retains its appeal and invariably is the favored model presented in textbooks.

STRATIGRAPHIC AND SEDIMENTOLOGIC APPROACHES TO EVAPORITES

Stratigraphic and sedimentologic approaches may yield different interpretations. It is commonly overlooked that sedimentologic conclusions have stratigraphic implications, and vice versa. This interdependence is particularly relevant to the study of evaporite sequences because their stratigraphy is commonly difficult to determine. In most instances correlations cannot be checked biostratigraphically and must be evaluated from isolated subsurface data points. Stratigraphic relations invariably must be inferred, because only very rarely can they be observed directly at outcrop. Thus stratigraphic relationships are, to varying degrees (and depending upon the spacing of the subsurface control), ambiguous. As such they allow a variety of sedimentologic interpretations to be considered—some of which differ widely (deep-water deposition as opposed to brine-flat environments, for example). In contrast, unambiguous sedimentologic conclusions, if they can be made, should suggest less ambiguous stratigraphic relationships.

By using the Paradox Basin as an example, it is possible to illustrate the difficulties that occur when different approaches to the study of basinal evaporites are used. In one interpretation (figure 1.4), basin-central evaporites are considered to pass laterally into coeval basin-flanking carbonates (Peterson and Hite 1969; Hite 1970). Carbonates and the various evaporites are considered to be contempo-

---

Figure 1.4
North-south cross-section of Hermosa Group in Paradox Basin showing the inferred lateral passage of evaporitic cycles of the Paradox Formation into carbonate cycles of the southern basin flank, and the lateral persistence of black shale markers that define the cycles in both basin and basin flank. (Modified from Peterson and Hite 1969.)

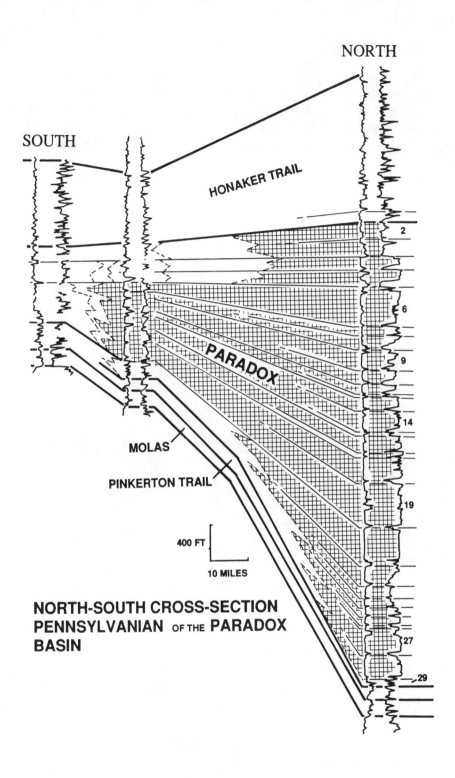

NORTH

SOUTH

HONAKER TRAIL

PARADOX

MOLAS

PINKERTON TRAIL

400 FT

10 MILES

**NORTH-SOUTH CROSS-SECTION
PENNSYLVANIAN** OF THE **PARADOX
BASIN**

raneous with one another, and this implies that only depositional models that allow such contemporaneity should be considered valid (figure 1.5). A stratified saline-basin model is such a possibility, and would imply that the evaporites accumulated in the deeper parts of the basin, perhaps in as much as several hundred feet of brine. This

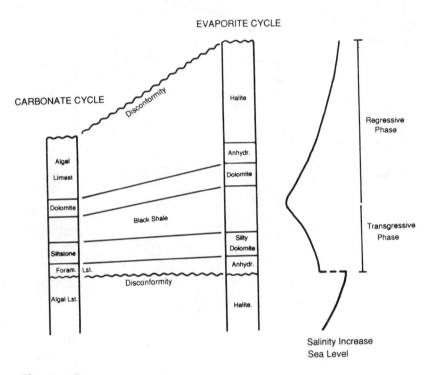

Figure 1.5
Diagrammatic correlation of lithofacies in idealized basin-central evaporite and basin-flanking carbonate cycles of the Paradox Formation, Paradox Basin, United States. Note that deposition of algal buildups on the basin flank are equated with deposition of anhydrite and halite in the basin center. Curve shows the relative sea level and salinity conditions during the cycles. Increasing salinity in the basin is correlated with sea-level falls, and salinity decreases with episodes of sea-level rise. In this interpretation the evaporites are deep-water deposits. (Hite and Buckner 1981.)

model has been subsequently expanded and developed to explain other features of the evaporite (Hite 1986).

In another interpretation (figure 1.6a), the very existence of halite in the basin is itself taken as sufficient evidence that the basin was essentially disconnected from the ocean (Kendall 1987a, using the arguments of Lucia 1972). Under such circumstances, evaporation commonly exceeds the rate of influx and the basin becomes desiccated (Maiklem 1971). In this scenario the basin is disconnected from the ocean by an emergent sill composed of the basin-flanking carbonates, and the greater part of these carbonates must have been deposited during earlier episodes, prior to basin desiccation. In this instance, carbonates and evaporites cannot have been contemporaneous, and a different stratigraphic packaging of the sediments must be considered.

It is important to restate the important fact that stratigraphic models (i.e., a series of correlations) cannot be verified by observation in evaporite systems; they are based upon assumed correlations of discrete subsurface data, and this data is commonly open to different interpretations. If the correlation is inappropriately drawn, then the stratigraphic model it defines fails, and so does any sedimentologic conclusion or inference based upon it. Conversely, if sedimentologic information is available, then this can be used independently to create a depositional model, and *this has stratigraphic implications*. In the Paradox Basin example, if it can be shown that the Paradox evaporites exhibit features that indicate deposition on a salt pan or shallow playa lake environment (implying a desiccated basin floor), then the second sedimentologic model prevails, together with its stratigraphic implications—namely, that evaporites and carbonates were not deposited contemporaneously. Sedimentologic studies of evaporites are thus of great significance in determining (or at least in constraining) the stratigraphy of evaporite-containing basins.

Most of the major controversies dealing with the environmental conditions under which basin-center evaporites formed can be attributed, I believe, to the two different approaches applied to these sediments and to an overreliance upon inferred stratigraphic relationships. The very different interpretations given to probably the most discussed evaporite basins—the Silurian Salina Group of the

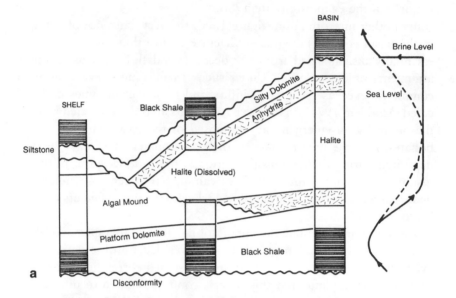

a

b

Michigan Basin and the Elk Point evaporites of western Canada—can be assembled into two main groups: those based primarily upon stratigraphic data, and those based upon sedimentologic evidence. It is contended that interpretations of the first group are, from their very nature, inherently weaker than those based upon correct sedimentologic evidence and reasoning. Nevertheless, depositional models that ignore stratigraphic data are probably also doomed to fail as valid interpretations of basin history. It is also evident that if mistakes are made in evaluating evaporite depositional conditions, then our stratigraphic interpretations may be similarly invalid.

An example of this is provided by the stratigraphic correlations and depositional models suggested for the relationships between the Niagaran carbonate buildups and surrounding Salina evaporites in the Michigan Basin. Some authors have relied significantly upon the presence within the buildups (or their flanking beds) of intervals containing displacive anhydrite nodules (Felber 1964; Mesolella et al. 1974:50; Sarg 1982). These intervals are correlated with the basal (A-1) evaporite unit of the interreef succession (figure 1.7a). This correlation has profound effects on the possible genetic and stratigraphic relationships that can be deduced between the reef and offreef (evaporitic) successions (see Mesollela et al. 1974 for a succinct treatment of the various stratigraphic models). Yet if the nodular anhydrites

---

Figure 1.6

(a) Revised correlation of idealized evaporite and carbonate cycles in the Paradox Formation (Kendall 1987a, 1987b) with dissolved halite on basin flanks restored. Evaporites exhibit an onlap relationship to underlying carbonates and were deposited during episodes when the basin became desiccated, and the earlier-formed basin-flanking carbonates became exposed. Carbonates and evaporites accumulated at different times in the basin, and during episodes when the basin had very different paleogeographies. Evaporites accumulated in playa environments—an interpretation supported by sedimentological evidence. (b) Lower Desert Creek cycle unfossiliferous carbonates with cm-bedding with erosive bases, lithoclastic intervals, desiccation cracking (not shown), and multiple levels containing sediment-casted, hopper-halite molds—all features consistent with deposition in a playa environment. Relative position of this sample is indicated by an asterisk along the right-hand column ("Basin").

are not depositional or early diagenetic emplacements, but instead are late diagenetic replacement nodules similar to those described by Machel (1986) from Upper Devonian reefs in Alberta (figure 1.7b), then their genetic significance is considerably lessened and the inferred stratigraphic relationships (together with the genetic connotations associated with this correlation) need not be correct. The importance of correctly identifying the type of anhydrite present within the carbonate buildup sequences cannot be overemphasized.

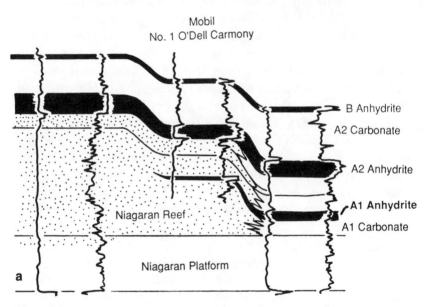

Mobil
No. 1 O'Dell Carmony

B Anhydrite

A2 Carbonate

A2 Anhydrite

A1 Anhydrite

A1 Carbonate

Niagaran Reef

Niagaran Platform

a

Figure 1.7
(a) Reef to offreef correlation, Silurian of Michigan Basin (after Sarg 1982). Note how the identification of the anhydrite interval in the reef-flanking well (Mobil O'Dell Carmoney) as a sedimentary unit, rather than as a late-diagenetic replacement, is critical to the establishment of a correct reef-offreef, carbonate-evaporite stratigraphy. (b) Late-diagenetic replacive nodular anhydrite within pressure-solution seams mimicking displacive sabkha anhydrite; Nisku (Upper Devonian) of Alberta, Canada (photo courtesy of H. Machel). If the anhydrite within the Mobil O'Dell Carmoney well were of this type, the suggested stratigraphic correlations would not necessarily be correct.

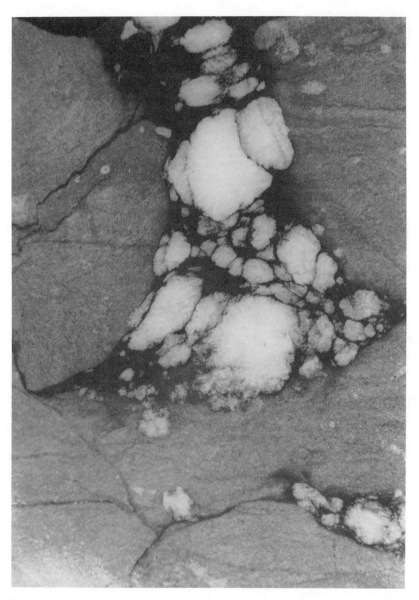

b

PROBLEMS OF STRATIGRAPHIC CORRELATION WITHIN EVAPORITE SEQUENCES

Why are correlations in evaporites, and between evaporites and other facies, inherently difficult? Apart from the question of the data we must work with (discussed above), there are at least five additional factors that affect our attempts at stratigraphic resolution within evaporite sequences.

1. *Lack of biostratigraphic control.* Although not devoid of life, evaporitic (and associated mesosaline) environments tend to be inhabited by specialized organisms that have low biostratigraphic potential. Thus there is commonly a lack of biostratigraphical control that can be employed to resolve correlation problems. Commonly we are able to bracket the age of the evaporite sequence by biostratigraphically dating horizons above and below the evaporite, but intra-evaporite subdivision is difficult, if not impossible.

2. *Rapidity of evaporite deposition.* Evaporites usually accumulate at relatively fast rates, so that even if biostratigraphically useful fossils are present, they are incapable of resolving the stratigraphy to the detail required to discriminate between different stratigraphic alternatives. Schreiber and Hsü (1980) compiled rates of optimum deposition established for evaporites in Recent environments (table 1.1).

Table 1.1
Rates of Deposition of Evaporites from Marine Water and Marine-fed Marginal Deposits

Sediment Type	Areas of Formation	Observed Rates of Deposition
Sulfates and Carbonate (sulfates usually anhydrite)	Sabkha	1 m/1000 yr vertically; 1 km progradation/1000 yr
Sulfates	Subaqueous (Solar Ponds)	1-40 m/1000 yr
Halite	Subaqueous (Solar Ponds)	10–100 m/1000 yr

Source: Schreiber and Hsü 1980.

These rates are matched by those of ancient deposits which have been estimated by age determination of beds above and below the evaporite. More than 2 km of salt in the Messinian of the Mediterranean accumulated in less than 2 million years, and more than 300 m of evaporites in the Middle Devonian Muskeg–Prairie· Evaporite Formations of western Canada were deposited in the time interval represented by only part of a single conodont zone—possibly only 500,000 years. Such depositional rates require that the deposition of many saline giants was initiated within preexisting deep basins—even if isostatic subsidence (the result of loading by the evaporites) is considered.

In the Paradox Basin, twenty-nine cycles containing halite (and perhaps forty cycles in all) have been recognized (Hite 1970) within the Desmoinesian. This corresponds to an average duration for each halite-bearing cycle of only 40,000 years, and much of this interval would have been spent in the episodes of nonevaporite deposition. The stratigraphic resolution required to discriminate between the different stratigraphic and depositional models (noted above) is that *within* each depositional cycle. This resolution is beyond the scope of present biostratigraphic methods.

In a similar fashion, in the Michigan Basin and its environs, biostratigraphy is seemingly powerless to resolve the stratigraphy of the Salina Group and its lateral equivalents in sufficient detail. Shaver (1977) found no biostratigraphic evidence for depositional breaks in carbonate (and reef) deposition on the flanks of the Michigan Basin and concluded that reef growth and evaporite deposition in the basin center were synchronous. Kahle (1978), on the other hand, showed that episodes of exposure (which may be inferred to be coincident with episodes of evaporative drawdown in the basin) had interrupted reef growth. However, the time required for such drawdown and for basin-center evaporite deposition was so short in geologic terms that no biostratigraphic breaks could be recognized in the flanking carbonate sequences.

3. *Complexity of stratigraphic relationships.* Stratigraphic relationships between evaporites and enclosing facies (carbonates or clastics) are usually complex, not layer-cake in character. Because evaporites may accumulate very rapidly, it is common that they have only very thin or no stratigraphic correlatives. This conclusion may also be

reached by considering the degree of disconnection from the ocean that is required to allow evaporites to form (Lucia 1972). In stratigraphic terms this disconnection implies that the evaporites pass laterally into disconformities that represent the emergent rims of the evaporite basin. Facies changes are consequently abrupt, and, because of the limited nature of the subsurface control, these lateral changes are only rarely sampled in wells, let alone by wells with core. The nature of these abrupt lateral contacts is therefore obscure and may be controversial. They may be interpreted as a very rapid interfingering of facies, or (less commonly) they may be identified as disconformable contacts between sequences of different age and paleogeographic setting.

   4. *Removal of evaporites in the subsurface.* Commonly, whether we recognize it or not, we use similarities in the thickness of units (or consistent rates of change in such thicknesses) to assess stratigraphic correlations (the "slice" method of stratigraphic correlation relies entirely upon this procedure). This procedure is often complicated in evaporite sequences by three related processes: removal of soluble components by ground- or formation-waters, which may result in the complete elimination of stratigraphic intervals; removal of part of a saline interval, as when carnallite-bearing horizons are converted to sylvite-halite rocks, causing loss of volume and consequent reduction in stratal thickness; and the fact that when the first two processes occur concurrently with deposition of overlying evaporites, the depositional surface becomes depressed, and thickened evaporite sections are deposited in these depressed areas. These processes may occur in localized areas that have abrupt boundaries. Stratigraphic correlation across such areas may be complicated, especially if there have been multiple episodes of dissolution, diagenetic leaching and/ or thickness enhancement, and multiple episodes of evaporite deposition. Anomalous thickening of evaporites caused by alteration or removal of underlying salts has been recorded from many basins, most notably the Michigan (Mesollela et al. 1974), the Elk Point (McCamis and Griffith 1967; Wardlaw 1968; Corrigan 1975), and the Paradox (Kendall 1987b).

   Removal of evaporites by later dissolution has other implications that have relevance to stratigraphic correlation and depositional models. Complete and unidentified removal of halite from a basin leaves part of the geologic history unrecognized. Kendall (1987b) identified major amounts of halite dissolution in the Paradox Basin, some of which occurred early enough to have influenced deposition of overlying strata. The previous nonrecognition of this salt dissolution was to a large extent responsible for the incorrect interpretation that the basin-central evaporites and the flanking carbonate sections were coeval. In reality, "flanking-carbonate" sequences (including phylloid-algal mounds) were originally overlain by "basinal" evaporites that have subsequently been removed.

Furthermore, since removal of soluble salts commonly begins at the margins of basins and proceeds inward, the final distribution of the more saline evaporites in a basin may bear little relationship to the original distribution. Removal of the marginal parts of a halite interval that has overstepped an earlier anhydrite unit, for example, will convert an onlapping situation into an apparent offlapping one (figure 1.8). In such situations the pattern of evaporite deposition may be converted into a "bull's-eye" pattern, which Hsü (1972) and Schmalz (1969) concluded represents deposition in a basin with negligible influx or reflux during evaporite deposition (influx across a sill imposes a "tear-drop" pattern). Hsü (1972) used the Middle Devonian of western Canada as an example of what he called an "off-centered bull's-eye pattern" (figure 1.9). However, the evidence for this claim is not definitive. Much of the area identified as the outer anhydrite envelope to the north, east, and south of the central halite and potash region (particularly the area in the south identified as the Swift Current Platform) has had halite removed from it by subsurface dissolution. Whatever pattern of evaporite distribution originally occurred in the Elk Point Basin has been substantially altered by later events.

5. *Vertically exaggerated cross sections.* Finally, some of our stratigraphic problems are of our own making. They arise because of the complexities of the stratigraphic relationships and because we choose to represent them with cross sections that have significant vertical exaggeration. There is a strong tendency when correlating between well control to make correlation lines approximately horizontal and as parallel to the stratigraphic datum used as is possible—giving rise to "layer-cake" stratigraphy. Stated differently, there is a natural reluctance to draw correlations that are steeply dipping, if other pos-

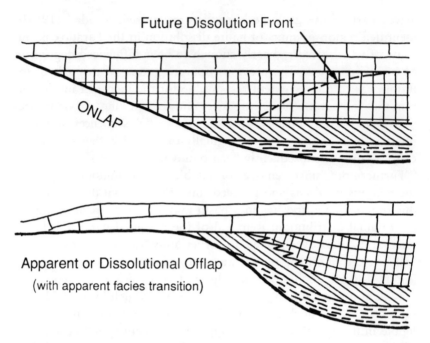

Future Dissolution Front

ONLAP

Apparent or Dissolutional Offlap
(with apparent facies transition)

Figure 1.8
Creation of an apparent offlapping stratigraphic pattern from an on-lapping one by marginal dissolution of evaporites.

sibilities are available. Busson (1980) showed (figure 1.10) that such tendencies are particularly enhanced when vertically exaggerated sections are used. Correlations that appear "wrong" on exaggerated sections (because they appear too steep) appear to be much more reasonable when seen in sections with lesser degress of vertical exaggeration. "Acceptable" correlations are more likely to show facies transitions between evaporitic and nonevaporitic sections and to impose age equivalency between them. On the other hand, if evaporites do accumulate at relatively fast rates, then thick evaporite sections must have formed within preexisting depressions, or we have to find a special cause for rapid, short-term subsidence. Correctly drawn correlations in such areas as strong depositional relief must of necessity dip. If we impose less steeply dipping correlations onto our cross sections, this will impose false age equivalency to evaporites and other sedimentary facies that they laterally juxtapose. This leads in turn to incorrect depositional models, or it will cause conflicts with sedimentologic evidence acquired independently.

An illustration of this situation is possibly supplied by the work of Smosna et al. (1977) upon the Silurian of West Virginia. (It must be emphasized that Smosna and his colleagues have not been chosen here out of any animosity but simply because their excellent paper is one of a very few in which factual stratigraphic data is presented.) These authors, using a widely spaced well control, concluded that the Salina evaporites are age equivalents of parts of the Tonoloway Limestone. Their paleogeographic reconstruction thus involves an evaporite basin flanked by a contemporaneous carbonate shelf (figure 1.11). This reconstruction implies that at least the earlier Salina

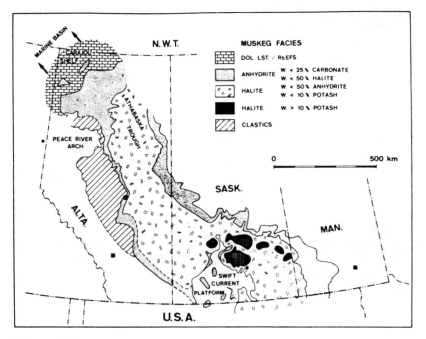

Figure 1.9
Evaporite facies of Middle Devonian, Upper Elk Point Group in western Canada (after Klingspor 1969). Interpreted by Hsü (1972) as an off-centered bull's-eye pattern. The facies distribution today, however, is at least in part the result of marginal dissolution of halite in Saskatchewan, Manitoba, and northeastern Alberta.

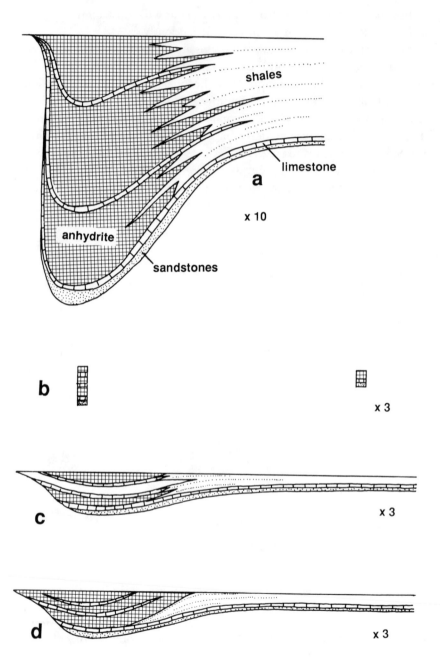

evaporites are deep-water deposits—a conclusion disputed by most sedimentologists who have examined the evaporites. This impasse appears to have been caused by the assumption (enhanced by using vertically exaggerated cross sections for correlation) that lateral juxtaposition of evaporites and carbonates implies age equivalency. An alternative explanation is that the evaporites may correlate with disconformities within the Tonoloway succession, so that the evaporites never accumulated at times when shelf carbonates were forming, and vice versa.

With few exceptions it is most unlikely that *subaqueous* evaporites will interfinger laterally with units of carbonates, clastics, or other evaporite facies of equivalent thickness—even in situations where this appears to be suggested in areas of close well control. This conclusion is reached by considering the very different amounts of the different evaporite minerals that can be precipitated from a given batch of brine, and hence the very different potential rates of evaporite accumulation. In the Paradox Basin the lateral equivalency of sequences of evaporites and carbonates having similar thicknesses is suggested by the occurrence of numerous black shale intervals that extend from basin center to basin flanks (figure 1.4). Shales define stratigraphic packages (sedimentary cycles) that are composed predominantly of evaporites (halite and potash salts) at the basin center and of carbonates toward the basin flanks. This situation does not necessarily imply (as does figure 1.4) that the evaporites and carbonates in these packages are age equivalent. This is only implied if, within each cycle, correlations are drawn that are nearly parallel to the cycle bound-

---

Figure 1.10
Diagrams illustrating effect of vertical exaggeration on correlations between basin-central thick evaporites and marginal clastic or carbonate sequences (modified from Busson 1980). (a) "Conventional" correlation (with significant vertical exaggeration) equating evaporites with marginal deposits. (b) Original data drawn with lesser amounts of vertical exaggeration. (c) and (d) Alternative correlations, with (c) corresponding to the conventional view (a), and (d) depicting an absence of any facies equivalency between the evaporites and the marginal deposits. With lesser amounts of vertical exaggeration, alternative (d) appears less unreasonable.

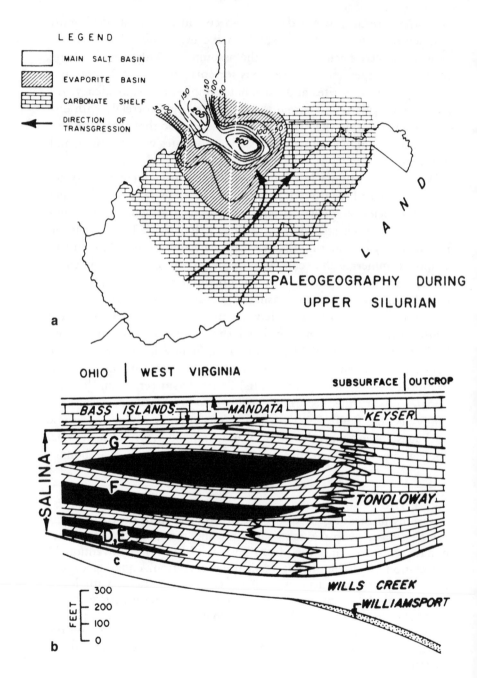

LEGEND

☐ MAIN SALT BASIN
▨ EVAPORITE BASIN
▦ CARBONATE SHELF
← DIRECTION OF TRANSGRESSION

PALEOGEOGRAPHY DURING
UPPER SILURIAN

a

OHIO | WEST VIRGINIA

SUBSURFACE | OUTCROP

BASS ISLANDS — MANDATA

KEYSER

SALINA

G

F

TONOLOWAY

D,E

c

WILLS CREEK

WILLIAMSPORT

FEET
300
200
100
0

b

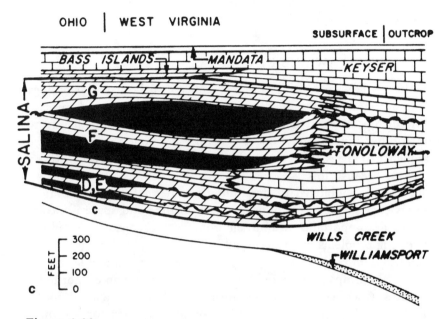

Figure 1.11
Correlation of evaporite and carbonate strata, Silurian of West Virginia. (a) Paleogeography during Upper Silurian with coeval evaporite basin and carbonate shelf (modified from Smosna et al. 1977). (b) East-west cross-section of Upper Silurian–Lower Devonian strata of northern West Virginia, showing correlation of Salina halite beds (black) with parts of the Tonoloway Limestone (Smosna et al. 1977). (c) Suggested alternative correlation with Salina halite units corresponding with depositional breaks in the Tonoloway Limestone.

aries. However, correlations within each cycle are not necessarily layer cake, and although they are spatial equivalents, the evaporites and carbonates may have accumulated at different times and in different depositional environments (figure 1.6). In this interpretation, one time line at least—that separating episodes of carbonate and evaporite deposition—must be drawn with a steeper dip relative to the cycle boundaries.

WALTHER'S LAW AND EVAPORITES

Walther's Law is considered a key concept in both sedimentology and stratigraphy, despite being constantly misquoted and misused (Middleton 1973). It states that only facies that develop adjacent to each other can be superimposed. It does not state that the vertical sequence of facies always reproduces the horizontal sequence. Nevertheless, because this is commonly the case, studies of vertical sequences are employed to determine the lateral facies distribution. Knowledge of facies relationships in both time and space is clearly of greatest importance to both environmental reconstruction and to resolving problems of chrono- and lithostratigraphy. The genetic relationship implied by Walther's Law between vertical and lateral facies relationships is commonly not the situation within evaporite sequences, where even the correct form of Walther's Law need not be applicable. This clearly is a disaster for studies of evaporites where lateral stratigraphic relations can only be inferred.

Walther's Law is a useful concept in situations where the change in facies over a given location is a simple consequence of facies migration in the environment caused by depositional up- and out-building (as in the succession of facies or subfacies within a point-bar deposit), or when the change is a simple product of sea-level change (a prograding barrier sand body, for example). In evaporite systems, which by their very nature are delicate balances between rates of inflow, outflow, and evaporation, even very slight changes in external conditions—such as changes in sea level, integrity of the sill, climate change, etc.—can result in major changes in the depositional environment. Facies tracts present at one stage in the basin history may be destroyed and their place taken by others that did not exist prior to the change. In this fashion Walther's Law is violated, since facies are superimposed which never developed adjacent to each other. If interpretations are drawn from such superimpositions of facies using Walther's Law, very inaccurate stratigraphic correlations and/or environmental syntheses may be drawn. Two examples will suffice.

A core taken in the lowermost Devonian strata of the Cold Lake subbasin of northern Alberta sampled a redbed-halite sequence. In this core different facies may be identified and related to different depositional environments by making analogy with Recent playas or continental sabkhas (figure 1.12; Kendall 1984, figure 8). At the base, coarse-grained and pebbly arkosic sandstones with small- and large-scale cross-bedding and clay drapes closely resemble modern wadi or ephemeral stream deposits. They pass upward into massive to finely laminated, dolomitic and anhydritic, silty and arenaceous mudstones that contain increasing amounts of displacive hopper-halite crystals in their upper parts and grade insensibly into a chaotic halite-mudstone (haselgebirge) facies (figure 1.12). These sediments may be identified as the deposits of dry and saline mud flats, respectively (Hardie et al. 1978). Overlying bedded halite probably formed in an ephemeral playa lake.

It is tempting, since each part of the cored sequence can be identified with part of a generalized playa model (Hardie et al. 1978; Kendall 1984), to apply Walther's Law, and suggest that on the basis of the observed succession, the different facies were arranged in a similar lateral sequence—with marginal wadi deposits giving way successively toward the basin center to dry and saline mud flats and to a central playa lake (figure 1.13). However, the well from which the core was obtained is situated in the basin center, and it is difficult to devise a depositional history that would allow marginal facies to form initially and be replaced successively by more "basin-central" facies as time progressed. In fact, correlation of the cored well with others in the subbasin strongly suggests that the sequence found in the cored well depicts a series of changes that probably affected the entire basin more or less simultaneously. Inspection of figure 1.14 shows that at one time the bedded halite (representing the ephemeral playa lake environment) apparently occupied almost the entire depositional basin. During much of the period when the underlying mudstones were deposited (in dry and saline mud flats), no bedded halite was being deposited in a central playa lake. Reliance upon

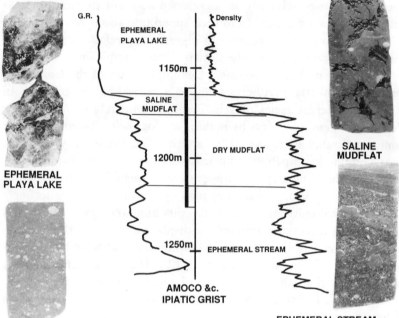

Figure 1.12
Well log and parts of core from Amoco Ipiatic Grist well, Alberta, showing transition between Lower Lotzberg Salt and Basal Red Beds. Red anhydritic and dolomitic mudstones pass upwards with increasing amounts of displacive halite hopper crystals into chaotic halite-mudstones ("haselgebirge"), marking a change from dry to saline mudflat subenvironments. Using Walther's Law, the entire cored sequence might suggest that a lateral change from alluvial fan–ephemeral stream to playa lake environments existed.

Walther's Law generates false and misleading genetic and stratigraphic interrelationships between the different lithofacies.

The second example—the Upper Ordovician Red River Formation of the Williston Basin (figure 1.15)—has some similarities with the first but is more controversial. In the central part of the Williston Basin, sedimentary cycles that are developed in the upper part of the Red River (the Herald Formation of Kendall 1976) are most complete, with each terminating in a thin but widely developed bed of anhydrite. The cycles are relatively simple, being composed of a basal marine carbonate unit, commonly fossiliferous and extensively bioturbated, surmounted abruptly by essentially unfossiliferous, finely laminated carbonates (commonly microcrystalline dolomites) capped by similarly laminated or thin-bedded dolomitic anhydrite, sometimes with such distorted bedding that the anhydrite resembles nodular anhydrite (Kendall 1976, plate 29; Kendall 1984, figure 26). Misidentification of the anhydrite intervals as displacive nodules and the underlying laminated dolomites as algal mat deposits have led many authors to apply Walther's Law and conclude that the Red River (Herald) depositional sequences represent shallowing-upward cycles with laminated dolomites representing intertidal deposits, and anhydrites representing supratidal deposits (Asquith et al. 1978; Asquith 1979; Carroll 1978; Clement 1985; Roehl 1967; Ruzyla and Friedman 1985; Thomas 1968). Furthermore, some of these authors

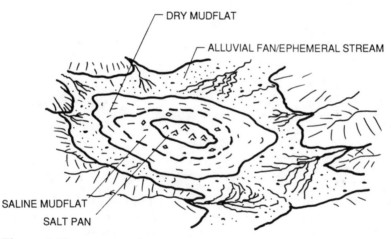

Figure 1.13
Diagram of concentrically arranged facies tracts in continental playa or inland sabkha. The centrally located ephemeral playa lake (or salt pan) is surrounded successively by saline and dry mudflats and sand flats with marginal alluvial fan and ephemeral stream (wadi) deposits. This type of facies distribution suggested by a noncritical application of Walther's Law to the Amoco Ipiatic core, does not appear to have been present during much of the Lower Elk Point history of the Cold Lake subbasin.

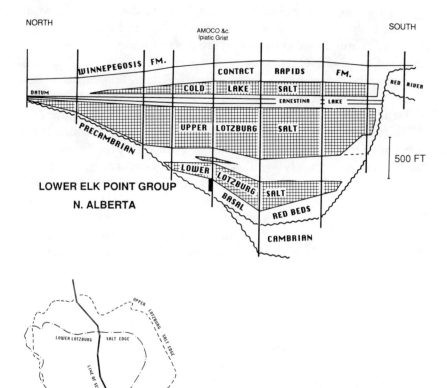

Figure 1.14
Cross-section through Lower Elk Point Group in Cold Lake subbasin, Alberta. Datum is the Ernestina Lake Formation—an apparently
short-lived marine incursion into an otherwise continental basin. When
the upper part of the Lower Lotzberg Salt was forming, almost the
entire depositional area was occupied by a salt-precipitating environment (playa lake), and much of the Basal Red Beds have no halite
equivalent. A similar but even more abrupt change in conditions from
dry and saline mudflats to playa lake halite is present at the base of
the Upper Lotzberg Salt—a halite-precipitating environment that again
filled almost the entire basin. (Modified from Hamilton 1971, with
the addition of the Amoco well.)

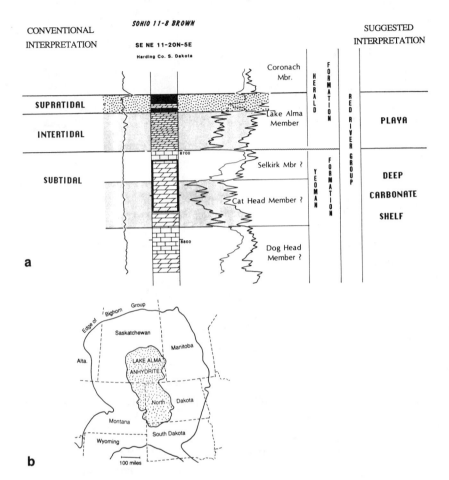

Figure 1.15
(a) Lowermost sedimentary cycle developed in Red River Group (Lake Alma Member of Herald Formation and underlying Yeoman Formation, equivalent to the Red River "C" cycle of some authors), Williston Basin. In some interpretations the burrow-mottled carbonates of the Yeoman are considered to be subtidal deposits; the microcrystalline laminated dolomites of the lower Lake Alma to be intertidal algal mat sediments; and the anhydrites of the upper Lake Alma (black) to be supratidal deposits. This is despite the necessity of invoking a basin-wide 50-ft intertidal sequence, and confinement of the supratidal anhydrite facies to the basin center (b). (Modified from Kendall 1984.)

have suggested hydrocarbon exploration strategies based upon this interpretation.

Kendall (1976), Kohm and Louden (1978), and Longman et al. (1983) reevaluated the lithologic evidence for the shallowing-up-wards (subtidal to supratidal) interpretation of the Red River cycles and concluded that all parts of the sequences are of subaqueous origin. Kendall (1976) also argued that the stratigraphic evidence—the basin-central restriction of uppermost cycle members, and the excessive thickness of the units identified as intertidal throughout the basin—is at variance with a sabkha interpretation for the anhydrites. (Sabkha anhydrites would be expected to be better developed at basin margins than in the basin center and should be underlain by widely distributed intertidal sequences no thicker than the tidal range at the time of deposition; yet the basal Herald cycle contains an "intertidal" laminated dolomite unit consistantly thicker than 50 ft throughout the central part of the basin; see Kendall 1976, figure 23.) Cycles have been reinterpreted as reflecting a progressive increase in water salinity with time—the "brining-upwards cycles" of Longman et al. (1983)—but in each case the salinity change is regarded as being progressive, with each salinity stage (facies) passing gradationally into the next as the basin becomes more hypersaline.

Reevaluation of the Red River lithofacies (partially reported in Kendall 1984, 1985) suggests that the different parts of the Red River cycles may not be genetically related to one another. Kendall (1985) provided evidence and regional arguments to suggest that the Lower Red River (the Yeoman Formation of Kendall 1976), which forms the marine carbonate member of the lowest Red River cycle, is a deep shelfal deposit, whereas the lithologies, sedimentary structures, lack of a marine fauna, lack of any marine equivalents, and the basin-central facies pattern are consistent with an interpretation of the Red River laminated dolomites and anhydrites as playa deposits (Kendall 1985). Comparisons between photographs of Red River sediments (Kendall 1976; Clement 1985; Neese 1985) and those of playa deposits from the Green River Formation (especially those of Smoot 1983) are particularly instructive. If these two environmental interpretations (deep shelf and inland playa) are correct, then Walther's Law cannot be applied to the Red River sequences. The absence of any transitional shallow shelf/coastal deposits indicates that the cycles must contain disconformable contacts between the marine and playa components. The presence of lithoclastic intervals at the contact in some cycles (Kendall 1976) provides some supporting evidence for the existence of these sedimentary breaks. At no time were marine and playa sediments deposited simultaneously—they represent discrete episodes in the Ordovician history of the Williston Basin. One possible scenario that could explain the Red River depositional sequences is that they reflect rapid sea-level falls accompanying Gondwana glacial episodes. Deep shelfal environments, representing high sea-level stands, may have been abruptly replaced by playas when sea levels fell below an upstanding rim that marked the edges of the ancestral Ordovician Williston Basin. Exposure of the rim isolated the basin from the ocean, caused evaporative drawdown, and created a vast inland playa.

Walther's Law therefore must not be applied indiscriminately to evaporite sequences. The speed with which evaporite depositional environments can change in response to relatively slight changes in external factors suggests that major (and many minor) lithologic boundaries in evaporite sequences may be more or less isochronous across the basin. Vertical sequences in evaporites reflect basin-wide changes more than they indicate lateral facies relationships. The same general conclusion will be reached from a different direction in the following section.

THE SILLED-BASIN MODEL FOR EVAPORITE DEPOSITION

The model most commonly used to explain the origin of basin-filling evaporites is that of the silled basin. The model predicts or explains many of the basic stratigraphic features of evaporitic sequences, but (as will be argued here) it is untenable. It was devised to explain the major compositional features of large evaporite bodies, in particular their thickness; the presence of thick units of the same mineralogy; and the variation in the proportion of different salts from that expected from simple seawater evaporation (with the more saline salts being underrepresented).

In essence, the model, a synthesis of those devised by Ochsenius (1877), King (1947), and Scruton (1953), suggests the presence of a shallow sill or bar across the mouth of the basin, partially separating it from the sea (figure 1.3). The bar functions to prevent the unrestricted access of seawater into the basin (thus allowing evaporation to concentrate seawater into a brine) and to inhibit the outflow of the evaporation-concentrated brines. In order to explain the absence, or paucity, of the more saline components in evaporitic successions, the model allows these more soluble phases to be exported from the basin as a dense concentrated brine. This brine escapes (refluxes) from the basin by overflowing the sill or percolating through it. Seawater entering over the sill forms a surface layer above the outflowing bottom brines. Inflow occurs because of the negative water balance existing in the basin. Losses by evaporation create a slope on the water surface declining into the basin, and this induces a basinward flow of seawater. As King (1947) noted, unless the chemocline between the inflowing surface seawater and the outflowing dense brines lies below wave base, mixing of the two waters will occur and reflux of the more soluble components will be inefficient or prevented. In the model, as the inflowing seawater moves across the evaporite basin it suffers evaporation and becomes progressively denser, and a lateral density gradient is generated. When the surface-brine density equals the density of the bottom brines (in the most distal parts of the basin), mixing occurs. In order to maintain a constant volume of bottom brine, some of it must reflux over or through the sill at the same rate as bottom brine is supplied by concentration of the surface waters. Because all parts of the brine column will be affected by the inward flow, reflux must overcome the pressure field induced by this "slight" hydraulic head between the sea and the interior of the evaporite basin; the topographic restraint imposed by the sill; bottom friction; and frictional losses (drag) imposed by the incoming flow. Scruton (1953) believed that the last-mentioned factor alone may be sufficient to create a hydrodynamic barrier preventing brine reflux over the sill.

During the progressive brine concentration, inferred to have occurred across the basin, the more soluble components of the brine successively become concentrated to the point where their solubility products are exceeded and they are precipitated. This produces a lateral change in the mineralogy of the evaporite precipitate across the basin—with the most soluble components precipitated in the more distal parts of the basin. The silled-basin model therefore predicts the presence of a lateral facies change across evaporite basins and furthermore implies that these different facies are coeval. The fact that some evaporite basins do display a marked change in facies, with increasing amounts of more soluble components away from the entrance to the basin (the Middle Devonian Elk Point Basin of western Canada being a near-perfect example; Grayston et al. 1964) provided powerful support for the model.

The silled-basin model (and other models that involve a free connection between the evaporite basin and the sea) is, however, subject to many problems—the chief amongst them being, first, the size of any possible connection or inlet and corollaries of this; and second, the hydrostatic conditions that occur in saline basins.

Inlet Size and Basin Restriction

Lucia (1972) and Kinsman (1974) have stressed the fact that the relatively slow rates of evaporation that can occur (less than 5 m per year, and probably in most cases less than 2 m per year; Shaw 1977) impose stringent limits on the size of any evaporite basin inlet. Lucia calculated that the ratio between the surface area of a basin and the cross-sectional area of the inlet (a ratio that is, in effect, a measure of the balance between the rates of water loss and gain in the basin) must be at least 10^6 before gypsum can be precipitated, and for halite to precipitate, this ratio must exceed 10^8 (figure 1.16). These estimates presuppose that all water entering the basin does so by surface inflow through the inlet. If in addition a basin receives seawater by seepage inflow through a porous barrier, or is diluted by groundwater influx, surface runoff, or rainfall, the above-mentioned ratios must be correspondingly larger. The calculated size of any inlet that can lead to a basin precipitating halite is so small, even for large basins, that it must be concluded that halite precipitation requires complete surface disconnection of the evaporite basin from the sea (Lucia 1972). Any change in the size or shape of the inlet by erosion (by the inflow), constriction by deposition, or changes induced by even the slightest degree of tectonic movement (including subsidence) or by sea-level changes will markedly affect the influx rate. This will alter the fine balance that must exist between the rates of inflow, reflux, and evaporation that are required to generate the thick units of near-constant evaporite facies that exist in many evaporite basins

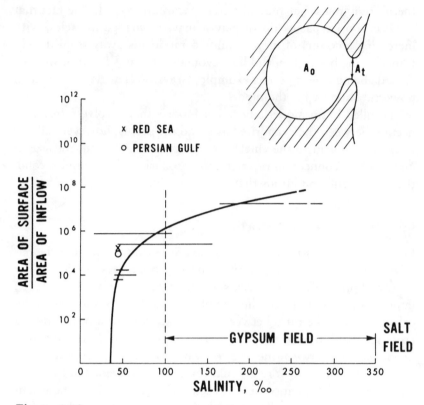

Figure 1.16
Relationships between the ratios of the areas of an inlet and an evaporite basin, and the maximum salinity developed in the basin (from Lucia 1972).

and that were the impetus for developing the silled-basin reflux model in the first place. If there is surface disconnection, there is no opportunity for any surface reflux of the concentrated brines out of the basin.

Furthermore, disconnected basins can be fed seawater only by seepage influx through permeable barriers, and these rates of inflow are very much smaller than those for surface influx. So much so, in fact, that in most instances, they are unable to offset water losses due to evaporation. Maiklem (1971) made a quantitative estimate of water influx through a barrier and of water losses in a Middle Devonian evaporite basin in western Canada (figure 1.17a). His main conclusion was that seepage through the barrier would have been insufficient to offset evaporative losses in the basin, and that the basin would have dried up (figure 1.17b). This was true even when extraordinarily large values for the barrier permeability—73 darcies!—were used. The loss of brine volume from the basin and the consequent lowering of its brine level is termed "evaporative drawdown." It is a feature that should affect all surface-disconnected evaporite basins. The fact that the majority of evaporites appear to have formed in shallow-water or brine-pan environments, even those inferred to have formed on the floors of preexisting deep basins, is a strong indication that evaporative drawdown occurred frequently in the past and that evaporite basins were essentially disconnected from the sea.

The occurrence in some basins of evaporites that appear to have formed in deep-water settings suggests that in some circumstances, evaporite basins do not desiccate to dryness. Maintenance of a brine column in the basin will require either aperiodic or seasonal inputs of seawater into the basin over the barrier as a supplement to the seepage inflow; a reduction of the evaporation rate; or seepage inflow occurring around the entire periphery of the basin or over much of the basin floor. Since much of this inflow is unlikely to be seawater, this third possibility suggests that some—perhaps many—of the evaporites in large basins may be of nonmarine (continental) origin, a possibility also suggested by other evidence (Hardie 1984). Episodic surface inflow would remove some of the difficulties inherent in the idea of a continuous feed of seawater into the basin. It might also provide an explanation for the near-ubiquitous occurrence of rhythmic bedding in deep-water evaporites. Each rhythm would mark an episode of surface inflow, followed by static evaporation of the basin-contained brines. The fact that at least some genetic units appear to have components present in proportions similar to that of seawater (Hardie 1984) may support this possibility.

Kinsman (1976) and Taylor (1980) have suggested that high air humidities may be responsible for reducing the evaporation losses of a brine. If brine and air are at the same temperature, evaporation can occur only when the vapor pressure of the brine (which varies with its chemical activity, itself a function of the brine's ionic strength or salinity) exceeds the partial pressure of water vapor in the air (i.e.,

PRESQUILE BARRIER REEF

a

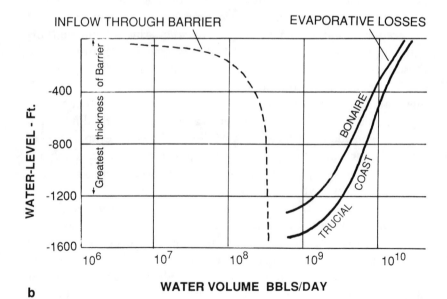

b

its relative humidity). Thus if the atmospheric humidity over an evaporite basin exceeds 76%, brines in the basin cannot be concentrated by evaporation to the point where halite is precipitated, when the brine has a water activity of 0.76. As values of brine activities and air humidities converge, the evaporation rate approaches zero. Taylor (1980) suggested that in a situation where climatic conditions did not allow evaporation to concentrate the basin inflow above a specified concentration, then the brine level in the basin would fall only to the level where the increased inflow (caused by the increased hydraulic head between the ocean and the lowered brine level in the basin) balances evaporation losses up to that specified concentration. Both Kinsman (1976) and Taylor (1980) used this scenario to explain the long-term deposition of a single evaporite mineral facies. In fact, this mechanism will, alone, produce very little evaporite deposition. Once the brine in the basin reaches the specified concentration, it cannot be further concentrated by evaporation, and no further precipitation would occur. Continued evaporite deposition would only occur by the evaporation of the inflow, and this would generate more brine incapable of being concentrated any further. In a short space of time the basin would completely fill with such a brine and evaporite deposition would cease. Continued deposition can only occur if there is continued influx. For this to occur, space must be made available in the basin, and, in the absence of evaporative losses, this space must be generated by some type of brine reflux. Further development of

Figure 1.17
Comparison between the rates of seepage inflow and the rates of evaporative water loss from the Elk Point Basin of western Canada (after Maiklem 1971): (a) inferred original size of Elk Point Basin and its northern barrier (Presqu'ile Barrier Reef); (b) graph of water level in basin vs. volume inflow and water loss in Elk Point Basin. Inflow curve represents the condition for the barrier with a permeability of nearly 73 darcys. Water-loss curves were calculated for climatic conditions similar to those of Bonaire and the Trucial Coast. Rates of seepage influx are always lower than the rates of water loss from the basin. Isolated basins will therefore exhibit a tendency toward desiccation.

evaporative models should concentrate upon possible reflux mechanisms within desiccated basins.

Hydrostatic Conditions in Evaporite Basins

Shaw (1977) argues that the silled-basin model is hydrostatically unsound. He chose to examine the model at the stage where halite was first beginning to precipitate in the most distal parts of the basin. He followed King (1947) in supposing that for efficient reflux of dense brines to occur, the depth at which the boundary between inflowing and outflowing waters occurred must lie deeper than wave base. Shaw chose a depth of 100 m, but the exact depth is immaterial to the main argument. Shaw also assumed that if halite were precipitating, then the influx must have been reduced (by evaporation) to approximately one-tenth of its original thickness in these distal areas (in actual fact if the basin inlet is of very small size the influx spreads out over the width of the basin, and the depth of the concentrated brine will be proportionately less).

An equilibrium level must occur, set by the sill depth, at which hydrostatic pressures should be equal. Departures from this equilibrium will cause flow of water or brine within the basin. Removal of 90% of the water column at the distal parts of the basin clearly would create an enormous hydrostatic imbalance, and this would be offset by the addition of a column of dense bottom brine (figure 1.18). Although the different parts of the basin would now be in hydrostatic balance, there would be a large elevation difference between the brine level at the distal part of the basin and the surface of the inflow at the sill (this is Scruton's "hydraulic head"). If the inflow were 10 m deep, Shaw's calculations would suggest that a surface elevation difference of over 1.5 m would occur across the basin. Of even greater importance (and unnoted by Shaw) is the elevation difference that would occur between the influx water layer and the concentrated brine. For a 10 m influx, the interface would lie at 9.47 m above the sill elevation at the distal part of the basin, but at only 1 m above the same datum at the sill—a difference of nearly 8.5 meters! The slopes on these boundaries would be highly unstable. Liquids are incapable of sustaining any shear stress, so that even if such slopes at the surface momentarily existed, water would rush from the sill into

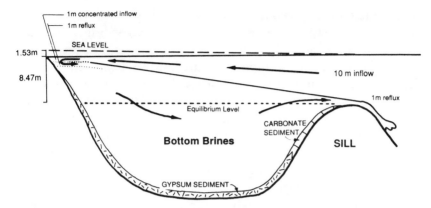

Figure 1.18

Hydrostatic conditions in a silled evaporite basin (modified from Shaw 1977). At the stage where the basin is just beginning to precipitate halite, the influx has been reduced to one-tenth of the original depth at the distal end of the basin. In order to achieve hydrostatic balance within the basin, the distal areas must be balanced by addition of bottom brines. The difference in water depths in different parts of the basin that are in hydrostatic balance would necessitate (for a 10 m influx) an elevation difference of the water surface greater than 1.5 m across the basin and an inclined surface between brines of different density (with an elevation difference across the basin of just less than 8.5 m).

the basin until an essentially level surface was restored. Similarly, brine would move rapidly toward the sill until an essentially horizontal interface with the inflow water body had been attained. It must be concluded that lateral salinity gradients of the magnitude required to allow evaporite deposition in part of a silled basin are not possible. The present-day Red Sea, a silled, deep basin in an arid setting, exhibits only a very small salinity variation. Surface salinity ranges from 36‰ at the sill to just over 40‰ in the Gulf of Suez— a lateral salinity gradient of only 1‰ per 400 km (Neumann and McGill 1962).

The main implication of Shaw's (1977) analysis of the silled, deep-water model for evaporite deposition is not, as Shaw supposed, that deep-water evaporites are incapable of being formed, but that a deep-water evaporite basin is incapable of precipitating different mineral facies simultaneously in different parts of the basin. This is precluded because of the large differences in the densities of brines that are capable of precipitating the different evaporite minerals. In deep-water basins, the density or elevation heads are large enough to ensure that the water in a basin will be more or less laterally homogeneous.

The same conclusion as reached independently by Hardie et al. (1978) in their analysis of the conditions within perennial saline lakes. These lakes probably do not differ significantly from some ancient evaporite basins, except possibly for scale and in the nature of the waters that enter the basin. Hardie et al. (1978) suggested that surface waters are concentrated by evaporation, become denser, and sink before saturation with respect to a given evaporite mineral is reached. This surface brine mixes with bottom waters (or displaces them) so as to increase the overall brine density in the basin. Eventually the entire brine body becomes dense enough so that the evaporating surface waters remain at the surface sufficiently long to become supersaturated with respect to the evaporite mineral. Evaporite crystals, nucleated at the surface, slowly settle through the bottom waters and dissolve, and the entire bottom brine body increases in density. At the point where the entire brine body becomes saturated with respect to the evaporite mineral, net accumulation occurs on the floor of the basin. Continued evaporative concentration of the surface waters slowly increases the overall concentration of the brine body until first the surface layer and then the entire brine body becomes saturated with respect to the next most soluble mineral phase, and precipitation and net accumulation of that mineral begins. The entire brine body changes salinity in concert, and changes in mineral precipitation should affect the entire basin more or less simultaneously.

This last conclusion has important sedimentologic and/or stratigraphic significance. If basin-central evaporites exhibit pronounced lateral facies changes, then it may be concluded that either the facies changes are not real but are the result of miscorrelation, or else the evaporites are not deep-water deposits but must have been deposited in shallow-water or brine-flat settings, where rapid and extreme lateral changes in brine composition and density are possible. Frictional drag imparted by the bottom on the shallow (or nonexistent) brine column will inhibit lateral brine mixing.

Rapid lateral variations in evaporite mineralogy in depositional cycles are recorded by Matthews (1977) from closely spaced cores in the Devonian Detroit River Group of the Michigan Basin. These are most unlikely, however, to be deep-water deposits as Matthews suggests. The close spacing of cores makes it most unlikely that any significant miscorrelation of the cores has been made. So the rapid facies changes must be considered to be real. If the evaporites are deep-water deposits, this would imply that the basin would have had to have exhibited very rapid and extreme changes in salinity—commonly changing from carbonate to halite supersaturation over distances of only a few miles. Unfortunately, the absence of any sedimentologic detail in the descriptions of these Detroit River cores precludes any documentation of the inferred shallow-water or brine-flat depositional environments implied by the presence of the rapid lateral facies changes.

The laminated carbonates, anhydrites, and halites of the Permian Castile Formation of the Delaware Basin (West Texas and New Mexico) are probably the best-known and most often-quoted example of inferred deep-water evaporites. Units of brecciated anhydrite occur in the western part of the basin and correlate with stratigraphic units of halite in the east (figure 1.19). Most breccias appear to have formed by the subsurface dissolution of halite, and their presence thus marks the former presence of this facies in the west. With only a single exception, every halite bed in the eastern part of the basin has an equivalent bed of breccia in a well located only 32 km from the western basin edge (Anderson et al. 1972). Halite beds are therefore considered to have originally had a basin-wide distribution.

Of the greatest importance to us here is the fact that a correlation of anhydrite laminae between different wells indicates that halite began to accumulate in the Delaware Basin almost simultaneously. In addition, beds of laminated anhydrite also occur within the halite beds, have basin-wide distribution, and appear to be composed of the same number of laminae. This also suggests that major mineral changes occurred isochronously over the entire basin. Anderson et al. (1972) are less certain about the timing at the end of halite deposition but suggest that this also was probably synchronous.

Individual laminae that are inferred to be composed of material precipitated from evaporating surface waters (pelagic or cumulate laminae) must also, from their lateral persistance and the persistance

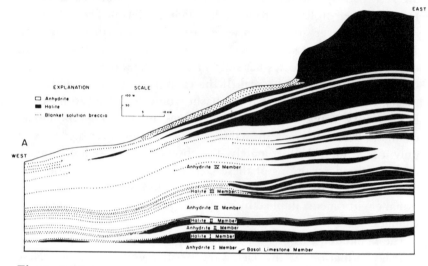

Figure 1.19
Correlation of halite beds and laterally equivalent halite-solution breccias in the Castile and Lower Salado formations across the Delaware Basin (from Anderson et al. 1972). Castile stratigraphy suggests the evaporite basin simultaneously switched from anhydrite to halite precipitation over the whole basin, a feature supporting the view that deep-water evaporite basins are filled with brines with laterally uniform salinities.

of groupings of laminae, represent isochronous events. They may represent seasonal precipitates, annual events, or aperiodic deposits, but they must have been deposited in the same time episode over the entire basin. Thus the surface brines in the Delaware Basin (and in any other basin where this type of deposit is found) simultaneously began to precipitate carbonate over the whole basin, followed by a simultaneous change to calcium sulfate (or even to halite) precipitation.

The lateral persistance of pelagic or cumulate evaporite laminae across entire depositional basins provides considerable support for their deep-water origin and for the simultaneous change in the salinity of surface (and bottom) brines over the entire basin. Lateral facies changes are thus indicative of shallow-water, not deep-water, settings.

EVAPORITES AND SEA-LEVEL CHANGES

The following section has benefited markedly from discussions with Brian Logan of the University of Western Australia and is largely based upon the present author's attempts to apply to ancient evaporite sequences many of the ideas generated by Logan in his studies of the present-day salt basin of Lake MacLeod.

Much of current stratigraphic effort is concerned with the effects of sea-level changes on stratigraphic sequences. Sea-level changes have always been seen as important to the interpretation of evaporite sequences, especially if the silled-basin model was employed. Clearly, changes in sea level would have the most profound effects on the size of the inlet and consequently upon the amount of influxing and refluxing brines. This variation would be expressed in profound changes in basin brine salinity and therefore in the type of evaporite mineral deposited. In extreme cases, flooding of the sill would cause evaporite deposition to cease altogether, whereas sea-level falls would cause greater restriction and the deposition of more saline evaporites. Hite (1970), for example, in his interpretation of the evaporite cycles in the Paradox Basin, correlated the amount of reflux that could occur over the sill, the salinity within the evaporite basin, and the type of evaporite deposition, with changes in sea level (figure 1.20). At lowest stands of sea level, the basin sill was at its shallowest; reflux did not occur or was inhibited to the greatest extent; the salinity within the basin consequently increased; and the most saline evaporites were precipitated (halite or potash salts). With rising sea level, the sill became deeper, allowing greater amounts of brine reflux, and this export of brine from the basin progressively caused a reduction in the brine salinity in the basin, and a consequent change in evaporite deposition to less saline minerals (calcium sulfate and eventually, carbonate).

However, if evaporite basins (especially those that contain halite) are considered to have been disconnected from the ocean during evaporite deposition (as discussed above), then there can be no simple relationship between sea level and the type of evaporite deposited in

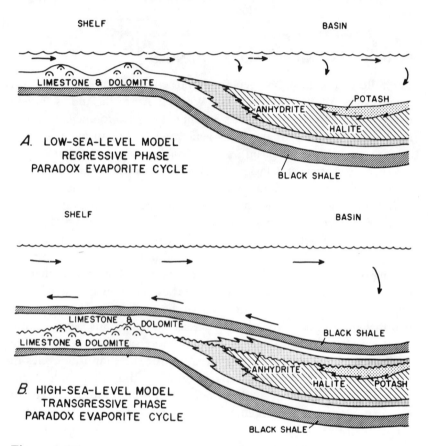

Figure 1.20

Interpretation of Paradox Formation sedimentary cycles according to Hite (1970). At low sea-level stands (a), the shelf is shallow, preventing surface reflux, and brines in the Paradox evaporite basin progressively increase in salinity. Uniform flow of seawater into the basin promotes algal reef growth on the flanking shelf coeval with halite deposition in the basin. At high sea-level stands (b), the shelf is deeper, allowing surface reflux and causing a progressive decrease in basin salinity. Outflow of hot saline brine across the shelf inhibits limestone deposition and causes deposition of black shales across both basin and shelf.

the basin. The amount of surface reflux is not a possible control, as Hite (1970) suggested, because no surface reflux occurs.

The type of evaporite mineral that is precipitated at any stage will be controlled by the salinity of the brine in the basin. In the absence of any mixing with any other water type, mineralogy will be controlled by the evaporation rate and the length of time the water has spent in the basin being affected by evaporation (i.e., the residence time of the brine in the basin). In turn, the brine residence time is affected primarily by the absolute and the relative rates of water inflow and brine reflux out of the basin. Increases in the absolute and relative rates of water inflow into a basin will increase the brine residence time, thus allowing evaporation to concentrate the inflowing water to higher salinities. Conversely, decreases in the rates of water influx (relative to rates of brine reflux) may significantly decrease the residence time of brine in the basin, so that evaporation is only capable of increasing brine salinity to a small degree before the brine is exported out of the basin. Decreases in brine residence time will be recorded by changes in evaporite deposition to minerals of lower solubility.

One of the most obvious methods of changing the absolute rates of water influx into an isolated and desiccated evaporite basin is a change in sea level. With the sea level rising (but not to high enough levels to allow drowning of the basin rim, which would cause evaporite deposition to end), the hydraulic head between the ocean and the depressed brine level is increased, resulting in greater amounts of water seepage into the basin through the permeable sill or barrier. In the absence of any increased rates of brine reflux, this increased influx will increase the brine residence times in the basin and result in the development of higher brine salinities. In this setting, therefore, sea-level rises may be marked by deposition of evaporites of increased solubility—exactly the reverse of the conventional interpretation that used surface reflux across a sill. Lowered sea level would cause reduction in the absolute influx rate, decreased brine residence times, and deposition of evaporites of lower solubility. Again, this is exactly the opposite of what conventional wisdom suggests.

The different interpretation of the relationship between deposition and sea-level changes just discussed may help explain some puzzling aspects of evaporite basin stratigraphy. An example is provided by

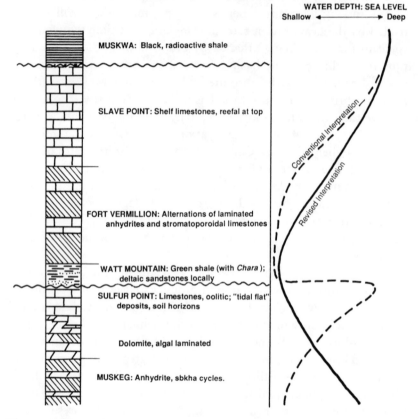

Figure 1.21
Hypothetical well section through Middle Devonian strata of northern Alberta with "conventional" and "revised" sea-level curves. Convential curve suggests that the upward decrease in basin salinity seen in upper Muskeg and Sulphur Point formations reflects a sea-level rise. Revised curve suggests that this same sequence records a sea-level fall that culminated with deposition of the nonmarine Watt Mountain clastics.

the relationships, shown in figure 1.21, between the Middle Devonian Muskeg evaporites (Elk Point Basin of northern Alberta) and the overlying, essentially nonmarine clastic interval of the Watt Mountain Formation and its relations with overlying anhydrites and open-marine carbonates (Fort Vermillion and Slave Point Formations). The markedly cyclic anhydrites of the uppermost Muskeg Formation pass transitionally upwards into carbonates of the Sulphur Point Formation, which southward pass gradationally into Muskeg anhydrites. The vertical sequence from anhydrites into carbonates would be interpreted conventionally as reflecting a sea-level rise that allowed more exchange to occur between the Elk Point Basin and the open ocean to the north. However, this interpretation would necessitate a subsequent rapid sea-level fall to account for the overlying nonmarine Watt Mountain.

The Watt Mountain–Fort Vermillion (anhydrites)–Slave Point (open-marine carbonates) sequence would require yet another sea-level rise. A much simpler picture is suggested, however, when it is considered that, in a disconnected and desiccated evaporite basin, sea-level falls may be accompanied by changes in evaporite deposition from more to less saline mineralogy. The Muskeg to Sulphur Point transition may be interpreted to mark such a sea-level fall, which climaxed with the nonmarine clastic deposits of the Watt Mountain. The following sea-level rise initially restored part of the evaporite basin, leading to deposition of the Fort Vermillion anhydrites, but eventually the sea-level rise drowned the seaward barrier of the Elk Point Basin, and the depositional site ceased to be one of evaporite deposition.

In the evaporites of the Paradox Basin, the upper parts of sedimentary cycles (figure 1.20) are composed of an upward sequence beginning with halite (with or without potash salts), passing through anhydrite and silty dolomites, and abruptly terminated by overlying units of black shale. These upper ("transgressive") parts of cycles were interpreted by Hite (1970) to have formed during periods of sea-level rise—a reasonable suggestion, because they terminate with black shales that are considered to be the deepest-water deposits of the entire depositional cycle. But, as we have seen, rises in sea level should result in increased residence times for brine in the basin, increasing basin salinity, and deposition of a sequence displaying an upward increase in the solubility of the evaporite minerals precipitated.

Kendall (1987a, 1987b) has suggested that the observed upward decrease in the solubility of phases precipitated in the upper parts of Paradox evaporite cycles may have been caused by two related effects that resulted from rapid filling of the basin topography with evaporites. Rapid aggradation in the Paradox Basin (caused by the very fast rates of evaporite deposition) reduced the hydraulic head existing between the open ocean and the depressed brine-level in the basin, and this offset or even reversed the tendency for the hydraulic head to be augmented by sea-level rise. Of greater importance, however, were the changes in the rate of brine seepage–reflux that occurred as a result of the sediment accumulation (figure 1.22). Initially, evaporite deposition would have been confined to the more central parts of the basin, where brine reflux would have been severely limited by the presence of underlying units of black shale. Brine residence times would have been long, and brines would have become concentrated to high salinities. With continued deposition the depositional area expanded onto the basin flanks where, particularly on the northeastern flank, the black shale aquitards are absent and the section gains permeable coarse clastics derived from the Uncompahgre Uplift. It is suggested that increasing amounts of brine reflux out of the basin through these porous marginal deposits would have progressively reduced the residence times of brine within the entire basin. As the salinity of brines in the basin dropped, the more soluble phases ceased being precipitated, and an upward sequence from halite to carbonates was formed.

The occurrence of desiccation cracking, sediment-casted halite molds, and bedding indicative of repeated flooding events in the carbonate units at the top of Paradox sedimentary cycles (figure 1.6b; Kendall 1987a, 1987b) indicates that these deposits were laid down in a still-desiccated basin rather than in an increasingly deepening water body as Hite (1970) supposed. The knife-sharp boundaries that exist between these playa (or inland sabkha) carbonates and the overlying deep-water black shales must record the sudden drownings of the Paradox Basin. They occurred as sea levels rose above the level of the sill or barrier that formerly isolated the basin from the open sea.

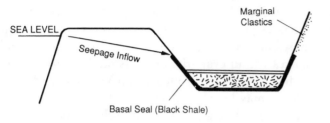

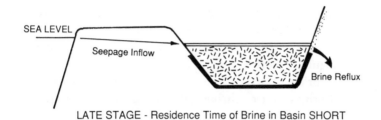

Figure 1.22
Alternative explanation of Paradox Basin depositional cycles. At times of evaporite deposition in the basin, flanking carbonates no longer form. During the initial stages (a), evaporite deposition is confined to the basin center where brines are confined by underlying impermeable black shales. Brine residence times are long, and brines reach high salinities. At later stages (b), the site of evaporite deposition has expanded onto the basin flanks, where brines reflux out of the basin through porous marginal facies. Brine residence times are progressively decreased, so that brines are unable to be concentrated as much as during earlier stages. The progressively less saline basin brines precipitate an upward succession of less saline mineralogy (halite to anhydrite to carbonate).

CONCLUSIONS

Stratigraphic studies of evaporite basins have been somewhat limited in the past because of the nature of the data and because of the application of depositional models that are inappropriate. Such models either do not appreciate the extreme degree of basin restriction required before evaporites can form, or fail to recognize that the large density contrasts that occur between brines capable of precipitating different mineral phases preclude coincident deposition of these different mineral phases in different parts of a deep, brine-filled basin. Lateral facies changes within evaporite deposits imply deposition in shallow water or brine flats—environments in which lateral variations in brine salinity may be maintained by bottom friction.

It is believed that almost all large evaporite deposits (especially those that contain halite) were formed within basins that were disconnected from the open sea. In such basins seawater influx would be confined to seepage through a permeable barrier (figure 1.22a). Rates of seepage influx are significantly lower than possible rates of evaporation even in the most arid of climates. Unless the seawater influx is augmented by other sources of water (rainfall, surface inflow, continental brine influx), the basin will become desiccated. Reflux of basinal brines over or through the barrier in such basins will be negligible. The absence or reduced quantity of the more soluble components in evaporite sequences (as compared with theoretical sequences produced by static seawater evaporation) requires that reflux must have occurred even in the most isolated of basins. Since this is unlikely to have occurred through the barrier, reflux must occur through the basin floor or laterally in marginal areas of the basin (figure 1.22).

The type of evaporite deposited in a basin at any given time will depend upon the salinity of the basinal brine, and this, in turn, will be governed by the time it resides in the basin. Brine residence times are determined by both the absolute and the relative rates of water influx and brine reflux into and out of the basin. Sea-level changes may affect the absolute and relative rates of water influx, and thus cause changes in the evaporite facies deposited in the basin, but their effects may be offset or superceded by the consequences of sediment aggradation in the basin.

The type of depositional model invoked to explain an evaporite unit has significant stratigraphic implications. Similarly, concepts about the stratigraphic packaging of evaporite and associated nonevaporite deposits have genetic significance. It is believed that for many evaporite basins, this interdependence of stratigraphic and genetic interpretations has gone unrecognized and that many inconsistencies remain. Application of stratigraphic concepts and appropriate depositional models is believed to be a fruitful line of approach in the interpretation of such basins.

REFERENCES

Anderson, R. Y., W. E. Dean, D. W. Kirkland, and M. I. Snider. 1972. Permian Castile varved evaporite sequence, west Texas and New Mexico. *Bull. Geol. Soc. Amer.* 83:59–86.

Asquith, G. B. 1979. *Subsurface Carbonate Depositional Models—A Concise Review.* Tulsa, Okla.: The Petroleum Corp.

Asquith, G. B., R. L. Parker, C. R. Gibson, and J. R. Root. 1978. Depositional history of the Ordovician Red River C and D zones, Big Muddy Creek Field, Roosevelt County, Montana. In *The Economic Geology of the Williston Basin: Montana, North Dakota, South Dakota, Saskatchewan, Manitoba,* pp. 71–76. D. Rehrig, chairperson, Williston Basin Symposium, Montana Geol. Soc. 24th Annual Conference. Billings, Mont.

Briggs, L. I., Jr. 1957. Quantitative aspects of evaporite deposition. *Michigan Academician* 42:115–123.

Briggs, L. I. 1958. Evaporite facies. *Jour. Sedim. Petrology* 28:46–56.

Busson, G. 1980. Large evaporite pans in a clastic environment: How they are hollowed out, how they are filled. *Bull. Cent. Rech. Explor.—Prod. Elf-Aquitaine* 4:557–588.

Carroll, W. K. 1978. Depositional and paragenetic controls on porosity development, upper Red River Formation, North Dakota. In *The Economic Geology of the Williston Basin: Montana, North Dakota, South Dakota, Saskatchewan, Manitoba,* pp. 79–94. D. Rehrig, chairperson, Williston Basin Symposium, Montana Geol. Soc. 24th Annual Conference. Billings, Mont.

Clement, J. H. 1985. Depositional sequences and characteristics of Ordovician Red River reservoirs, Pennel Field, Williston Basin, Montana. In P. O. Roehl and P. W. Choquette, eds., *Carbonate Petroleum Reservoirs,* pp. 74–84. New York: Springer-Verlag.

Corrigan, A. F. 1975. The evolution of a cratonic basin from carbonate to evaporitic deposition and the resulting stratigraphic and diagenetic changes, Upper Elk Point Subgroup, northeastern Alberta. Ph.D. diss., University of Calgary, Calgary.

Curtis, R., G. Evans, D. J. J. Kinsman, and D. J. Shearman. 1963. Association of dolomite and anhydrite in the recent sediments of the Persian Gulf. *Nature* 197:679–680.

Felber, B. E. 1964. Silurian reefs of southeastern Michigan. Ph.D. diss., Northwestern University, Evanston, Ill.

Grayston, L. D., D. F. Sherwin, and J. F. Allan. 1964. Middle Devonian. In R. G. McCrossan and R. P. Glaister, eds., *Geological History of Western Canada*, pp. 49–59. Calgary: Alberta Soc. Petrol. Geologists.

Hamilton, W. N. 1971. *Salt in East-Central Alberta. Bull. Research Council Alberta* vol. 29.

Hardie, L. A. 1978. Evaporites, rifting and the role of $CaCl_2$ hydrothermal brines. *Geol. Soc. Amer. Abstracts with Programs* 10(7):416.

Hardie, L. A. 1984. Evaporites: Marine or non-marine? *Amer. Jour. Sci.* 284:193–240.

Hardie, L. A., J. P. Smoot, and H. P. Eugster. 1978. Saline lakes and their deposits: A sedimentological approach. In A. Matter and M. E. Tucker, eds., *Modern and Ancient Lake Deposits*, pp. 7–41. Intern. Assoc. Sedimentologists Spec. Publ. no. 2.

Hite, R. J. 1970. Shelf carbonate sedimentation controlled by salinity in the Paradox Basin, Colorado and Utah. In J. L. Rau and L. F. Delwig, eds., *Third Symposium on Salt*, 1:48–66. Cleveland: Northern Ohio Geol. Soc.

Hite, R. J. 1985. The sulfate problem in marine evaporites. In B. C. Schreiber and H. L. Harner, eds., *Sixth Symposium on Salt*, 1:217–230. Alexandria, Va.: Salt Institute.

Hite, R. J., and D. H. Buckner. 1981. Stratigraphic correlations, facies concepts, and cyclicity in Pennsylvanian rocks of the Paradox Basin. In D. L. Wiegand, ed., *Geology of the Paradox Basin*, pp. 147–157. Rocky Mountain Assoc. Geologists 1981 Field Conference. Denver.

Hsü, K. J. 1972. Origin of saline giants: A critical review after the discovery of the Mediterranean evaporite. *Earth Science Reviews* 8:371–396.

Kahle, C. F. 1978. Patch-reef development and effects of repeated subaerial exposure in Silurian shelf carbonates, Maumee, Ohio. In *The North-Central Section: Geological Society of America, Field Excursions*, pp. 63–115. Boulder, Colo.

Kendall, A. C. 1976. *The Ordovician carbonate succession (Bighorn Group) of southeastern Saskatchewan.* Dept. Mineral Resources, Saskatchewan Geol. Survey, Rept. no. 180. Regina.

Kendall, A. C. 1984. Origin and geometry of Red River dolomite reservoirs, western Williston Basin: Discussion. *Bull. Amer. Assoc. Petroleum Geologists* 68:776–779.

Kendall, A. C. 1985. Depositional and diagenetic alteration of Yeoman (Lower Red River) carbonates from Harding Co., North Dakota. In M. W. Longman, K. W. Shanley, R. F. Lindsay, and D. E. Eby, eds., *Rocky Mountain Carbonate Reservoirs—A Core Workshop*, pp. 51–93. SEPM Core Workshop no. 7. Tulsa, Okla.

Kendall, A. C. 1987a. Depositional model for carbonate-evaporite cyclicity: Middle Pennsylvanian of Paradox Basin (abstract). *Bull. Amer. Assoc. Petroleum Geologists* 71:576.

Kendall, A. C. 1987b. Early salt dissolution: Pennsylvanian of Paradox basin, Colorado and Utah. *SEPM Annual Midyear Meeting Abstracts* 4:41.

King, R. H. 1947. Sedimentation in Permian Castile Sea. *Bull. Amer. Assoc. Petroleum Geologists* 31:470–477.

Kinsman, D. J. J. 1974. Evaporite deposits of continental margins. In A. L. Coogan, ed., *Fourth Symposium in Salt*, 1:255–259. Cleveland: Northern Ohio Geol. Soc.

Kinsman, D. J. J. 1976. Evaporites: Relative humidity control on primary mineral facies. *Jour. Sedim. Petrology* 46:273–279.

Klingspor, A. M. 1969. Middle Devonian Muskeg evaporites of western Canada. *Bull. Amer. Assoc. Petroleum Geologists* 53:927–948.

Kohm, J. A. and R. O. Louden. 1978. Ordovician Red River of eastern Montana and western North Dakota: Relationships between lithofacies and production. In *The Economic Geology of the Williston Basin: Montana, North Dakota, South Dakota, Saskatchewan, Manitoba*, pp. 99–117. D. Rehrig, chairperson, Williston Basin Symposium, Montana Geol. Soc. 24th Annual Conference. Billings, Mont.

Krumbein, W. C. and L. L. Sloss. 1963. *Stratigraphy and Sedimentation*. 2d. ed. San Francisco: Freeman.

Logan, B. W. 1987. *The MacLeod Evaporite Basin, Western Australia*. AAPG Memoir no. 44. Tulsa, Okla.

Longman, M. W., T. G. Fertel, and J. S. Glennie. 1983. Origin and geometry of Red River dolomite reservoirs, western Williston Basin. *Bull. Amer. Assoc. Petroleum Geologists* 67:744–771.

Lucia, F. J. 1972. Recognition of evaporite-carbonate shoreline sedimentation. In J. K. Rigby and W. K. Hamblin, eds., *Recognition of Ancient Sedimentary Environments*, pp. 160–191. Soc. Econ. Pal. Miner. Spec. Publ. no. 16. Tulsa, Okla.

McCamis, J. G. and L. S. Griffith. 1967. Middle Devonian facies relationships, Zama area, Alberta. *Bull. Canadian Petrol. Geol.* 15:434–467.

Machel, H.-G. 1986. Early lithification, dolomitization, and anhydritization of Upper Devonian Nisku buildups, subsurface of Alberta, Canada. In J. H. Schoeder and B. H. Purser, eds., *Reef Diagenesis*, pp. 336–356. Berlin: Springer-Verlag.

Maiklem, W. R. 1971. Evaporative drawdown—a mechanism for water-level lowering and diagenesis in the Elk Point Basin. *Bull. Canadian Petrol. Geol.* 19:487–501.

Matthews, R. D. 1977. Evaporite cycles and lithofacies in Lucas Formation, Detroit River Group, Devonian, Midland, Michigan. In J. H. Fisher, ed., *Reefs and Evaporites—Concepts and Depositional Models*, pp. 73–91. Amer. Assoc. Petrol. Geologists Studies in Geology no. 5. Tulsa, Okla.

Mesolella, K. J., J. D. Robinson, L. M. McCormick, and A. R. Ormiston. 1974. Cyclic deposition of Silurian carbonates and evaporites in Michigan Basin. *Bull. Amer. Assoc. Petroleum Geologists* 58:34–62.

Middleton, G. V. 1973. Johannes Walther's Law of the correlation of facies. *Bull. Geol. Soc. Amer.* 84:979–988.

Neese, D. G. 1985. Depositional environment and diagenesis of the

Red River Formation, "C" interval, Divide County, North Dakota and Sheridan County, Montana. In M. W. Longman, K. W. Shanley, R. F. Lindsay, and D. E. Eby, eds., *Rocky Mountain Carbonate Reservoirs—A Core Workshop*, pp. 95–124. SEPM Core Workshop no. 7.

Neumann, A. C. and D. A. McGill. 1962. Circulation of the Red Sea in early summer. *Deep Sea Research* 8:223–235.

Ochsenius, K. 1877. *Die Bildung der Steinsalzlager und ihrer Mutterlaugensalze unter spezieller Berucksichtigung der Floze von Douglashall in der Engln'schen Mulde.* Halle/Salle: C.E.M. Pfeffer Verlag.

Peterson, J. A. and R. J. Hite. 1969. Pennsylvanian evaporite-carbonate cycles and their relationship to petroleum occurrence. *Bull Amer. Assoc. Petroleum Geologists* 53:884–908.

Roehl, P. O. 1967. Stony Mountain (Ordovician) and Interlake (Silurian) facies analogs of Recent low-energy marine and subaerial carbonates, Bahamas. *Bull. Amer. Assoc. Petroleum Geologists* 51:1979–2032.

Ruzyla, K. and G. M. Friedman. 1985. Factors controlling porosity in dolomite reservoirs of the Ordovician Red River Formation, Cabin Creek Field, Montana. In P. O. Roehl and P. W. Choquette, eds., *Carbonate Petroleum Reservoirs*, pp. 41–58. New York: Springer-Verlag.

Sarg, J. F. 1982. Off-reef Salina deposition (Silurian), southern Michigan Basin, implications for reef genesis. In C. R. Handford, R. G. Loucks, and G. R. Davies, eds., *Depositional and Diagenetic Spectra of Evaporites— A Core Workshop*, pp. 354–372. SEPM Core Workshop no. 3. Tulsa, Okla.

Schreiber, B. C. and K. J. Hsü. 1980. Evaporites. In G. D. Hobson, ed., *Developments in Petroleum Geology*, 2:87–138. London: Applied Science Publishers.

Schmalz, R. F. 1969. Deepwater evaporite deposition—a genetic model. *Bull. Amer. Assoc. Petroleum Geologists* 53:798–823.

Scruton, P. C. 1953. Deposition of evaporites. *Bull. Amer. Assoc. Petroleum Geologists* 37:2498–2512.

Shaver, R. H. 1977. Silurian reef geometry—new dimension to explore. *Jour. Sedim. Petrology* 47:1409–1424.

Shaw, A. B. 1977. A review of some aspects of evaporite deposition. *Mountain Geol.* 14:1–16.

Shearman, D. J. 1966. Origin of marine evaporites by diagenesis. *Trans. Inst. Mining Metallurgy* 75:B208–B215.

Sloss, L. L. 1953. The significance of evaporites. *Jour. Sedim. Petrology* 23:143–161.

Smoot, J. P. 1983. Depositional subenvironments in an arid closed basin: The Wilkins Peak Member of the Green River Formation (Eocene), Wyoming, U.S.A. *Sedimentology* 30:801–827.

Smosna, R. A., D. G. Patchen, S. M. Warshauer, and W. J. Perry, Jr. 1977. Relationships between depositional environments, Tonoloway Limestone, and distribution of evaporites in the Salina Formation, West Virginia. In J. H. Fisher, ed., *Reefs and Evaporites—Concepts and Depositional Models,* pp. 125–143. Amer. Assoc. Petrol. Geologists Studies in Geology no. 5. Tulsa, Okla.

Taylor, J. C. M. 1980. Origin of the Werraanhydrit in the U.K. southern North Sea—a reappraisal. In H. Füchtbauer and T. Peryt, eds., *The Zechstein Basin with Emphasis on Carbonate Sequences,* pp. 91–113. Vol. 9 of *Contr. Sedimentology.* Stuttgart: E. Schweizerbart'sche Verlagsbuchhanlung.

Taylor, S. R. 1983. A stable isotope study of the Mercia Mudstones (Keuper Marl) and associated sulphate horizons in the English Midlands. *Sedimentology* 30:11–31.

Thomas, G. E. 1968. *Notes on Textural and Reservoir Variations of Ordovician Micro-dolomites, Lake Alma–Beaubier Producing Area, Southern Saskatchewan.* Saskatchewan Dept. Mineral Resources and Saskatchewan Geol. Soc. Core Seminar on the Pre-Mississippian rocks of Saskatchewan, Oct. 3–4, 1968. Regina.

Wardlaw, N. C. 1968. Carnallite-sylvite relationships in the Middle Devonian Prairie Evaporite Formation, Saskatchewan. *Bull. Geol. Soc. America* 79:1273–1294.

2

Peritidal Evaporites and Their Sedimentary Assemblages

CHRISTOPHER G. ST. C. KENDALL
JOHN K. WARREN

In 1985 Warren and Kendall described the peritidal evaporite carbonate assemblage as having two depositional settings: subaerial salt
flats, or sabkhas, and subaqueous coastal salinas. They showed that
the major distinction between the sediments of these two evaporite
settings is that sabkha sediments tend to accumulate on the topographically higher areas, whereas salinas accumulate sediment in topographic lows. Both evaporative sequences can cap tidal flat/la-
goonal carbonates, but the anhydrites of the sabkhas tend to be nodular
and enclosed within a carbonate matrix that is derived from the lower
part of the tidal flat cycle and is carried up into the supratidal part
of the cycle by storm washover. In contrast, the evaporites of salinas

Christopher G. St. C. Kendall is a Professor in the Departments of Geology and of
Marine Sciences at the University of South Carolina, Columbia.

John K. Warren is an Assistant Professor in the Department of Geological Sciences,
University of Texas at Austin.

tend to accumulate in standing bodies of water and exhibit a laminar fabric which contains little or none of the carbonate matrix so common to sabkha sediments. In addition, the evaporites associated with the salinas tend to be thicker, and quite commonly they include not only calcium sulfate, but halite. Significantly, both types of sedimentary settings and sedimentary suites are associated with hydrocarbons, and act as seals. The salina evaporites are probably the more important of these trapping mechanisms, because they form more continuous and complete barriers to migrating fluids. However, as can be seen in the descriptions of the evaporites from Abu Dhabi that follow, sabkha evaporites can be locally extremely continuous, and so in some cases probably form just as good a seal over hydrocarbons as do playas.

Large portions of the text of the article that follows draw upon a paper on Abu Dhabi by Butler et al. 1982; a paper on South Australian salinas by Warren (1988); and a paper by Ward et al. (1986) on the Permian Basin. It focuses on two Recent examples, the sabkhas of Abu Dhabi, and the salinas of coastal Australia, and it shows how the same sedimentary associations can be seen in the upper Permian Carlsbad Group from the margin of the Delaware Basin of Texas, one of the many ancient examples where salina and sabkha sediments occur. It should be noted that this comparison between the modern and the ancient suffers from lack of compatibility in terms of scale or areal extent. The ancient evaporitic sabkhas and salinas encompassed areas two to three times larger than their modern equivalents.

Recent examples of sabkhas are quite numerous. In addition to occurring around the present Arabian or Persian Gulf, they are found along the margins of the Red Sea and along the margins of the Mediterranean, including the coast of Libya, Tunisia, portions of Egypt, and sections of the Eastern Mediterranean, including Turkey, Greece, and Israel. They also form around Shark Bay in western Australia and around the margins of South Australia. They are associated with the coastal portions of Baja California and have been found associated with clastics in the Laguna Madré in Texas.

Salinas also are common in the Recent. In addition to Lake MacLeod of western Australia, better-known examples include the Coorong and Lake Marion of South Australia, and some are found along of the shores of Tunisia and the coastal fringe of the Sinai.

Ancient examples of sabkha and salina sedimentary association are also numerous. For instance, not only do they occur in the Carlsbad Group from the Delaware and Midland Basins, but they also occur in the Devonian Elk Point Basin of western Canada; the Pennsylvanian of the Williston and the Paradox Basin; the Jurassic Smackover, where the Buckner evaporites act as an updip seal along the Gulf Coast of the United States; the Jurassic Kimmeridgian Arab Formation of Saudi Arabia, where the Hith anhydrite seals the carbonates; the lower Cretaceous Ferry Lake anhydrite, a widespread subaqueous deposit which acts as a seal in portions of the Gulf Coast of the United States; and where the shallow-water Oligocene Asmari shelf limestone is associated with the overlying Fars evaporites.

ABU DHABI

The Abu Dhabi carbonate evaporite complex is associated with an assemblage of carbonate sediments and clastics that are common to other parts of the Arabian Gulf. This enclosed sea is an area of high net evaporation. Its major sources of lower-salinity water are the Straits of Hormuz and the Shatt Al Arab Delta. This arm of the sea is an asymmetric basin that is deeper to the east, along a belt that lies adjacent to and parallels the Iranian shoreline, shoaling abruptly towards Iran but more gently towards the Arabian Shield.

To the west a gently dipping shelf or ramp occurs to just seaward of the complex of islands and carbonate and evaporitic sediments that mark the shoreline of Saudi Arabia, the Emirates, Qatar, and Kuwait. In addition to carbonates and evaporites, the Arabian Gulf is fed clastics from the north by the Shatt Al Arab where the confluence of the Tigris, Euphrates, and Karoon rivers form an enormous wedge of deltaic sediments which are spreading into this predominantly carbonate province. Similarly, along the eastern shore of the Arabian or Persian Gulf, seasonal streams from the Zagros Mountains are feeding alluvial fans into the predominantly carbonate province of the Iranian shoreline.

The sedimentary facies that accumulate in the Emirates are tied to the elevated salinities in the waters that adjoin them (Kinsman 1964). Here, over a basement of Tertiary and Quaternary sediments, a thin veneer of Recent sediments that have built locally to sea level form coastal terraces and prograding supratidal coastal salt flat sequences, known as sabkhas. Locally, the Tertiary crops out and forms a series of low hills against which Pleistocene sediments accumulated as cross-bedded, wind-blown carbonates, deposited during the last major glacial eustatic low stand in the Arabian Gulf. These latter sediments are known locally as "Miliolites," so named from the abundance of the foraminifera which occur in them. Locally, these cross-bedded eolianites form hills and line the inner margins of the present-day salt flats, or sabkhas. They are sometimes truncated so that their surface lies close to the present water table, and their festoon cross-beds are exposed at these wind-deflated surfaces. They underlie much of the Holocene carbonate/evaporite complex.

The Holocene Sediments of Abu Dhabi

Offshore from Abu Dhabi and in the deeper tidal channels, cutting through the barrier island lagoons, skeletal debris accumulates (figure 2.1). This skeletal debris has higher and higher percentages of lime mud as the water deepens offshore (Houbolt 1957). In contrast, close to the shallow shoreline of the Emirates, the lime mud content of the sediments becomes less, and in water of depths between 20 and 30 ft is composed of nearly pure pelecypod sands. To the west of Abu Dhabi, the shallow-water carbonates just landward of this molluscan debris consist of coral reefs and coralgal sands, whereas to the east, oolite shoals accumulate. These form on the tidal deltas developed by the tidal channels draining the lagoons in the vicinity of Abu Dhabi Island. The channels debouch from between the barrier islands to form these intertidal deltas (Figure 2.1). To the lee of the reefs and of the oolites shoals, grapestones accumulate. They are particularly common in the shallower waters of the more open Khor Al Bazam Lagoon, whereas pelleted lime muds are accumulating in the protected barrier island lagoons to the east. Lining the inner shores of the lagoons of Abu Dhabi are algal mats and mangrove swamps, and prograding behind the shoreline are supratidal salt flats, or sabkhas, in which the evaporite minerals are accumulating (figure 2.1).

Coral Reefs. Coralgal shoals and coral reefs form along the northern and windward side of the barrier rimming the seaward margin of the

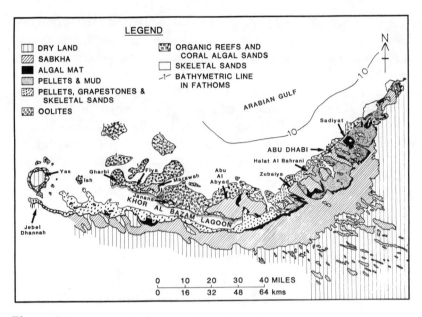

Figure 2.1
Map of the coastal carbonate facies of Abu Dhabi, the United Arab
Emirates.

Khor Al Bazam Lagoon; to the north of the peninsula of Zubaiya
that protrudes out from between the island of Abu Al Abyad and
Halat Al Barani; and in front of the center of the barrier islands of
eastern Abu Dhabi. Reefs, particularly of *Acropora,* also line some
of the tidal channels draining the barrier island lagoons, as for instance between Halat Al Barani and Abu Dhabi Island, and between
Abu Dhabi and Sadiyat. The reefs to the north of the Zubaiya Peninsula accumulate along the seaward edge of a coastal terrace which
is itself made up of bioclastic sands derived from this and earlier
Pleistocene reefs. These sands spill across the coastal terrace into the
lagoon that fronts the Zubaiya Peninsula. This coastal terrace has
karstic solution depressions, or "blue holes" in it, which probably
formed during Pleistocene sea-level lows and which are now being
filled by Holocene sediment. Similar reefs and associations can be
seen seaward of Abu Al Abyad, Janana, and Marawah (Kendall and
Skipwith 1969; Purser and Evans 1973). Sediment from these reefs

is carried by storms and tidal currents across the barrier to spill over into the Khor Al Bazam in a series of shallow-water flood deltas fed by axial flood channels.

Oolite Shoals. In the eastern portion of the Abu Dhabi coast, barrier islands, which protect tidal lagoons, are cut by tidal channels that form ebb deltas between and to the north of the barrier islands. These ebb deltas are the site of the accumulation of oolite shoals. These oolitic grains are finer along the seaward margin of the shoals and become coarser southward in a landward direction. The better-studied oolite areas occur on either side of Abu Dhabi Island between Halat Al Barani and Zubaiya (Kinsman 1964), and between Abu Dhabi and the island of Sadiyat (Purser and Evans 1973). The oolite shoals are very similar to those of the Bahamas and, as can be seen in aerial photographs taken over a period of fifteen years, maintain their general configuration between major storms. The oolites are very sharply confined to the shoals, so that within half a mile offshore from an oolite shoal the ooid content of the sediments is sharply depleted (Purser and Evans 1973). Large megaripples form on the shoals and have a wave length of between 20 and 40 m. The height of the individual megaripple varies from about 2 to 3 m. The oolites from the shoals are being carried by waves and by wind onto the barrier islands and are forming eolian dunes.

Grapestone and Pellet Shoals. Grapestones and pellets accumulate in the quieter waters behind the barrier islands and on the coastal terraces on the south side of the Khor Al Bazam (Kendall and Skipwith 1969; figure 2.1). Most of these grapestones are formed in intertidal water, where largely aragonitic cementation is binding the bioclastic and pelleted sand of these coastal terraces to form crusts, which are then broken into fragmented aggregates by storms induced by the strong northern Shamal winds of this area. Locally within this grapestone area, foraminifera tests and other bioclastic remains are being infested by blue-green algae. These algae are altering the tests through a selective process of micritization induced by solution and reprecipitation within the mucilageous envelopes secreted by the algae (Kendall and Skipwith 1969). It is thought that diurnal photosynthesis is driving this process. As a result, the original magnesium calcite foraminifera tests are being converted to micritic aragonite.

Lime Mud. The protected lagoons of the area around Abu Dhabi are the site of the accumulation of lime mud and pelleted carbonates (Kinsman 1964; Kendall and Skipwith 1969; and Purser and Evans 1973; figure 2.1). Here, in some cases, cementation is taking place within the pellets so that they form small hardened grains.

Mangrove Swamps. Locally, at the landward side of these pelleted lime mud accumulations, mangroves grow at the mouths of the tidal creeks draining portions of the coast, particularly in the vicinity of Abu Al Abyad and immediately to the east of the Zubaiya Peninsula. These mangrove swamps protect algal flats in their lee. These algal mats form over the higher portions of the tidal flat. In some cases these creek areas are enclosed by eolian dunes, which are sometimes formed of gypsum. The mangrove species associated with these swamps is largely the black mangrove, *Avicennia marina*. Locally, along the tidal creeks draining the mangrove swamps, crabs are extensively re-working sediment to cause the high level of bioturbation seen locally in the lime muds of the tidal flats.

Algal Flats. Locally, in more restricted and protected areas particularly along the eastern portion of the Khor Al Bazam, algal flats are accreting and prograding along the upper part of the tidal flat (Kendall and Skipwith 1968; figure 2.1). These algal flats stretch for some 48 km along the coast and vary between 1 and 4 km in width. The present algal surface of these flats caps an algal peat whose preservation can probably be linked to the pickling effects of the high salinities of the area and the reducing environment over which the algae are accumulating. The algal flats growing out into the Khor Al Bazam are not the first Holocene sediments to accumulate here. Instead, in its early stages, the Khor Al Bazam appears to have been a more open lagoon and subject to higher wave energies. The result is that this body of water was lined by beach ridge cheniers formed of cerithid gastropod debris. Subsequently, the Khor Al Bazam filled and became more restricted. At the same time, lime muds started to accumulate in the lagoon, while in the intertidal zone algal mats began to grow. These algal stromatolites have since prograded 5 to 6 km seaward.

As Kendall and Skipwith (1968) describe them, the algal peats show a variety of sedimentary structures. For instance, along the seaward edge of the algal flats, the surface is more pustular and looks like a flat surface of black cinders. This seaward margin of the algal flat is an area where wave and tidal scour occurs, producing washout features. If the cindery zone of pustular algae is traced landward, the surface becomes smoother but is locally broken into a series of desiccation polygons which are coated by leatherlike dark green layers of algae. Where tidal creeks drain the algal flat, these meandering bodies of water are being obstructed by further growth of algal stromatolites. Here these form layers of peat whose pink surface looks like huge lily pads several meters across, which themselves may be desiccated to form large polygons 50 cm to 1 m in diameter. These same polygons can be identified in trenches cut in the supratidal salt flats, now well landward of the present algal flat. Landward of the polygonal zone, the green to black algal flat surface is crenulated, and on its landward margin this surface covers a mush of gypsum crystals. In the supratidal, the algal mat becomes a horizontal layer, since the algae are no longer growing vigorously and the now pink surface no longer expands and becomes crenulated.

In addition to the accumulation of algal peat, the algal flats are the site of the entrapped lime mud which is probably being both precipitated within the algal layers and washed onto them, particularly along the landward edge of the algal flat within the lower portion of the crinkled zone. Landward, gypsum is being precipitated in the form of the mush mentioned earlier, which ranges in thickness from a few centimeters to as much as 10 cm. The lime muds overlying and caught up in this mush show evidence of dolomitization.

Sabkha. The flat algal surface in the upper algal flat marks the landward edge of the sabkha. When one digs a trench in this sabkha, starting at the surface, the sedimentary sequence begins with a halite layer, which passes down in sequence into storm washover carbonate, and a gypsum mush or anhydrite that has replaced the gypsum mush. Below, an algal peat usually occurs. This peat frequently contains large gypsum crystals, 5 to 10 cm across. Beneath are lower tidal-flat muds and sands. This vertical sabkha sequence, which stretches back across the sabkha to the beach ridges, also reflects the lateral sedimentary sequence from land to sea. The major difference landward is that the sediments overlying the gypsum mush or anhydrite are covered by an accumulation of thicker and thicker storm washover carbonates and clastics derived from the tidal flat. This washover sediment probably includes gypsum crystals which were eroded from the upper algal flats and probably form the nucleus to the cumulus cloud–like nodules of anhydrite found just within and beneath the sabkha surface. The halite surface of the upper algal flat and the adjacent sabkha is disrupted into a series of expansion polygons.

Locally, within the sabkha sedimentary sequence just below the algal peat at the top of what was previously the intertidal zone, a carbonate crust occurs that is cemented by calcium carbonate and sometimes also by gypsum. This forms a seal between the underlying more marine sediments with their marine waters, and the overlying supratidal salt flats with their waters with elevated salinity. On the most landward side of the sabkha, particularly within the sabkhas close to Abu Dhabi, the anhydrite is being replaced by gypsum. This is in response to the influx of the fresher continental waters from the Arabian interior which enter the coastal system and freshen it, thus causing alteration of what was once anhydrite back to gypsum again.

The Abu Dhabi Sabkha

In the vicinity of Abu Dhabi Island (figures 2.2 and 2.3; Butler et al. 1982), the supratidal salt flat is around 13 km wide and covers approximately 100 sq km. The Quaternary sediments of the coastal flats are around 12 m in thickness. They overlie Miocene sediments and consist of about 9 m of Pleistocene eolianites, with up to 2.7 m of Recent lagoonal subtidal and intertidal carbonates (figure 2.4). The overlying supratidal sediments are about a meter in thickness (figures 2.4 and 2.5).

The sediments forming these salt flats have at least three sources: storm tide transport from the tidal flats and lagoons; wind transport from the interior of the Arabian Peninsula; and transport by occasional rain floods from outwash fans (Butler 1969; Patterson and

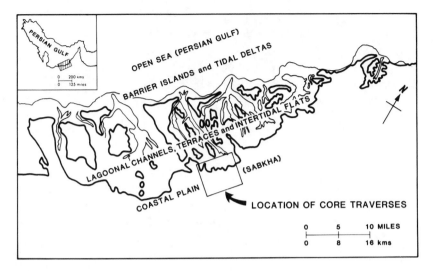

Figure 2.2
Location of sabkha to the south of Abu Dhabi Island (after Butler et al. 1982). These traverses are detailed on figure 2.3.

Kinsman 1981). The evaporite minerals that occur within these salt flats (Curtis et al. 1963; Butler 1969, 1970; and Butler et al. 1982) are the subject of this part of the paper. Based on subsurface sediment type, the coastal salt flats can be subdivided laterally into distinct outer and inner environments. The outer environment, which is 10 km wide and lies inland from the lagoons, is underlain by shallow-water and tidal-flat carbonate sediments. The inner environment, 3 km wide at the landward margin of the salt flat, is underlain by Pleistocene eolianites and locally variegated detrital clays and silts of Tertiary age.

At an elevation of 0.6–1.0 m above the upper intertidal zone, the surface of the coastal salt flats has only minimal changes in relief over large distances. Its seaward slope is between 1/1,000 and 1/3,000 (figure 2.6; Butler 1969; Patterson and Kinsman 1981). Exceptions occur along the axes of former lagoonal tidal channels, which contain the youngest carbonate sediments of this coastal sabkha. Locally, the surface is interrupted by rectangular monadnocks of cemented Pleistocene eolianite which pierce the Holocene sediments. In the vicinity of the monadnocks, the supratidal sediments are essentially detrital

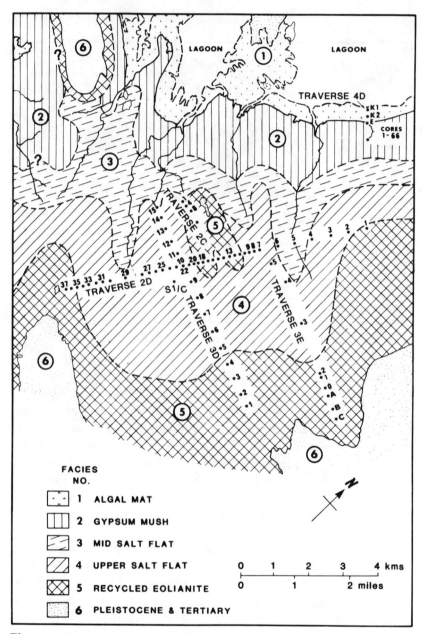

Figure 2.3
Map of the distribution of facies belts of the Abu Dhabi sabkha and the core locations discussed in text (Butler et al. 1982). See figure 2.2 for location of this map.

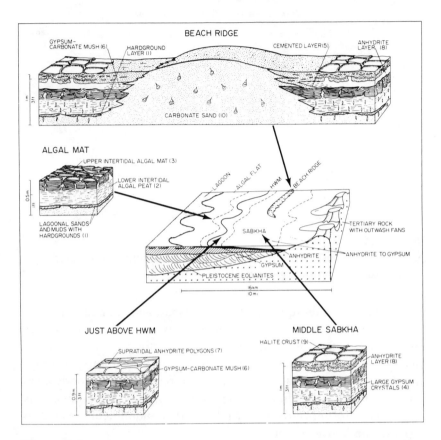

Figure 2.4
Schematic block diagrams showing sediment and evaporite distribution in Abu Dhabi sabkha, Arabian Gulf. All peripheral diagrams are keyed to central block of sabkha. HWM = high-water mark. (1) Lagoonal carbonate sands and/or muds with carbonate hardgrounds. (2) Vaguely laminated lower tidal-flat carbonate-rich algal peat. (3) Upper tidal-flat algal mat formed into polygons. (4) Large gypsum crystals (many lenticular). (5) Cemented carbonate layer. (6) High tidal-flat to supratidal mush of gypsum and carbonate. (7) Supratidal anhydrite polygons with windblown carbonate and quartz. (8) Anhydrite layer replacing gypsum mush and forming diapiric structures. (9) Halite crust, formed into compressional polygons. (10) Deflated beach ridge of cerithid coquina and carbonate sand. (Warren and Kendall 1985.)

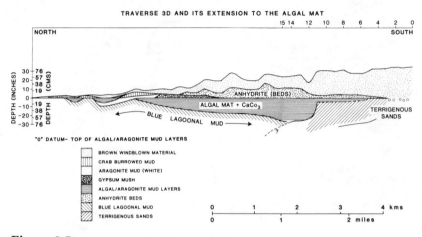

Figure 2.5
North-south cross section of supratidal sediments along Traverse 3-D; datum is top of algal peat layer (Butler et al. 1982).

Pleistocene eolianites (quartz, calcite, feldspar, and abundant miliolid foraminifera) eroded from the monadnocks and redistributed over the salt-flat surface by winds and by marine floodwaters. The combined effect on the surface sediments of wind deflation (Patterson and Kinsman 1981) and, to a lesser extent, reworking by floodwaters is responsible for the almost planar surface of the coastal flats. This surface lies just above the water table and is undoubtedly related to it (Patterson and Kinsman 1981).

Hydrology. The climate along the coast of Abu Dhabi favors the precipitation of evaporites. The sea evaporates at about 124 cm per year (Privett 1959), while salt-flat groundwater evaporates at about 6 cm per year (Patterson and Kinsman 1981). Rainfall is low, averaging 3.8 cm yearly. Temperature at the salt-flat surface can vary diurnally as widely as from 10° C to 53° C, as measured in May 1967 (Butler 1969).

As a consequence of the climatic extremes, interstitial fluids over most of the coastal flats approach saturation or are saturated with respect to sodium chloride. Butler (1969) and Patterson and Kinsman (1981) found surface chloride values of 4.0 to 4.7 moles per kg over much of the Abu Dhabi sabkha. Water is added to the sediments at

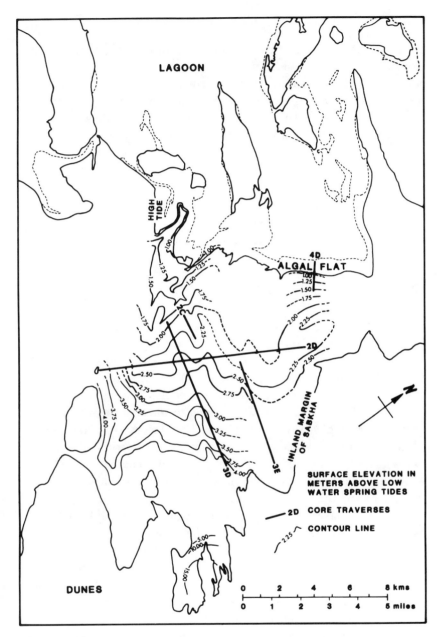

LAGOON

HIGH
TIDE

4D
ALGAL FLAT

INLAND MARGIN
OF SABKHA

N

SURFACE ELEVATION IN
METERS ABOVE LOW
WATER SPRING TIDES

2D   CORE TRAVERSES

2.25   CONTOUR LINE

DUNES

| 0 | 2 | 4 | 6 | 8 kms |

| 0 | 1 | 2 | 3 | 4 | 5 miles |

Figure 2.6
Map showing relief on the surface of the Abu Dhabi sabkha (after
Patterson and Kinsman 1981). The location of the core traverses are
also shown (Butler et al. 1982).

a rate that is fast enough to prevent precipitation of bittern minerals from the interstitial fluids. Water, originating from the lagoon and continental groundwater, produces two gross diagenetic environments (Butler 1969): a marine environment, confined to the seaward portion of the coastal flats and including approximately the upper 0.6 m to 1.2 m of the sabkha sediments; and a continental environment, confined to the landward portion of the coastal flats and generally restricted to the subsurface. This overlay of the marine environment on the continental environment is dynamic (Butler 1969; Patterson and Kinsman 1981); in the past when the coastal flats were narrower, the marine waters played a more important role, but with progradation of the coast, continental waters have become increasingly important.

Lagoon waters are added at the sediment surface by a process which Butler (1969) termed "flood recharge." During times of strong onshore winds or storms (the Shamals), sheets of lagoon waters (>2 cm deep) are driven across the salt-flat surface. Occasional torrential rains add to the volume of the floodwaters (Patterson and Kinsman 1981). There is a direct correlation between the frequency and area of flooding, and the distribution and nature of the evaporite minerals recognized in surface cores and shallow trenches. Hsü and Schneider (1973) ascribe the origin of the brines in this nearshore area to evaporative pumping, a process that pulls the brines from the lagoon through the tidal flat sediments. In contrast, Patterson and Kinsman (1981), like Butler et al. (1982), believe flood recharge coupled with water under a continental influence from a landward direction determine the nature of these brines.

The surface layers of these coastal salt flats are variably cemented by halite. In consequence, floodwaters become rapidly saturated with sodium chloride. The saturated floodwaters both sink into the sediments and drain back into the lagoons. As a result, the ionic balance in brines and subsequent diagenesis is greatly altered by the flood recharge process, especially in areas of former lagoon tidal channels.

Continental groundwaters infiltrate salt flat sediments from the landward margin of the coastal plain and from the intra-salt-flat monadnocks. Locally, these waters move vertically from the underlying uncemented Pleistocene sand aquifer into the overlying carbonate sediments. More commonly, however, the upward movement of continental groundwaters into the salt flat is restricted by the cemented layer of carbonate sand mentioned earlier. This layer is 8 cm or so thick here and is cemented by gypsum and carbonate. As in other locales, it is developed at the top of the subsurface lagoonal sediments throughout the coastal flats at a depth of approximately 2.1 m. It is probably a continuation of a similar layer found at the surface of the lower tidal flat of the inner Khor Al Bazam Lagoon (Kendall and Skipwith 1969). These authors believe this crust is forming through the precipitation of carbonate from the evaporating tidal waters.

Evaporites. Evaporites within sediments of the Abu Dhabi coastal flats occur in a variety of modes. The anhydrite found here is finely crystalline, has the consistency of cream cheese, is thixotropic, and occurs in loosely consolidated sediments. Anhydrite pseudomorphs after discoidal gypsum, rare in the salt flat, are restricted to crystals exposed at the sediment surface, where Butler (1969) reports the anhydrite as having the characteristic discoidal shape of gypsum crystals. In thin section, these can be seen to be a late infill of solution casts after gypsum. Commonly such pseudomorphic outlines are subsequently obliterated through the effects of the compaction of the surrounding carbonate matrix.

Butler (1969) concluded that the anhydrite of the Abu Dhabi coastal flats is predominantly a secondary mineral of polygenetic origin. He explains the dehydration of gypsum to anhydrite by a combination of three mechanisms: direct dehydration to anhydrite; a solution-precipitation process; and stepwise dehydration via intermediary bassanite. Moiola and Glover (1965) recorded similar mechanisms of anhydrite formation from Clayton Playa, Nevada. Stratigraphic and textural relationships between gypsum and anhydrite and the chemistry of interstitial brines are consistent with the several origins for the anhydrite. The bulk of the sabkha anhydrite replaces metastable gypsum; as the parent gypsum is polygenetic (formed as a primary precipitate, as a by-product of dolomitization of preexisting aragonite, and as a by-product of aragonite dissolution) the daughter anhydrite is also polygenetic. Some anhydrite is formed by dolomitization and aragonite dissolution without the formation of intermediary gypsum. The SO_4^{2-} ion involved in the formation of gypsum (and thus anhydrite) comes from brines derived from seawater and from continental groundwaters.

In the coastal salt flats, anhydrite nodules range from 0.05 mm to 30 cm in diameter, with the larger nodules being aggregates of smaller nodules. Nodule size is probably related in part to the size of the initial gypsum crystals, small nodules having formed from small gypsum crystals and large ones from larger gypsum crystals. Small nodules contain pelleted aragonite muds, foraminifera tests, and frosted angular quartz grains. These contaminants are conspicuously absent from larger nodules. If they were originally present, they have either been forced out of the nodule during growth or were actually replaced by anhydrite. Nodule shape is also variable: some nodules are ellipsoidal with their longest dimension oriented either horizontally or vertically, whereas others are spherical, pelletlike, doughnut-shaped, contorted, or have mammilated surfaces. Nodules are formed of anhydrite laths varying in length within any nodule from 2 microns to 100 microns. The arrangement of laths is typically felted, but parallel or radial orientations are not uncommon.

Solitary nodules are rare; more frequently, nodules are aggregated to form anhydrite layers that are 3 mm to 2.4 m thick. All layers of anhydrite nodules show the characteristic chicken-wire texture that is recognized in ancient nodular anhydrites. Nodules are generally separated by films of aragonite mud, Holocene calcian dolomite/aragonite mud, or detrital eolianite (quartz, calcite, and feldspar). Subsequent local replacement of aragonite mud films by anhydrite laths forms a pseudo-chicken-wire texture. In such a texture the anhydrite laths are oriented with long axes tangential to nodule boundaries.

Distribution of Evaporites. Understanding the occurrence and diagenesis of the upper intertidal and supratidal sediments of the Abu Dhabi sabkha is aided by the recognition of the distinctive facies belts (figure 2.3). The seaward portion of the coastal flats can be subdivided on the basis of flood frequency (Butler 1969) into five belts of distinctive facies which approximately parallel the intertidal area: algal mat facies (upper intertidal); gypsum mush facies (lower supratidal); mid-salt-flat facies (mid-supratidal); upper salt-flat facies (upper supratidal); and recycled eolianite facies (upper supratidal). Anomalies in this generalized pattern occur over the ancient tidal channels, where localized flooding commonly penetrates deeper into the central part of the salt flat.

Algal Mat Facies (Upper Intertidal). The algal mat facies forms a belt that is up to 0.8 km wide (figure 2.3) and is characterized by laminated mats of blue-green algae interlayered with aragonite mud. The algal mats are underlain by lagoonal and lower intertidal aragonite muds (see figures 2.4, 2.5, and many of the core photographs). Diagenetic features include:

1. Precipitation of aragonite mud.
2. Precipitation of gypsum crystals that are up to 2 mm long, lens-shaped, and flattened parallel to the c-axis (Cores K1, K2, and 1 of Traverse 4-D in figure 2.7; figure 2.4).
3. Localized cementation of surface sediments by aragonite, magnesite, and dolomite to form crusts.

Gypsum Mush Facies (Lower Supratidal). The gypsum mush facies is characterized by a gypsum mush which is emplaced above and below a crinkled algal mat and also beneath a flat algal mat (figures 2.4, 2.5 and 2.8). Cores 6, 16, 21, 26 and 36 of Traverse 4-D (figures 2.7 and 2.9) were taken in the gypsum mush facies belt, which is up to 2.5 km wide (figure 2.3). The gypsum mush, which reaches 30 cm in thickness, can be subdivided into two horizons on the basis of the associated sediment. The units are a lower horizon of isolated gypsum crystals that are dispersed through aragonite muds containing algal filaments, and an upper horizon of a mush of gypsum crystals (≤ 2 mm diameter) growing in Pleistocene storm-washover sediments and early diagenetic aragonite. Diagenetic features of the gypsum mush include:

1. Interstitial precipitation of aragonite.
2. Precipitated gypsum crystals up to 1.5 cm long, and solution of gypsum.
3. Local surficial alteration of gypsum to anhydrite via bassinite.
4. Local re-solution of anhydrite.
5. Camel and human footprints in this facies are commonly underlain by surface casts of anhydrite, possibly where local gypsum mush has inverted.

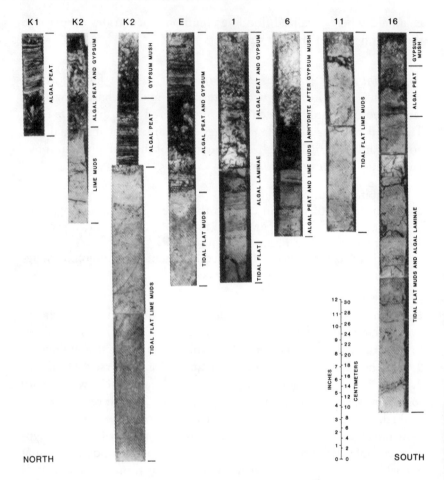

TRAVERSE 4D

Figure 2.7
Photographs of Cores K_1, K_2, E, 1, 6, 11, and 16 from Traverse 4-D. (See figure 2.3 for locations of the cores.) Cores are tied to the sediment surface (Butler et al. 1982).

Mid-salt-flat Facies (Mid-supratidal). The surface sediments of mid-salt-flat facies, which form a belt up to 1.6 km wide (figure 2.3), are characterized by a flat-to-crumpled layer of halite forming on the surface (figures 2.4 and 2.10). Diagenetic processes in the facies include:

1. Alteration of pockets of gypsum crystals in surface detrital sands to anhydrite nodules and layers (Cores 2/2 of Traverse 2-D, figure 2.11).
2. Partial replacement of preexisting gypsum mush by anhydrite (Core S_1/C of Traverse 3-D, figure 2.12).
3. Precipitation of gypsum as displacive and pore-filling crystals below the layer of gypsum mush (Cores 5, 7, and 21 of Traverse 2-D, figures 2.11, 2.13, and 2.14).
4. Alteration of the pore-filling gypsum to anhydrite.
5. Extensive local dolomitization ($\approx 80\%$) of preexisting aragonitic muds to form protodolomite.
6. Solution of gypsum crystals.

Primary structures include polygonal anhydrite layers (figure 2.4). The anhydrite layers are related to the isolated blebs and small nodules of anhydrite that are found above the gypsum mush toward the seaward margin of the flats. Traced landward, the small anhydrite nodules increase in size and number and eventually form a surface layer of interlocking polygons (figure 2.15). Even further landward,

Figure 2.8
Surface trench located near the seaward margin of supratidal zone. Upper portion of trench exposes bed of gypsum mush overlying algal peat and underlying a poorly preserved algal surface (Butler et al. 1982).

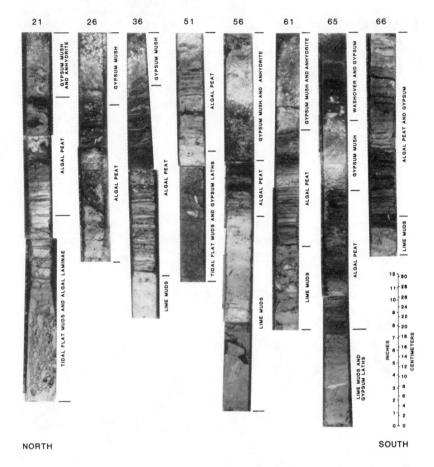

TRAVERSE 4D

Figure 2.9
Photographs of Cores 21, 26, 36, 51, 56, 61, 65, and 66 from Traverse 4-D. (See figure 2.3 for locations of the cores.) Cores are tied to the sediment surface (Butler et al. 1982).

where the surface layer is overlain by reworked eolianite, the polygons are festooned in section and range in diameter from approximately 30 cm to 1 m (figure 2.16). The upper edges of the margins of the polygons are characteristically planed off, a result of erosive

Figure 2.10
Salt polygons, 30 to 50 cm in diameter, covering the surface of mid-
to upper salt flat (Butler et al. 1982).

flooding water that moved across the sabkha. Additional festoons
may stack up over the original anhydrite layers (figure 2.17). Slumping and compaction, related to alteration of the underlying gypsum
mush to anhydrite, locally distorts the original layering. The festoon
layers of anhydrite influence subsequent development of anhydrite
structures further landward in the upper salt-flat region (Facies 4).

Upper Salt-flat Facies (Upper Supratidal). The upper salt-flat facies
belt, up to 4.8 km wide (figure 2.3), is flooded only once every four
to five years. Diagenesis in the belt is controlled by continental
groundwaters mixing with the marine-derived groundwaters. Seven
diagenetic features are recognized:

1. Addition of near-surface nodular anhydrite layers (Cores 4 and 5
   of Traverse 3-D, figure 2.12).
2. The complete replacement of gypsum mush by chicken-wire anhydrite, forming a bed that is up to 30 cm thick and covers an
   area of approximately 10 sq km (Cores 7, 20, and 30 of Traverse
   2-D; figures 2.13, 2.14, and 2.18). The bed forms ptygmatic folds,

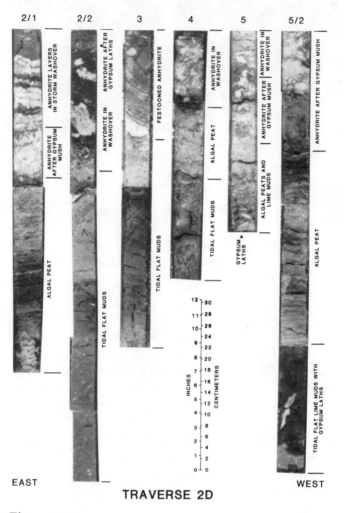

EAST

WEST

TRAVERSE 2D

Figure 2.11
Photographs of Cores 2, 3, 4, and 5 from Traverse 2-D. (See figure 2.3 for locations of cores.) Cores are tied to sediment surface (Butler et al. 1982).

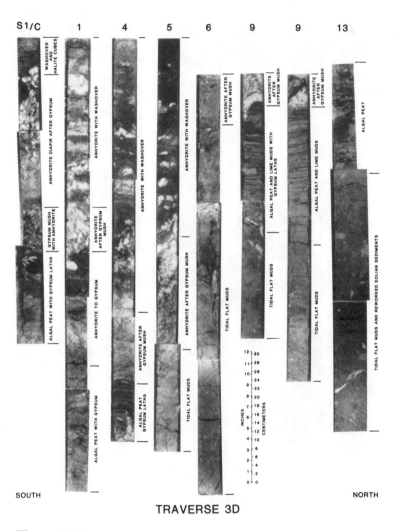

TRAVERSE 3D

Figure 2.12
Photographs of Core S_1/C and Cores 1, 4, 5, 6, 9, and 13 from Traverse 3-D. (See figure 2.3 for locations of the cores.) Cores are tied to sediment surface (Butler et al. 1982).

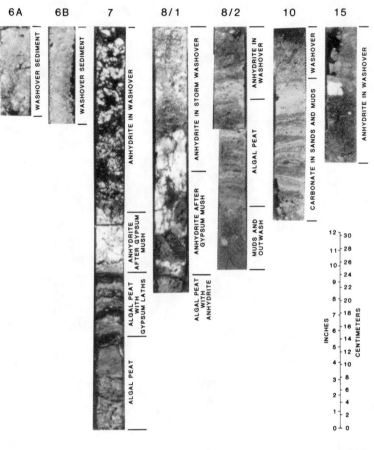

TRAVERSE 2D

Figure 2.13
Photographs of Cores 6, 7, 8, 10 and 15 from Traverse 2-D. (See figure 2.3 for locations of cores.) Cores are tied to sediment surface (Butler et al. 1982).

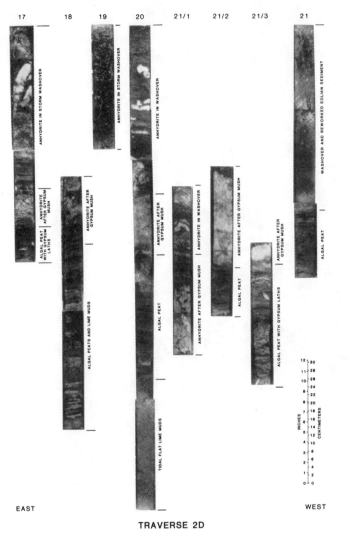

TRAVERSE 2D

Figure 2.14
Photographs of Cores 17, 18, 19, 20, and 21 from Traverse 2-D. (See figure 2.3 for locations of cores.) Cores are tied to sediment surface (Butler et al. 1982).

Figure 2.15
Polygonal saucers of anhydrite exposed at salt-flat surface (Butler et al. 1982).

disharmonic folds (thrust folds), and diapirlike structures (figures 2.4, 2.19, 2.20, and 2.21).

3. The local alteration within algal laminae of the subsurface algal mat of small, ≤2 mm long, discoidal gypsum crystals to anhydrite nodules that are >1 mm in diameter (Core 17 of Traverse 2-D; figure 2.14).

4. The development of abundant, large (≈6 cm, but occasionally up to 25 cm long), lenticular gypsum crystals (selenite) within the algal mat and underlying lagoonal sediment (figure 2.4, numerous cores, and figure 2.22).

Figure 2.16
Surface trench from upper salt flat, showing cross section of anhydrite polygonal saucers. Note anhydrite layer which replaced gypsum mush. Doug Shearman is the scale (Butler et al. 1982).

5. The formation of Holocene calcian-dolomite in trace amounts ($\approx 10\%$).
6. The filling of internal molds in gastropod shells by anhydrite (figure 2.23).
7. The local occurrence of halite in the form of hopper crystals ($<6\,\text{cm}^3$) and as a cement (Core S_1/C of Traverse 3-D, figure 2.12; Core 30 of Traverse 2-D, figure 2.18; and Core 4 of Traverse 3-E, figure 2.24).

The local cementation of supratidal evaporite and algal mat sediments by pore-filling halite and hopper crystals is related to the upward movement of continental waters. Cementation occurs within circular or ellipsoidal plugs that are up to 17 m in diameter and 1 m deep. These areas are coincident with holes in the subsurface cemented layer of carbonate sand. Structures include diapirlike features that are developed within the layer of secondary anhydrite (2 above) that overlies the subsurface algal mat of this facies (figures 2.5 and 2.19). At some localities the diapirlike features are interpreted to be primary because a gypsum mush occurs between adjacent festoon layers of anhydrite. The diapirlike shapes have been enhanced by additional anhydrite growth and compaction. The diapirs appear to be true piercement structure in areas of the sabkha where there is an influx of continental groundwaters that may have promoted anhydrite formation.

Other structures include contorted and folded anhydrite layers, which occur throughout this facies. Contorted anhydrite layers, in vertical section resembling convolutions of intestines, have been termed "enterolithic veins" (Hahn 1912) or "ptygmatic" structures. In the coastal salt flats, these contorted layers frequently occur within the upper 30 to 60 cm of detrital eolianite (figure 2.21) and on a smaller scale in chicken-wire anhydrite. West (1965) described similar structures from the Purbeckian of southern England, and Holliday (1967) recognized them in the Carboniferous of Spitsbergen.

Throughout the sabkha, thin vertical tubes that are infilled with carbonate sediment are common in the anhydrite (figure 2.25, and Core 8/1 of Traverse 2-D, figure 2.13). The tubes appear to be pri-

Figure 2.17
Stacked or "festooned" anhydrite polygons exposed in shallow trench with incipient development of anhydrite diapirs. Note 6-inch ruler for scale, and algal peat at base of trench (Butler et al. 1982).

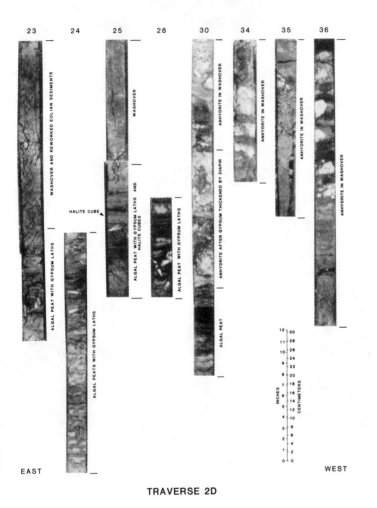

Figure 2.18
Photographs of Cores 23, 24, 25, 28, 30, 34, 35, and 36 from Traverse 2-D. (See figure 2.3 for locations of the cores.) Cores are tied to sediment surface (Butler et al. 1982).

Figure 2.19
Surface trench from upper salt flat. Basal sediments exposed are an algal peat with scattered large prismatic crystals of gypsum. White diapiric layer is anhydrite that replaced gypsum, and upper layer is storm washover and windblown carbonates and clastics with anhydrite nodules (Butler et al. 1982).

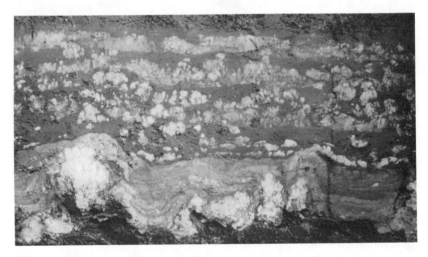

Figure 2.20
Anhydrite layers shown in surface trench located between upper salt flat and recycled eolianite (Butler et al. 1982).

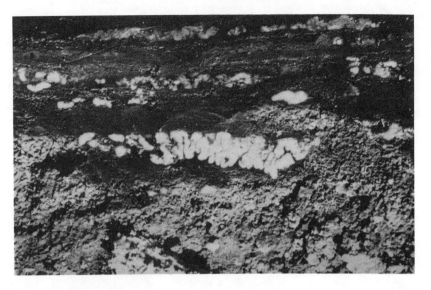

Figure 2.21
Surface trench showing ptygmatic layering of anhydrite layers (Butler et al. 1982).

mary or very early diagenetic features that are formed within the supratidal environment. Though we do not understand their origin, the identification of similar vertical veins in ancient nodular anhydrites would suggest a supratidal setting. Although the anhydrite that is found in Pleistocene eolianites at the landward margin of the coastal salt flat (Facies 5) has the characteristic nodular texture, we have not seen vertical sand tubes in these deposits. Deformational structures are also not found in areas like the landward margin of the coastal salt flats, where gypsum is forming at the expense of anhydrite.

Recycled Eolianite Facies (Upper Supratidal). The recycled eolianite facies, marking the landward margin of the salt flats, has a maximum width of approximately 3 km (figure 2.3). Sediments are essentially continental in origin and consist of Pleistocene eolianites overlain by up to 60 cm of recycled Holocene eolianites, shell lag material, and bedded skeletal debris (Cores 1 and 2 of Traverse 3E, figure 2.24). The groundwater regime is entirely continental, and four diagenetic and sedimentary features are recognized:

Figure 2.22
Gypsum crystal collected at algal flat/lagoon facies transition (Butler et al. 1982).

Figure 2.23
Anhydrite pseudomorphing gastropods (Butler et al. 1982).

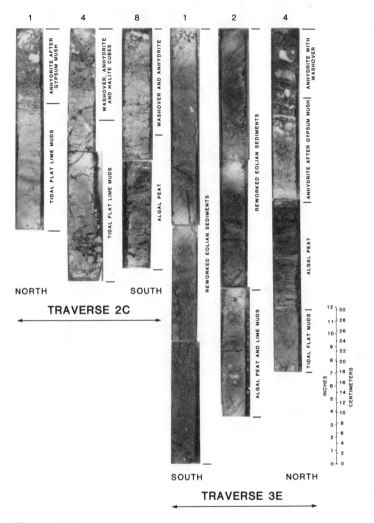

Figure 2.24
Photographs of Cores 1, 4, and 8 from Traverse 2-C, and Cores 1, 2, and 4 from Traverse 3-E. (See figure 2.3 for locations of cores.) Cores are tied to sediment surface (Butler et al. 1982).

Figure 2.25
Anhydrite layer exposed in surface trench. Note the numerous vertical carbonate-filled tubes that penetrate the layer (Butler et al. 1982).

1. Beds, up to 2 m thick, that are formed of alternating layers (≈5 cm thick) of anhydrite nodules and detrital sands, and thick beds, up to 2.4 m thick, of anhydrite nodules.
2. Alteration of nodular anhydrite layers to gypsum (figures 2.26 and 2.27; Core 1 of Traverse 3-D, figure 2.12). Secondary gypsum crystals are flattened parallel to the a-c plane, are up to 7 cm long, and are zoned with anhydrite laths.
3. Interstitial precipitation of gypsum forming cemented layers of sediment.
4. Locally, surface halite crusts containing polyhalite and sylvite.

Figure 2.26
Trench showing anhydrite layer in recycled eolianite facies. The white anhydrite to the left is inverting to gypsum on the right (Butler et al. 1982).

Figure 2.27
Surface trench exposing gypsum in recycled eolianite facies. The gypsum has almost completely replaced the anhydrite (Butler et al. 1982).

Stratigraphic Relationships. Commonly, wherever there is an upper intertidal algal mat exposed on the surface of the coastal salt flats (Facies 1), a layer of gypsum crystal mush up to 30 cm thick and of variable width occurs along its landward margin (Facies 2; see also figures 2.3 and 2.5). In the sabkha subsurface south of Abu Dhabi, buried algal mats are found up to 9 km inland from the present shoreline. By analogy with the present surface algal mat and gypsum mush relationship, the inland subsurface algal mat should be overlain by a gypsum crystal mush. Instead, in the upper salt-flat region (Facies 4), the subsurface algal mat is overlain by a layer of chicken-wire anhydrite up to 30 cm thick, containing the same vertical distribution of sediments as observed in the gypsum mush that lies seaward (figures 2.3 and 2.5). Trenches across the sabkha show the progressive replacement, from the top down, of a gypsum mush by anhydrite that has a chicken-wire texture. Porosity (determined by neutron activation; courtesy of Union Oil Research, California) of the gypsum mush ranges from 15% to 35%, with an average of 30%. The layer of secondary chicken-wire anhydrite, on the other hand, has an almost uniform porosity of approximately 20%.

Locally, within the middle salt-flat region (Facies 3), gypsum crystals (<2 mm long) within the algal laminations are replaced by anhydrite to form thin (<0.5 cm) laminae of small (<1 mm) anhydrite nodules which alternate with organic layers. On compaction, such laminae might easily resemble the familiar varved anhydrites commonly recognized in ancient examples and considered to have formed from standing bodies of water. Also near the middle salt-flat region, the layer of secondary chicken-wire anhydrite is overlain by 30 to 60 cm of detrital eolianite and storm-washover sediment (Facies 4). These sediments contain thin (<3.2 cm thick) layers of nodular anhydrite, and often exhibit ptygmatic structure. The near-surface layers of nodular anhydrite that postdate the layer of secondary chicken-wire anhydrite can be explained by a flood recharge model. After a flood, the upper few centimeters of the salt flat in Facies 3 contain small (≈0.2 mm) discoidal and prismatic gypsum crystals (Core 2/2 of Traverse 2-D, figure 2.11). The floodwaters have a composition similar to the brines collected from evaporating pans that are saturated with sodium chloride. During flooding, the ionic strength of lagoon waters is increased by a solution of halite at the surface, thereby causing the further precipitation of Ca^{2+} and SO_4^{2-} ions as gypsum. Subsequent burial of the gypsum crystals and their alteration to anhydrite would yield the layers of anhydrite nodules that are observed in the mid-sabkha region. The same process may also explain the occurrence of layers of anhydrite nodules within the detrital sands at the landward margin of sabkhas (Facies 5).

Abu Dhabi Sabkha: Conclusions

The Abu Dhabi salt flats contain a suite of supratidal evaporites whose description can aid in the study of similar ancient deposits. This is important from an economic standpoint, because carbonate platform or shelf reservoirs are commonly sealed by updip supratidal evaporites. The generalized shoaling-upward cycle of the Abu Dhabi sabkha is recognized in many ancient sedimentary sequences. The cycle comprises, in ascending order, a carbonate mud or cross-bedded carbonate sand of lower tidal flat origin; an algal-laminated carbonate mud with gypsum rosettes; a dense bed of anhydrite that is 20 cm or so thick and locally diapiric; and a cap of interbedded windblown and storm-washover carbonates and clastics with thin ptygmatic and nodular anhydrite.

Evaporites of the Abu Dhabi salt flats form many structures that are similar to those recognized in ancient sabkha environments. The structures include ptygmatic (enterolithic) layers, polygonal layers, disharmonic fold layers, overthrust folds, diapirlike features, and sand veins. Some structures appear to be the result of the combined processes of anhydrite growth and compaction, whereas others are related to the supratidal origin of the anhydrite. Landward of the sabkha sediments, windblown quartz accumulates as huge dune fields forming the edge to this coastal complex.

The tidal-flat sediments from Abu Dhabi can be equated to the present accumulation of grapestone facies and pelleted lime muds in the Khor Al Bazam, while the coral reefs and oolite shoals are their seaward equivalents, forming the margin to the present-day prograding coastal sequence, which itself will eventually be capped by sabkhas.

The major characteristics of the sabkha evaporites that identify them as being of sabkha facies are the presence of the ptygmatic and chicken-wire anhydrite, the presence of a matrix of carbonates, the presence of vertical burrows in the anhydrite, and the lack of the finely laminated gypsums that are associated with playas.

SUBAQUEOUS EVAPORITES: SALINAS

Salinas adjacent to areas of carbonate accumulation are probably commoner in Australia than many other continental margins.

The South Australian salinas contain sequences of subaqueous gypsum up to 10 m thick and have formed in the last 6,000 years. The salinas occur as isolated lacustrine depressions within surrounding Quaternary calcareous coastal dunes (figures 2.28, 2.29). Inflow into

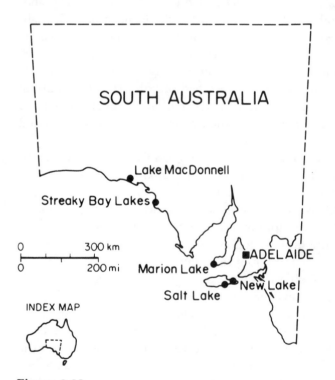

Figure 2.28
Location of South Australian gypsum-filled salinas (Warren and Kendall 1985).

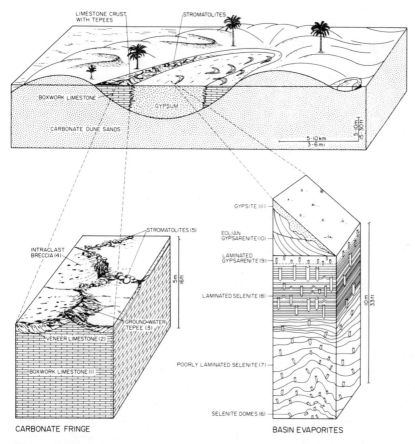

Figure 2.29
Schematic block diagrams showing sediment and evaporite distribution in South Australian salinas. (1) Boxwork limestone, showing
remnant algal or evaporite structures. (2) Veneer limestone, a hard
indurated crust in many places overthrust into tepees. (3) Groundwater tepee. (4) Intraclast breccia. (5) Stromatolites (domal forms
commonly cap tepees; larger laterally linked forms are more common
in deeper waters; stromatolites and tepees both occur in megapoly-
gonal pattern). (6) Selenite (coarse-grained gypsum) domes. (7) Poorly
layered selenite domes. (8) Laminated selenite. (9) Laminated gypsarenite (in some salinas this unit dominates and there is only minor
selenite). (10) Eolian gypsarenite. (11) Gypsite (silt-sized gypsum),
which forms when dune surface is stable. (Warren and Kendall 1985.)

the depressions is mainly from resurgence of marine-derived ground-waters that seep from the sea through the dune aquifer into the salina depressions. The driving force supplying the water to the basin is evaporative drawdown, moving seawater plus some meteoric groundwater into the basin edge whenever surface waters evaporate to levels below that of the surrounding ocean.

In plan view, the sediments in South Australian salinas show a bull's-eye pattern consisting of a carbonate rim surrounding a central zone of laminated gypsum (figure 2.29). Individual salinas measure up to 20 km × 12 km, with centers filled with as much as 10 m of laminated subaqueous gypsum. The carbonate unit of the lake edge is a unit distinct from the more central gypsum; the two are often separated by a sharp, often near-vertical contact. The $CaSO_4$ phase in the basin is always gypsum, not anhydrite, as temperatures are not high enough nor the basin arid enough for the formation of anhydrite. The South Australian coastal zone experiences Mediterranean-like weather, with cool, wet winters and hot, dry summers. Rainfall is 250 to 600 mm per year, and evaporation is 1,500 to 2,250 mm per year. Winter temperatures are 10° to 20° C and rarely fall below freezing; summer temperatures are often in the 30s, but can climb into the 40s.

Much of the salina gypsum is coarse-grained and bottom-nucleated, and pore-infilling gypsum is rare. The South Australian salinas are one of the few documented occurrences of natural, widespread, Holocene subaqueous gypsum deposition (Warren 1982a). The gypsum is usually more than 90% pure and is laid down as a characteristic shallowing-upward evaporite unit (figure 2.29). At the base, the succession is massive, poorly layered domes of coarse-grained gypsum (selenite). The elongate gypsum crystals in this zone show little or no preferred orientation, and carbonate impurities are distributed randomly through the gypsum (figure 2.30). Higher in the section, the degree of lamination increases as the dome amplitude decreases. Domes pass into horizontally laminated selenite where individual laminae appear to crosscut large, upwardly aligned gypsum crystals (figure 2.31). These large ("giant") crystals at first appear secondary, and their Miocene equivalents in the Sicilian Basin were once interpreted as secondary gypsum after anhydrite (Ogniben 1957). Studies of identical crystals in the South Australian salinas indicate

Figure 2.30
Massive domes of elongated crystals of gypsum with random orientation and random carbonate particles, New Lake, South Australia.

that the gypsum crystals (30–50 cm long) are primary, not secondary crystals. They grow with their long axes perpendicular to bedding, responding to the effects of crystal impingement and growth alignment. Carbonate laminae in the South Australian gypsums form by the precipitation and accumulation of aragonite peloids during the spring and early summer. The peloids are ostracode and brine shrimp fecal pellets mixed with the micritized remnants of algal tubules. At first the peloids mantle the underlying gypsum and are enclosed by

Figure 2.31
Laminated selenite cut by palisades of large crystals that are perpendicular to the depositional surface and contain layers of carbonate pelloids. New Lake, South Australia.

a cyanobacterial mat covering the gypsum. The peloids are subsequently encased by the upward poikilitic growth of the large gypsum crystals during the summer and fall. The giant gypsum crystals actually grow upward each season by millimeter-scale increments; each step of growth is in crystallographic continuity with its underlying parent crystal. Millemeter-laminated, coarse-grained gypsum beds are in turn overlain by a laminated accumulation of sand-size gypsarenite. The upper part of these horizontally laminated gypsarenites are often reworked into wave-oscillation ripples. The laminated and rippled gypsum is in turn overlain by a thin, massive, poorly bedded gypsarenite unit that formed under seasonally vadose or subaerial conditions. Capping the whole succession is a unit of cross-stratified eolian gypsum, and in areas stabilized by vegetation, a pedogenic crust of gypsite (silt-sized) has formed atop both lacustrine and eolian sediments (figure 2.32).

The carbonate unit forming about the edges of many South Australian salinas is composed of a boxwork limestone unit (4–5 m thick) overlain by a veneer fenestral limestone sheet less than 1 m thick (figure 2.29; Warren 1982b). The boxwork limestone is a diagenetic unit created by the cannibalization of earlier lacustrine carbonate by inflowing fresher, but still saline, marine-derived groundwaters. The "boxwork" contains algal structures, evaporite pseudomorphs, and other highly altered bedded fabrics dependent on the preexisting sediment types (figure 2.33). In some areas the limestone sheet above the boxwork limestone consists of subaerial algal tufas or subaqueous mm-laminated stromatolites and algal mats. In other areas this limestone consists of fenestral capillary crusts crosscut by tepees. The stromatolites grew as current-aligned domes up to 40 cm tall in the continuously subaqueous parts of the lake edge, or as mats in areas of very shallow and sometimes ephemeral surface water along the lake shoreline (Von der Borch et al. 1977; Warren 1982b). In some areas a little farther out in the lake, the stromatolite domes cap tepee

Figure 2.32
Cross-bedded and rippled gypsarenite. Marion Lake, South Australia.

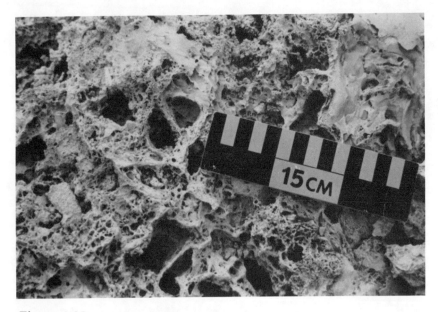

Figure 2.33
Boxwork carbonate. Lake Marion, South Australia.

structures. Like stromatolites, the tepee structures are confined to the capping carbonate unit and form an extensive fenestral limestone sheet up to 60 cm thick. Tepees are commonest in areas of resurging groundwater, where seasonal changes in the groundwater head encourage tepee growth (Warren 1982b). Most limestone sheets caught up in the tepee structures are capillary crusts composed of fenestral lime mudstones containing stromatactis-like voids, pisolites, laminar cements, and internal sediment.

Some of the South Australian salinas (namely, Marion Lake and the lakes nearby) were not always enclosed depressions fed by marine springs. Originally they were early Holocene marine embayments where open-marine skeletal grainstones and wackestones were deposited on the lagoon floor (Olliver and Warren 1976; Von der Borch et al. 1977). Today these sediments lie beneath the stromatolites and subaqueous gypsum of the lake fill. The lakes only became evaporitic when the narrow entrances to the embayments were blocked off from Southern Ocean waters by a buildup of sand spits and sea grass banks in the narrow entrances to the lagoons.

Lake MacLeod

In western Australia a very fine example of a salina is Lake MacLeod, which lies just north of Shark Bay on the coast (figure 2.34) and has been studied and described in some detail by Logan (1981). He has shown how a coastal lagoon, eroded into the underlying Pleistocene possibly by wind deflation during the most recent Pleistocene sea-level low, was isolated by prograding clastic deltaic sediments and carbonates, and has developed into a halite-, gypsum-, and carbonate-filled salina.

Its water surface now lies well below sea level and is maintained by the Indian Ocean water that penetrates the Pleistocene coastal dunes through fractures and caves, forming a series of springs. This water has almost normal marine salinity as it enters Lake MacLeod. Logan (1981) demonstrated that in the early stages of the development of

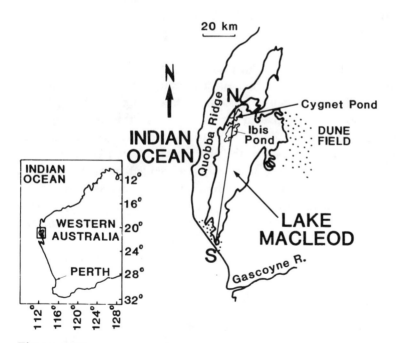

Figure 2.34
Location of Lake MacLeod in western Australia (Handford et al. 1984).

Lake MacLeod, these normal marine waters are believed to have filled this lake. Then, by a process of both evaporation and leakage downwards into the underlying sediments and spillover, it evolved a unique pattern of evaporite fill. The initial volume of water entering the system was removed through seepage and spillover with enough loss by evaporation to precipitate the initial carbonate fill on the floor of the lagoon (figure 2.35). As these carbonates collected, the seepage of water through the base of the lagoon was reduced, so that higher-salinity water, instead of being lost, evaporated further, and in consequence the concentration of the water was further elevated. Thus while some water of the saline body continued to escape by seepage, salinities were elevated enough for gypsum to now be precipitated across the floor of the deeper parts of the playa. Once this gypsum formed a continuous layer over these deeper areas, it became an even more effective lake floor seal than the carbonate. Logan (1981) has shown that at this point the rate of loss of water through seepage became less than the rate of loss of water through evaporation, and the lake was drawn down to a smaller body of water to the south. In response, the salinities of the playa were further elevated, and the lake began to precipitate halite, particularly at this southern end (figure 2.35). A longitudinal profile recording precipitation north–south down the lake would have shown carbonate being precipitated at its northern end, with a belt of calcium sulfate being precipitated close by to the south, while over most of the rest of the lake depression, halite was being precipitated. Eventually the lake became so filled with halite that the waters of the playa spread beyond its initial borders. Seepage through the floor of the extended parts of the playa was now sufficient to prevent the buildup of higher salinity waters, so ending the further precipitation of halites. In consequence, the lake is now the site of calcium carbonate precipitation to the north and calcium sulfate to the south, with halite having ceased to form, except by artificial means in the ponds of a salt plant to the south.

The character of the MacLeod fill is such that to the north a massive gypsum layer occurs which is 5 m and over thick and overlies a halite fill and the original carbonate lagoon floor (Handford et al. 1984; figure 2.35). To the south, halite extends over much of the playa, so that the vertical profile of the fill of the lake is of at least some 7 m or more of massive sodium chloride overlying the thin la-

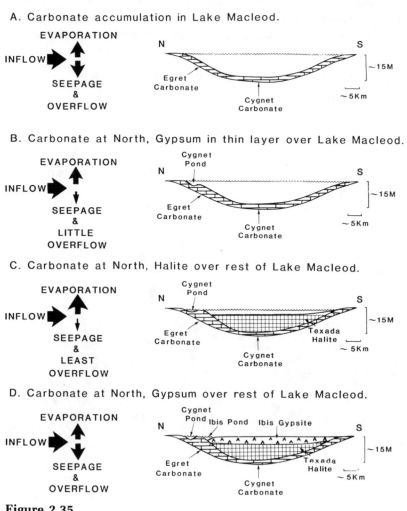

Figure 2.35
Generalized stratigraphic cross section of Lake MacLeod (after Handford et al. 1984) and its evolution through time (interpreted from Logan 1982). The cross section is located on figure 2.34.

goonal carbonates. Today the surface gypsum forms a mush of small prisms that lacks a marked carbonate matrix. By analogy, it is probable that Shark Bay, to the south, will evolve into similar playas by the accretion and progradation of the carbonate sills that already isolate the bay into such segments as Hamlin Pool or Depuch Loup.

PERMIAN CARBONATE/EVAPORITE SEQUENCES OF THE DELAWARE BASIN

The two peritidal sedimentary sequences, those of the sabkha and the playa, are extremely important as hydrocarbon seals in the ancient. A good example of these sediments occurs in the Delaware Basin Permian, both in the subsurface and in the rocks exposed in the Guadalupe Mountains of West Texas and New Mexico (figure 2.36). Ward et al. (1986) describe these in some detail, and their

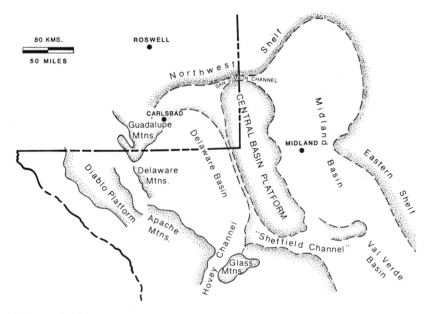

Figure 2.36
Map of the Permian Basin of West Texas and southeastern New Mexico, showing large-scale features delineated by subsurface studies, and mountain ranges exposing Guadalupian strata along the western margin of the Delaware Basin (Ward et al. 1986).

results are paraphrased through much of this section of this paper. They note that the Guadalupian Delaware Basin was rimmed by massive prograding carbonate silt banks and local reefs, the Capitan Limestone (figure 2.37). Seaward, during sea-level highs, the Delaware Basin was the site of the accumulation of basinal carbonates, while during sea-level lows, windblown clastics collected. Landward, the massive carbonate margin interfingers with interbedded heterogeneous shelf carbonates that in turn pass into mixed restricted carbonate tidal-flat dolomites and interbedded windblown quartz sands and silts, which themselves are interbedded with beds of gypsum and anhydrite updip. In the subsurface, these same carbonate evaporite rocks form classic hydrocarbon traps, with the evaporites plugging the porosity that occurs in the clastics and the updip carbonates of the inner shelf.

Thus, the Guadalupian-age sedimentary rocks of the Permian Basin of Texas and southeast New Mexico, seen in outcrops and wells, represent a classic example of the facies relationships that span a carbonate evaporite shelf (figure 2.37). As Ward et al. (1986) point out, the importance of clastics to the facies models for this region has been largely overlooked.

The Guadalupian formations in outcrop have long been recognized (e.g., Tyrrell 1969) as representing time-equivalent basin, shelf marginal (reef), and shelf (back-reef) deposits (figures 2.36, 2.37, and 2.38). The basinal deposits in the Delaware and Midland basins are thick, finely laminated siltstones and sandstones that thin towards the basin margin, and thin interbedded black to gray limestones that thicken toward the basin margin. The thesis of Ward et al. (1986) is that this limestone was deposited continuously through the upper Guadalupian, but that the sandstone was depositionally punctuated and actually drowned the volume of carbonates when clastic deposition was occurring. The siliciclastic basinal facies of the Delaware Mountain Group are divided by the carbonate stringers into, in ascending order, the Brushy Canyon, Cherry Canyon, and Bell Canyon formations (figure 2.38; King 1942; Harms 1974; Williamson 1977). The sandstones were probably transported from the shelf to the basin

FACIES BELTS & OUTCROP NOS.	BASIN (1, 2 & 3)	SHELF EDGE — APRON (4)	SHELF EDGE — CHEST OR REEF (5)	SHELF (BACK-REEF) — SAND FLATS (6)	SHELF (BACK-REEF) — BARRIER ISLANDS (7)	LAGOON (8)	COASTAL FLATS & PLAYAS (9)	FACIES BELTS & OUTCROP NOS.
SEDIMENTS	SILT AND SANDSTONE, LIME MUDSTONE AND CARBONATE DEBRIS	BRECCIAS, LIME MUD, GRAIN AND BOUND-STONE	LIME MUD, GRAIN AND BOUND-STONE.	LIMEY TO DOLOMITIC WACKE & GRAIN-STONES	DOLOMITIC, PISOLITIC, INTRACLASTIC GRAIN & WACKESTONE	DOLOMITIC MUDSTONES AND WACKESTONES	GYPSUM, AND/OR ANHYDRITE, QUARTZ SANDS & SILTS, DOLOMITIC MUDSTONES & WACKESTONES	SEDIMENTS
BEDDING AND SEDIMENTARY STRUCTURES	SANDSTONE CHANNEL FILLS, SILTSTONE DRAPES, LOW ENERGY BEDFORMS AND SOME GRADING. LIMESTONE-THIN LAMINATED AND GRADED BEDS. LIMESTONE DEBRIS, - CHAOTIC AND FINING UPWARD	IRREGULAR THICK BEDS OF GRADED, CROSS-BEDDED, CHAOTIC AND DENSE LIME-STONE	MASSIVE AND BIOHERMAL LIME-STONES WITH FRAC-TURES	WIDE-SPREAD HETERO-GENEOUS MEDIUM BEDS, CROSS-BEDDED TO DENSE	WIDESPREAD MEDIUM BEDS WITH FENESTRAE AND TEPEES	SHOALING CYCLES OF WIDESPREAD DOLOMITIC MUDSTONE AND WACKE-STONE BEDS CAPPED BY LAYERED AND LAMINATED MUDSTONES AND EVAPORITES	SHOALING CYCLES OF WIDESPREAD EVAPORITES QUARTZ SANDS & SILTS AND LAMINATED VUGGY DOLOMITES	BEDDING AND SEDIMENTARY STRUCTURES
POROSITY	INTERPARTICLE IN SILT AND SANDSTONE. 18-20% PRODUCTIVE, 17% NON-PRODUCTIVE	INTRASKELETAL, VUGGY AND FRACTURED. AVERAGE 4%		INTERPARTICLE, FENESTRAL AND FRACTURED. AVERAGE 8%		INTERCRYSTALLINE, FENESTRAL AND MOLDIC IN DOLOMITE. INTER-PARTICLE IN SANDS. 9-16% PRODUCTIVE. 9% NON-PRODUCTIVE	NONE	POROSITY
RESERVOIR POTENTIAL & SEALS	SANDSTONE RESER-VOIRS, SILTS AND LIMESTONES SOURCE & SEAL	WATER PRONE				STACKED RESERVOIRS IN SANDS AND CARBONATES. SEAL, UPDIP EVAPORITE	EVAPORITE PRONE, WITH PATCHY RESERVOIRS. EVAPORITE SEAL	RESERVOIR POTENTIAL & SEALS
FACIES BELTS & OUTCROP NOS.	BASIN (1, 2 & 3)	SHELF EDGE — APRON (4)	SHELF EDGE — CREST OR REEF (5)	SHELF (BACK-REEF) — SAND FLATS (6)	SHELF (BACK-REEF) — BARRIER ISLANDS (7)	LAGOON (8)	COASTAL FLATS & PLAYAS (9)	FACIES BELTS & OUTCROP NOS.
WIDTH OF FACIES BELTS	70 MILES (112 km)	1 MILE (1.6 km)	½ MILE (.8 km)	½ MILE (.8 km)	2 MILES (3.2 km)	2-8 MILES (3-13 km)	25 MILES (40 km)	WIDTH OF FACIES BELTS

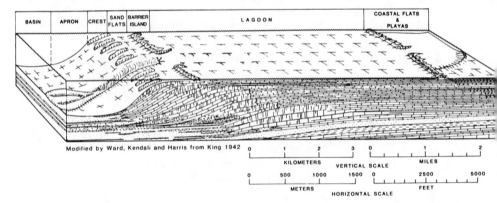

Modified by Ward, Kendall and Harris from King 1942

KILOMETERS   VERTICAL SCALE   MILES

METERS   HORIZONTAL SCALE   FEET

Figure 2.37
A generalized block diagram and summary of Guadalupian-age shelf, shelf margin, and basin facies, with representative outcrop localities numbered (Ward et al. 1986). Numbers refer to location on figure 2.39 map.

by gravity-driven currents (Harms 1974; Williamson 1977), whereas the limestones represent local slumps and turbidites of shelfal marginal and shelf carbonates (L. C. Babcock 1977). Some of the limestones are sheet-like and cover the whole basin, but most accumulated along the margins of the basin as wedge- or mound-shaped fans.

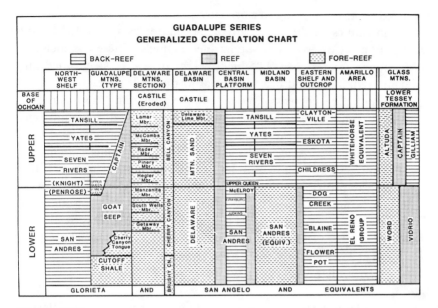

Figure 2.38
A generalized correlation chart of the Guadalupe series of the Permian Basin, correlating subsurface and outcrop shelf and basin deposits. Based on unpublished correlations by Y. B. Newson (Ward et al. 1986).

Basinal Clastics. The basinal clastics are subarkosic siltstones and sandstones, generally cemented by calcite (Jacka et al. 1968; Harms 1974; Williamson 1977, 1978). The siltstone commonly contains horizontal to slightly wavy laminae, 0.04 to 0.8 in (0.1–2.0 cm) thick, which are graded and become increasingly organic-rich upwards (Williamson 1977). The siltstone drapes underlying surfaces as uniform thick blankets with widespread lateral continuity, suggesting deposition from suspension. Unlike the siltstone, which extends across channels into interchannel areas, nearly all the sandstones are confined to channels. The sandstones, particularly in the Bell Canyon, are moderate to well-sorted, poorly cemented silty to very fine sand, with porosities in the subsurface as high as 27%.

Williamson (1977) believed that the sands were deposited by channelized, bottom-hugging density currents, and that the geometry of the sandstones in the Bell Canyon Formation is controlled by the configuration of erosional submarine channels. These channels are commonly greater than 0.25 mi (0.4 km) in width, steep-sided and flat-floored, and trend nearly normal to the shelf edge.

Thermohaline currents flowing off the shelf into the basin conceivably could have eroded channels and deposited the sandstones and siltstones within the Brushy Canyon Formation (Harms 1974). Williamson (1977) suggested that the flushing of saline lagoons adjacent to the steep shelf margin by storms or seasonal chilling of shelf waters was probably the source of dense water that drove density currents. Sediment could have been initially entrained by episodic mass wasting associated with unstable slopes. The sands may well have reached the break in slope during slight drops in sea level, and the sand-filled lagoons may then have been continually eroded by the density currents proposed by Williamson (1977).

Basinal Limestones. The carbonate tongues that divide the clastics include the Getaway, South Wells, Manzanita, Hegler, Pinery, Rader, McCombs, and Lamar limestones (figure 2.38). The relationship of these limestones to those of the shelf margin can be observed in numerous canyons (e.g., McKittrick Canyon, figure 2.39) that dissect the Capitan escarpment at nearly right angles. This relationship is one of a gradational change from dark, well-bedded basinal limestones with few fossils, to massive, fossiliferous and sometimes debris-filled carbonate aprons on the basin margin slope (figure 2.37).

In the Goat Seep shelf edge, significant channeling is recognized by Crawford (1981) in toe-of-slope facies exposed in Shirttail and Shummard canyons. Crawford (1981) interpreted these basinward trending channels as filled with shelf and shelf-edge sediments which slumped or were carried downslope in suspension through openings created by the slumps. Of the basinal limestone stratigraphically above those described by Crawford, the Lamar Limestone is one of the best known. This limestone, which equates to the Capitan, is wedge-shaped in outcrop, being more than 300 ft (90 m) thick near the basin edge and tapering to approximately 6 ft (2 m) in thickness at its most basinward exposure (L. C. Babcock 1977; figure 2.37). At its most ba-

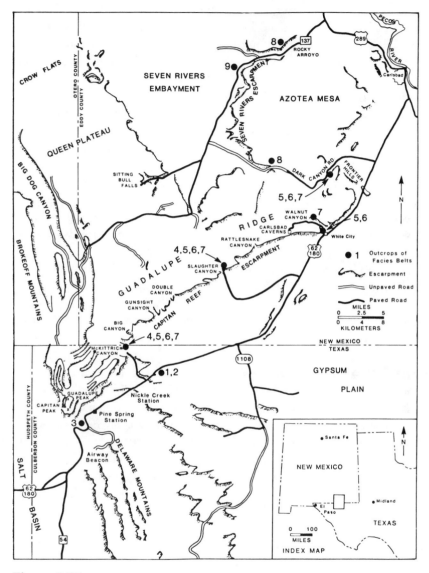

Figure 2.39
Map of Guadalupe Mountains showing localities of outcrops referred to in text and on figure 2.37 (Ward et al. 1986).

sinward position in the subsurface, the Lamar Limestone is represented by only a few feet of silty shale (Tyrrell 1969), devoid of fossils, with rare burrows, graded beds, or signs of active transport (L. C. Babcock 1977). The bedding of the basin margin zone is varied from evenly laminated, thin beds, to irregular beds containing carbonate breccia and massive moundlike buildups (L. C. Babcock 1977). In a similar position close to the basin margin, the Pinery Limestone displays downslope changes in bed form, including thinning of individual beds, increased bedding irregularity, diminution in the size of trough with wedging out of individual beds, an increase in gradational internal contacts, more undulous contacts, and high percentages of dark, fine-grained carbonate material (Koss 1977). In contrast, the Rader conglomerate horizon contains pebble- to boulder-sized lithoclasts deposited by subaqueous gravity flows, with sharp erosional contacts over deep troughs filled by grain-supported carbonate sands in truncated foresets (Koss 1977). However, the bedding in these basinal limestones becomes more regular basinward, where they contain more lime mud. All these different basinal limestones are interpreted to have been deposited in oxygen-depleted conditions, with the Lamar Limestone containing selectively silicified pelecypods and other benthic organisms.

Shelf Margin Deposits. The shelf margin (reef) deposits are confined to a narrow belt bordering the shelf (figure 2.37). They consist of massive light gray limestones (cut by occasional neptunian dikes, which are filled by carbonates and sand) and overlying steeply bedded, blocky rubble (fore reef facies) limestone. Shelf-edge carbonate deposits include the Goat Seep (King 1942, 1948; Newell et al. 1953; Boyd 1958; J. A. Babcock 1973, 1977; Hayes 1964; Crawford 1981; Yurewicz 1976, 1977; Schmidt 1977) and Capitan Reef complexes (Shumard 1858; Lloyd 1929; Newell et al. 1953).

The relationship of the Goat Seep Reef and Capitan Limestone to the basinal deposits of the Delaware Mountain Group is transitional. The basinal sediments thicken into the Goat Seep and Capitan apron as they become more fossiliferous and less well-bedded (figures 2.37 and 2.40). Locally, breccias consisting of shallow-water carbonates are common, but most of the sediment is a muddy lime siltstone. The

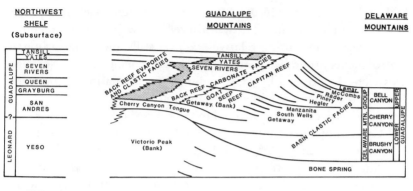

Figure 2.40
Diagrammatic cross section showing position of oil fields on North-west Shelf and Central Basin Platform. Almost without exception, the fields are situated within self dolomites or siltstones and are shelf-ward from the shelf edge (Ward et al. 1986).

character of the basin margin transition differs between the Goat Seep Reef, the lower and middle Capitan, and the upper Capitan apron.

The evolution of the shelf margin from a low to steep angle is ac-companied by a faunal change and an increased difference in the na-ture of the marine cementation (J. A. Babcock 1977; Yurewicz 1977; Mazzullo 1977; Schmidt 1977; Crawford 1981). Much of the fauna along the shelf margin were apparently filter feeders, suggesting that wave action on this break in slope was probably moderate. With the exception of the bound bioclastic material, most of the sediment is extremely fine-grained. Clearly, at times the margin oversteepened, for it was fractured and subsequently healed by a variety of marine cements, encrusting biota, and sediment. Fractures in the older Cap-itan are filled predominantly with detrital carbonate sediment (Ken-dall 1969), whereas fractures in the upper Capitan are locally filled by fibrous marine cements and algae (J. A. Babcock 1977). Both sets of fractures contain quartz sand, presumably carried there by the wind at low sea-level stands. Similar fractures are recorded in the Goat Seep by Crawford (1981).

The geometry of the Capitan margin observed on outcrop in North

McKittrick Canyon (figure 2.41, stages 1 through 5), is an apparent response to changes in relative sea level. These large-scale relationships suggest that during relative sea-level stillstands, the carbonate environments prograded seaward, while subsequent relative sea-level rises were outstripped by rapid sediment accumulation and progradation, producing a steplike configuration.

Similar geometries documented in the subsurface for the Goat Seep and Capitan by Silver and Todd (1969) suggest that they are due to either eustatic changes of sea level or, possibly, basinwide subsidence, with the few differences in geometry and facies association caused by the different local rates of carbonate accumulation. This widespread response to relative sea level is important, since it enables the geologist to understand the geometries of the Goat Seep and Capitan margin and shelf and predict similar occurrences elsewhere in the basin.

Shelf Sediments. Shelf deposits are variable, but occur in widespread sheetlike lenses (figure 2.37). In some cases, shelf carbonates contain sizable amounts of sand, as in the San Andres (Lee and Girty 1909; Needham and Bates 1943; Kottlowski et al. 1956), Grayburg (Dicky 1940; Moran 1954; Tait et al. 1962) and Queen (Crandall 1929; Moran 1954) formations. Similar but younger formations of shelf deposits include the Seven Rivers (Meinzer et al. 1926; Dickey 1940; Sarg 1977, 1981; Hurley 1978), Yates (Gester and Hawley 1929; Mear and Yarborough 1961) and Tansill (Deford and Riggs 1941; Neese and Schwartz 1977).

Seaward Shoals and Barrier Islands. The transition from the massive Capitan Limestone into coeval shelf deposits is abrupt and marked by a low-angle rollover (figure 2.41). In Slaughter Canyon (figure 2.39), Yurewicz (1977) showed that the shelf sediments are horizontal in their most landward position, while seaward, the dip of the bedding changes so that over a distance of approximately 1,000 ft (300 m), the shelf sediments descend some 100 ft (30 m). Shelf beds of the Goat Seep dip seaward too, but the angle of dip is less than that observed in the Capitan shelf equivalents. Where this shelf-to-basin transition is exposed, the bedded shelf grainstones and packstones with occasional quartz sandstone commonly end abruptly in

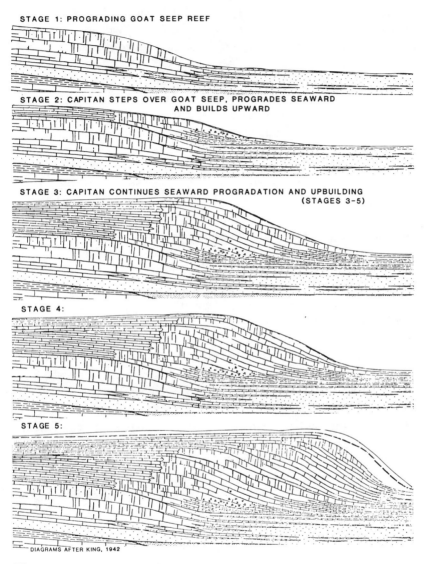

STAGE 1: PROGRADING GOAT SEEP REEF

STAGE 2: CAPITAN STEPS OVER GOAT SEEP, PROGRADES SEAWARD
AND BUILDS UPWARD

STAGE 3: CAPITAN CONTINUES SEAWARD PROGRADATION AND UPBUILDING
(STAGES 3-5)

STAGE 4:

STAGE 5:

DIAGRAMS AFTER KING, 1942

Figure 2.41
Schematic cross sections illustrating the vertical and lateral evolution
of the Capitan margin. Successive stages record the interplay between
carbonate accumulation and relative sea-level change. Such large-scale
changes can be seen on seismic profiles and have been documented
by core studies in the subsurface (Ward et al. 1986).

the massive Capitan. However, these grainstones and packstones are interbedded with lime siltstone and mudstone equivalents of the Capitan, which extend back into and lense out between grainstone and packstone beds.

Lying between the massive Capitan Limestone and the more widespread, reservoir-bearing cyclic and laminated shelf dolomites and sandstone of the Carlsbad Group is a narrow belt of well-bedded, thin limestones and dolomites, with occasional sandstones and abundant tepees. The limestones and dolomites are commonly grainstones consisting of silt-sized intraclasts, well-sorted fossils, *Mizzia*, and pisolites. The outer shelf equivalent of the Goat Seep shelf edge differs in that it lacks tepees, pisolites, and coated grains (Crawford 1981). The shelf margin muds and fossiliferous muds in the Carlsbad Group grade shelfward into subtidal and intertidal grainstones, which in turn grade into a pisolite and tepee barrier-island facies (Kendall 1969; Neese and Schwartz 1977; and Yurewicz 1977).

The width of the grainstone deposits, interpreted to have formed largely on subtidal flats, is extremely variable, and ranges from a few hundred feet to as much as 0.5 mi (0.8 km; Neese and Schwartz 1977). Bedding thickness in the outer-shelf facies varies from several inches to as much as 8 to 10 ft (2–3 m). Sedimentary structures observed in the subtidal to intertidal grainstones range from steep cross-beds to low-angle features that may be associated with the swash zone in front of barrier islands.

The pisolitic and tepee facies commonly extend to within 2 mi (3.2 km) of the break in slope, but the distance varies from 0.125–5.0 mi (0.4–8.0 km; Kendall 1969; Assereto and Kendall 1977). In contrast to the laterally adjacent grainstones (Yurewicz 1977), the pisolitic and tepee deposits contain evidence of exposure including fenestral cavities and desiccation cracking, in which early marine aragonitic cements accumulated in a supratidal-to-intertidal setting, probably within a line of barrier islands just back from the break in slope (Assereto and Kendall 1977). The quartz sands associated with the grainstones and tepee facies occur in even beds up to a meter or so thick and with few sedimentary structures.

Middle Shelf Lagoon. These shelf deposits interfinger with barrier facies downdip and downsection, but are in abrupt contact with evaporites updip (figure 2.37). These bedded sediments are formed by dense burrowed dolomites that are capped by algal laminated dolomites and are thought to be a shoaling-upward lagoon fill. Interbedded quartz siltstones and sandstones are not infrequent but become more common landward, where the carbonate facies may locally contain pelletal packstones and grapestone grainstones, suggesting a high-energy shoreline here (Sarg 1977; figure 2.37). A sparse fauna and flora including foraminifera, algae, and calcispheres suggts that conditions on the shelf were probably hypersaline to metahaline.

A modern analog to these facies is the Persian Gulf Khor Al Bazam described earlier, which evolved from a high-energy lagoon with well-developed carbonate beaches on its inner margin (Kendall and Skipwith 1969). As this lagoon filled, the character of the sediments along its inner margin changed to low-energy algal mat facies, and its center became muddy.

Shelf Salina and Sabkhas. The updip change from lagoonal dolomites to carbonates and evaporites with coarse quartz siltstones occurs within only about 500 to 650 ft (150–200 m; Sarg 1977). This facies transition is exposed in the Seven Rivers Embayment (figure 2.39) over an area about 3 miles (5 km) wide and extending about 50 mi (80 km) along facies strike. The evaporite facies consists predominantly of mosaic gypsum layers (3–20 ft, or 1–7 m thick), interbedded with dolomite and quartz siltstone deposited in a subaqueous environment (Sarg 1977). Contrary to Dunham (1972) and Sarg (1977), Ward et al. (1986) think that that the evaporites and siliciclastics represent deposition within playas and sabkhas formed over and within deflation flats behind an old shoreline topographically below the adjacent seawad lagoon, in a manner similar to what occurred in Lake MacLeod, north of Shark Bay in Australia (Handford et al. 1984; Logan 1981), or the coastal lagoons along the South Australia coast (Burne 1982; Warren 1982a, 1982b). Most of the gypsums were thus deposited in standing bodies of water, though Ward et al. (1986), believe that some of the evaporites were also deposited in the displacive mode found in coastal sabkhas of Abu Dhabi (Kendall 1969). The matrix-rich composition of these sediments favors sabkha, but geochemical data presented by Sarg (1977) argues against this: a classic unresolved geologic dilemma. Although these evaporite facies probably have both origins (sabkha and stand-ing body of water), they do form the major updip seal and trap for the hydrocarbons found in the subsurface of the Permian Basin. Lo-cally, the quartz sands associated with this facies, particularly in the Queen Formation, show evidence of local subaqueous channeling. Faunal associations and the interbedded carbonates suggest tidal channel fill.

The importance of reservoir development in Guadalupian shelf do-lomites, coupled with the laterally adjacent seal-forming evaporites, cannot be understated. Ward et al. (1986) show that hydrocarbon production occurs from within the shelf deposits of the San Andres, Grayburg, and Queen equivalents of the Goat Seep Reef, and of the Seven Rivers, Yates, and Tansill equivalents of the Capitan Reef (fig-ure 2.40). Porosities in the reservoirs are characteristically due to the presence of intercrystalline and moldic pores in bioturbated dolo-wackestones and dolopackstones. Moldic porosity formed from the dissolution of fusilinids and bivalve fragments. Porosities generally range between 7% and 15%, and permeabilities, although sometimes higher, are characteristically less than 10 md. Within a reservoir zone, the permeability may vary significantly, whereas the porosity remains relatively constant. This relationship in a clean dolomite is one that causes obvious exploration and exploitation problems, and unfor-tunately it is extremely difficult to predict ahead of the drill. The other significant reason for the loss of reservoir quality is plugging by evaporites. Anhydrite, gypsum, and halite form seals both updip of and capping the significant oil fields of Guadalupian age; these evaporites also occlude porosity in carbonates and sandstones, add-ing further to the sealing potential of the updip equivalents. These evaporites can also complicate reservoir distributions, because their occurrence across the shelf can be local.

Production from sandstone, siltstone, or dolomite with average po-rosities of 18% typifies the Queen, Seven Rivers, and Yates forma-tions (figure 2.42). During deposition of these formations, the Mid-land Basin, most of the Central Basin Platform, the Eastern Shelf, and the Northwest Shelf were sites of evaporite deposition. These shelves prograded toward the Delaware Basin: a minor amount to-ward the west, off the Central Basin Platform, where the slope was steeper and the shelf margin and associated shelf deposits stacked one

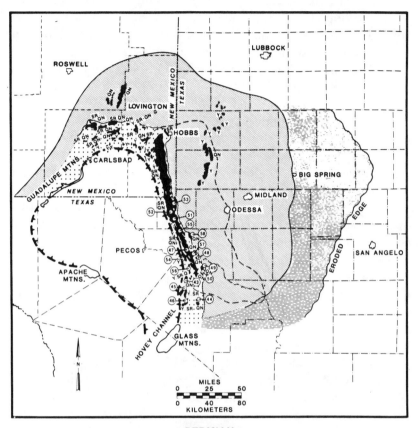

PERMIAN
UPPER GUADALUPE SERIES
YATES-SEVEN RIVERS-QUEEN FORMATIONS

⊤⊤ CAPITAN "REEF" FRONT (SUBSURFACE)

BACK REEF SANDSTONE
AND DOLOMITE

EVAPORITE FACIES – RED SANDSTONE,
RED SILT AND BEDDED ANHYDRITE.
POROSITY PLUGGED BY EVAPORITES.

⑤⑷ OILFIELDS LISTED IN TABLES

RED SANDSTONE, RED SHALE,
ANHYDRITE AND HALITE.

◗ OIL OR GAS FIELD

– – – OUTLINE CENTRAL BASIN PLATFORM

ʸ YATES PRODUCTION
ˢᴿ SEVEN RIVERS PRODUCTION
ᵠᴺ QUEEN PRODUCTION

Figure 2.42
Simplified geologic map of Yates, Seven Rivers, and Queen forma-
tions in the Permian Basin, showing the locations of producing fields
(Ward et al. 1986).

upon another; but a significant amount in southeastern New Mexico, where the slope was more gentle. The evaporite facies of the Queen Formation recognized in the Guadalupe Mountains occurs 15 to 18 mi (25–30 km) behind the Capitan Reef, whereas it lies 12 to 16 miles (15–25 km) behind the Capitan Reef along the western edge of the Central Basin Platform. The downdip limit of evaporative facies has not been defined exactly from surface geology in the Seven Rivers Formation, but it is present in the Seven Rivers Embayment approximately 8 mi (13 km) shelfward from the reef front along the western edge of the Central Basin Platform, a relationship Ward et al. (1986) claim was used by geologists in the 1940s to predict the downdip occurrence of the reef in the subsurface prior to drilling.

Fields producing from the Queen are a greater distance from the shelf margin than the Seven Rivers or Yates fields (figure 2.42). The fields along the northern margin of the Delaware Basin are designated as producing from the Queen, Seven Rivers, or Yates Formation, whereas along the western edge of the Central Basin Platform, the producing formation is undesignated. The Queen, Seven Rivers, and Yates produce with little or no break in production for approximately 90 mi (144 km) along the western edge of the Central Basin Platform (figure 2.42). The field boundaries are industrial and not geological. Five fields along this trend had estimated original primary reserves of 100 MM barrels of oil or greater, with the North Ward–Estes Field in Ward and Winkler counties, Texas, being the largest, with estimated original primary reserves of 300 MM barrels of oil. The production from the Queen, Seven Rivers, and Yates was in excess of 1.3 MM barrels of oil as of early 1980.

The Tansill Formation produces erratically along the Northwestern Shelf and the western edge of the Central Basin Platform, from small fields in back-reef dolomites. Hydrocarbon fields in the Tansill of southeastern New Mexico (figure 2.43) include Stateline, Sioux, Comanche, Seamow, and Parallel.

Permian Basin Example: Conclusions

Carbonate reservoir geometry and stratigraphy in the Permian Basin vary in response to paleogeographic setting. Facies are largely controlled by the proximity of the depositional basin to open-marine

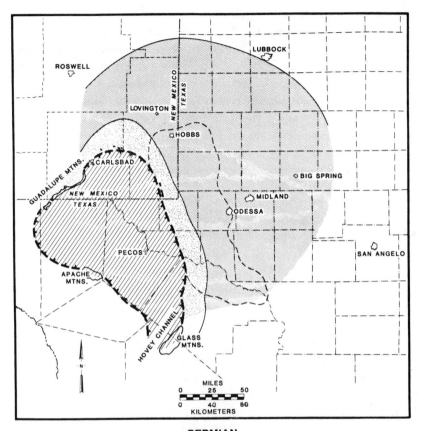

PERMIAN
UPPER GUADALUPE SERIES
TANSILL FORMATION

⊤ ⊤ CAPITAN "REEF" FRONT (SUBSURFACE)

▨ LAMAR LIMESTONE

☐ BACK REEF DOLOMITE

☐ BACK REEF ANHYDRITES AND EVAPORITES

– – – OUTLINE CENTRAL BASIN PLATFORM

⌐ OIL OR GAS FIELD

Figure 2.43
Simplified geologic map of Tansill Formation in the Permian Basin
(Ward et al. 1986).

circulation and the porosity of the depositional setting within the basin. Some of these carbonate facies extend into the basin as thin, widespread, dark limestones. The basin margin facies are formed by a massive crustal bank or reef, and an apron of sediment that was shed from the crest in thick irregular beds. This margin interfingers with thinner, regularly bedded, widespread, heterogeneous shelf carbonates that formed on subtidal sand flats and barrier islands characterized by pisolites and tepees. The major shelf carbonate facies is formed by cycles of dense dolomite and laminated dolomite probably deposited in a lagoon. These pinch out shelfward into widespread playa and sabkha evaporites, quartz sandstone and siltstones, and laminated dolomites.

CONCLUSION

Sabkhas form shoaling sequences in both modern and ancient sea-marginal settings. Most of the evaporites in a sabkha are deposited in the capillary zone as intrastratal displacive nodules and crystals. Idealized sabkha sequences are shoaling cycles with displacive evaporites in the upper part of the salt flat. Each cycle is capped by an erosion surface, the result of displacive crystal growth jacking up the surface of the wet mudflat into the vadose zone. Grainstone and sandy reservoirs in marine sabkhas are in the subtidal and intertidal sand bodies sealed by evaporites. Dolomitized sabkha mudstones can form hydrocarbon reservoirs with good intercrystalline porosity.

Subaqueous interpretations and models for ancient sea-marginal evaporites have been given a footing in the Holocene in the last decade with studies in Africa, Australia, the Mediterranean, and the Middle East. Our understanding of ancient sulfate textures from subaqueous settings is growing year by year as the complexities of the gypsum-anhydrite transition are unraveled. A large step forward was made when vertically aligned and elongate anhydrite nodules were recognized as ancient buried equivalents of subaqueous gypsum crystals (Warren and Kendall 1985). This knowledge, coupled with an understanding of the various vertical sequences laid down under different hydrological settings, is allowing more and more detailed environmental interpretations to be made of ancient evaporites. Stability of the water column and its salinity are paramount in controlling the type of succession. When conditions are stable, large bottom-nucleated crystals are deposited; under more changeable conditions of salinity and energy level, mechanically reworked accumulations of evaporites are deposited. The rapid deposition rate of subaqueous evaporites compared to rates of change in the water level in the basin means many evaporite sequences start off in relatively deep water depths (few to tens of meters) and quickly infill the basin to the water level. Any further evaporite deposition is in the capillary zone, and sabkha textures are the result.

Sabkha and subaqueous signatures are not mutually exclusive, and in the ancient, due to diagenetic alteration, these two facies may be difficult to distinguish using fabric alone. With the infill of an ancient subaqueous basin, the depositional signature will pass upsection from subaqueous to sabkha. Many ancient sabkhas probably held local brine ponds filled with subaqueous evaporites, and many large-scale evaporite basins filling with subaqueous salts were fringed and later covered by sabkhas. Changes in water level created alternations of sabkha and subaqueous signatures, a rise in water level changed deposition from sabkha to subaqueous, while a fall in brine level probably "sabkhaized" many subaqueous sequences. The next stage in the study of sea-marginal settings will be the detailed modeling of the hydrology in modern areas where the factors controlling dolomitization and evaporite plugging can be measured, understood, and then applied to understanding porosity distribution in ancient reservoirs.

ACKNOWLEDGMENTS

The authors express appreciation for the use of data, both published and unpublished, in this article. Christopher Kendall extends his thanks to his colleagues Godfrey Butler, Mitch Harris, and Bob Ward, with whom he has published papers on some of the examples given in this article. He also thanks geologists who have taken him into the field and shown their results. We would like to thank the members of the USC Basin Modeling Industrial Association, including Arco, Chevron, Cities Service, Lytton Industries (Core Labs and Western Geophysical), Marathon, Norsk Hydro, Phillips, Saga, Statoil, Sun, Texaco, and Union, who supported some of the drafting of the illustrations used in this paper. In particular we thank Brian Logan, who guided Christopher Kendall through Lake MacLeod and whose work we draw on to describe this playa. John Warren would like to thank the Owen Coates Foundation of the University of Texas for its financial support of this article, and ADNOC for its financial support of his studies in Abu Dhabi. We also thank James Sadd, who offered editorial criticism.

REFERENCES

Assereto, R. L. A. M. and C. G. St. C. Kendall. 1977. Nature, origin, and classification of peritidal teepee structures and related breccias. *Sedimentology* 24:153–210.

Babcock, J. A. 1973. The role of algae in the formation of the Capitan Limestone (Permian, Guadalupian), Guadalupe Mountains, West Texas–New Mexico. Ph.D. diss., University of Wisconsin, Madison.

Babcock, J. A. 1977. Calcareous algae, organic boundstones, and the genesis of the upper Capitan Limestone (Permian, Guadalupian), Guadalupe Mountains, West Texas and New Mexico. In *Upper Guadalupian Facies,* pp. 3–44. See Hileman and Mazzullo 1977.

Babcock, L. C. 1977. The paleontology of the Lamar Limestone. In *Upper Guadalupian Facies,* pp. 357–390. See Hileman and Mazzullo 1977.

Bornemanne, E., J. H. Doveton, and P. N. St. Clair. 1982. *Lithofacies Analysis of the Viola Limestone in South Central Kansas.* Kansas Geological Survey Petrophysical Series, no. 3. Wichita, Kans.

Boyd, D. W. 1958. *Permian Sedimentary Facies, Central Guadalupe Mountains, New Mexico.* New Mexico Bureau of Mines and Mineral Resources Bulletin, vol. 49. Albuquerque, N. Mex.

Budai, J. M., K. C. Lohman, and R. M. Owen. 1984. Burial dedolomite in the Mississippian Madison Limestone, Wyoming and Utah Thrust Belt. *Journal of Sedimentary Petrology* 54:276–288.

Butler, G. P. 1969. Modern evaporite deposition and geochemistry of co-existing brines: The sabkha, Trucial Coast, Arabian Gulf. *Journal of Sedimentary Petrology* 39:70–89.

Butler, G. P. 1970. Holocene gypsum and anhydrite of the Abu Dhabi sabkha, Trucial Coast: An alternative explanation of origin. In J. L. Rau and L. F. Delwig, eds., *Third Symposium on Salt* 1:120–152. Cleveland: N. Ohio Geological Society.

Butler, G. P., P. M. Harris, and C. G. St. C. Kendall. 1982. Recent evaporites from the Abu Dhabi coastal flats. In C. Robertson Handford, Robert G.

Loucks, and Graham R. Davies, eds., *Depositional and Diagenetic Spectra of Evaporites*, pp. 33–64. SEPM Core Workshop no. 3. Tulsa, Okla.

Burne, R. V. 1982. Interactions between saline continental groundwaters and peritidal carbonates, Spencer Gulf, South Australia (abs.). *Abstracts, 11th Intern. Congress on Sedimentology*, p. 157. Hamilton, Ontario.

Crandall, K. H. 1929. Permian stratigraphy of southeastern New Mexico and adjacent parts of western Texas. *AAPG Bulletin*, 13:927–944.

Crawford, G. A. 1981. Depositional history and diagenesis of the Goat Seep Dolomite (Permian, Guadalupian), Guadalupe Mountains, West Texas–New Mexico, Ph.D. diss., University of Wisconsin, Madison.

Curtis, R., G. Evans, D. J. J. Kinsman, and D. J. Shearman. 1963. Association of dolomite and anhydrite in the recent sediments of the Persian Gulf. *Nature* 197:679–680.

De Ford, R. K. and G. D. Riggs. 1941. Tansill Formation, West Texas and southeastern New Mexico. *AAPG Bulletin* 25:1713–1728.

Dicky, R. I. 1940. Geologic section from Fisher County through Andrews County, Texas, to Eddy County, New Mexico. In R. K. De Ford and E. R. Lloyd, eds., *West Texas–New Mexico Symposium*, part 1. *AAPG Bulletin* 24:37–51.

Dolton, G. L., A. B. Coury, S. E. Frezon, K. Robinson, K. L. Varnes, J. M. Wunder, and R. W. Allen. 1979. *Estimates of Undiscovered Oil and Gas, Permian Basin, West Texas and Southeast New Mexico*. U.S. Geological Survey, Open File Report no. 79-838. Washington, D.C.

Dunham, R. J. 1972. *Capitan Reef, New Mexico and Texas: Facts and Questions To Aid Interpretation and Group Discussion*. Permian Basin Section SEPM, Publication no. 72-14. Midland, Tex.: SEPM.

Esteban, M. and L. C. Pray. 1977. Origin of the pisolite facies of the shelf areas. In *Upper Guadalupian Facies*, pp. 479–486. See Hileman and Mazzullo 1977.

Evans, G., V. Schmidt, P. Bush, and H. Nelson. 1969. Stratigraphy and geologic history of the sabkha, Abu Dhabi, Persian Gulf. *Sedimentology* 12:145–159.

Frenzel, H. N. 1962. The Queen-Grayburg–San Andres problem solved. In G. Wilde et al., eds., *Permian of the Central Guadalupe Mountains, Eddy County, New Mexico: A Guidebook*, pp. 87–90. West Texas, Roswell, and Hobbs Geological Societies, Publication no. 62-48. Roswell, N. Mex.

Galley, J. E. 1958. Oil and geology in the Permian Basin of Texas and New Mexico. In L. G. Weeks, ed., *Habitat of Oil*, pp. 395–446. Tulsa, Okla.: AAPG.

Galley, J. E. 1971. Summary of petroleum resources in Paleozoic rocks of region 5, North, Central, and West Texas and eastern New Mexico. In I. H. Cram, ed., *Future Petroleum Provinces of the U.S.—Their Geology and Potential*, pp. 726–737. AAPG Memoir no. 15. Tulsa, Okla.

Gester, G. H. and H. J. Hawley. 1929. Yates Field, Pecos County, Texas.

In *Structure of Typical American Oil Fields* 2:480–499. AAPG Symposium. Tulsa, Okla.

Hahn, F. 1912. Untermeerische Gleitung bei Trenton Falls (Nordamerika) und ihr Verhaltnis zu ahnlichen Storungsbildern. *Neus JG. Min. Geol. Palaont.* 36:1–41.

Handford, C. R., A. C. Kendall, D. R. Prezbindowski, J. B. Dunham, and B. W. Logan. 1984. Salina-margin tepees, pisoliths, and aragonite cements, Lake MacLeod, western Australia: Their significance in interpreting ancient analogs. *Geology* 12:523–527.

Harms, J. C. 1974. Brushy Canyon Formation, Texas: A deep-water density current deposit. *GSA Bulletin* 85:1763–1784.

Hartman, J. K. and L. R. Woodard. 1971. Future petroleum resources in post-Mississippian strata of North, Central, and West Texas and eastern New Mexico. In I. H. Cram, ed., *Future Petroleum Provinces of the U.S.— Their Geology and Potential,* pp. 752–800. AAPG Memoir no. 15. Tulsa, Okla.

Hayes, P. T. 1964. *Geology of the Guadalupe Mountains, New Mexico.* U.S. Geological Survey Professional Paper no. 446. Washington, D.C.

Hileman, M. E. and S. J. Mazzullo, eds. 1977. *Upper Guadalupian Facies, Permian Reef Complex, Guadalupe Mountains, New Mexico and West Texas.* Permian Basin Section SEPM, Publication no. 77-16. Midland, Tex.: SEPM.

Holliday, D. W. 1967. Secondary gypsum in Middle Carboniferous rocks of Spitsbergen. *Geological Magazine,* 104:171–177.

Houbolt, J. J. H. C. 1957. Surface sediments of the Persian Gulf near the Qatar Peninsula. Ph.D. diss., University of Utrecht. Den Haag: Mouton and Co.

Hsü, K. J. and J. Schneider. 1973. Progress report on dolomization-hydrology of Abu Dhabi sabkhas, Arabian Gulf. In B. H. Purser, ed., *The Persian Gulf,* pp. 409–422. New York: Springer-Verlag.

Hurley, N. 1978. Facies mosaic of the Lower Seven Rivers Formation (Permian), North McKittrick Canyon, Guadalupe Mountains, New Mexico. Master's thesis, University of Wisconsin, Madison.

Jacka, A. D., R. H. Beck, L. C. St. Germain, and S. C. Harrison. 1968. Permian deep sea fans of the Delaware Mountain Group (Guadalupian), Delaware Basin. In B. Silver, ed., *Guadalupian Facies, Apache Mountain Area, West Texas: Symposium and Guidebook,* pp. 49–67. Permian Basin Section SEPM, Publication no. 68-11. Midland, Tex.: SEPM.

Jones, T. S. 1953. *Stratigraphy of the Permian Basin of West Texas.* 1–69. Midland, Tex.: West Texas Geological Society.

Kendall, C. G. St. C. 1969. An environmental reinterpretation of the Permian evaporite/carbonate shelf sediments of the Guadalupe Mountains. *GSA Bulletin* 80:2503–2525.

Kendall, C. G. St. C. and P. A. d'E. Skipwith. 1968. Recent algal mats of a Persian Gulf lagoon. *Journal of Sedimentary Petrology* 38:1040–1058.

Kendall, C. G. St. C. and P. A. d'E. Skipwith. 1969. Geomorphology of a recent shallow-water province; Khor al Bazam, Trucial Coast, Southwestern Persian Gulf. *GSA Bulletin* 80:865–892.

Kendall, C. G. St. C. and W. Schlater. 1981. Carbonates and relative changes in sea level. *Marine Geology*, 44:181–212.

King, P. B. 1942. Permian of West Texas and southeastern New Mexico. *AAPG Bulletin* 26:535–763.

—— 1948. *Geology of the Southern Guadalupe Mountains, Texas*. U.S. Geological Survey Professional Paper no. 215. Washington, D.C.

Kinsman, D. J. J. 1964. The recent carbonate sediments near Halat al Bahrani, Trucial Coast, Persian Gulf. In L. M. J. U. van Straaten, ed., *Deltaic and Shallow Marine Deposits*, pp. 189–192, vol. 1 of *Developments in Sedimentology*. Amsterdam: Elsevier.

Koss, G. M. 1977. Carbonate mass flow sequences of the Permian Delaware Basin, West Texas. In *Upper Guadalupian Facies*, pp. 391–408. see Hileman and Mazzullo 1977.

Kottlowski, F. E., R. H. Flower, M. L. Thompson, and R. W. Foster. 1956. *Stratigraphic Studies of the San Andres Mountains, New Mexico*. New Mexico Bureau of Mines and Mineral Resources Memoir no. 1. Albuquerque, N. Mex.

Lee, W. T. and G. H. Girty. 1909. *The Manzano Group of the Rio Grande Valley, New Mexico*. U.S. Geological Survey Bulletin no. 389. Washington, D.C.

Lloyd, E. R. 1929. Capitan Limestone and associated formations of New Mexico and Texas. *AAPG Bulletin* 13:645–658.

Logan, B. W. 1981. *Modern Carbonates and Evaporite Sediments of Shark Bay and Lake MacLeod, Western Australia*. Field Excursion Guidebook. Geol. Soc. Australia, 5th Australian Geol. Conv. Perth, Western Australia.

Longacre, S. A. 1980. Dolomite reservoirs from Permian biomicrites. In R. B. Halley and R. G. Loucks, eds., *Carbonate Reservoir Rocks*, pp. 105–117. SEPM Core Workshop no. 1. Tulsa, Okla.

Longacre, S. A. 1983. A subsurface example of a dolomitized middle Guadalupian (Permian) reef from West Texas. In P. M. Harris, ed., *Carbonate Buildups*, pp. 304–326. SEPM Core Workshop no. 4. Tulsa, Okla.

Mazzullo, S. J. 1977. Synsedimentary diagenesis of reefs. In *Upper Guadalupian Facies*, pp. 323–356. See Hileman and Mazzullo 1977.

Mazzullo, S. J. 1981. Facies and Burial Diagenesis of a Carbonate Reservoir: Chapman Deep (Atoka) Field, Delaware Basin, Texas. *AAPG Bulletin* 65:850–865.

Mear, C. E. and D. V. Yarborough. 1961. Yates Formation in Southern Permian Basin of West Texas. *AAPG Bulletin* 45:1545–1556.

Meinzer, O. E., B. C. Renick, and K. Bryan. 1926. *Geology of no. 3 Reservoir Site of the Carlsbad Irrigation Project, New Mexico, with Respect to Water-tightness.* U.S. Geological Survey, Water Supply Paper no. 580-A. Washington, D.C.

Moiola, R. J. and E. D. Glover. 1965. Recent anhydrite from Clayton Playa, Nevada. *American Mineralogy* 50:2063–2069.

Moore, C. H. and Y. Druckman. 1981. Burial diagenesis and porosity evolution, Upper Jurassic Smackover, Arkansas and Louisiana. *AAPG Bulletin* 65:597–628.

Moran, W. R. 1954. Proposed type sections for the Queen and Grayburg formations of Guadalupian age in the Guadalupe Mountains, Eddy County, New Mexico (abs.). *GSA Bulletin* 65:1288.

Needham, E. C. and R. L. Bates. 1943. Permian type sections in central New Mexico. *GSA Bulletin* 54:1653–1667.

Neese, D. G. and A. M. Schwartz. 1977. Facies mosaic of the upper Yates and lower Tansill Formations, Walnut and Rattlesnake canyons, Guadalupe Mountains, New Mexico. In *Upper Guadalupian Facies*, pp. 437–450. See Hileman and Mazzullo 1977.

Neumann, A. C., J. W. Kofoed, and G. H. Keller. 1977. Lithoherms in the straits of Florida. *Geology* 5:4–10.

Newell, N. D., J. K. Rigby, A. G. Fischer, A. J. Whiteman, J. E. Hickox, and J. S. Bradley. 1953. *The Permian Reef Complex of the Guadalupe Mountains Region, Texas and New Mexico.* San Francisco: Freeman.

Ogniben, L. 1957. Petrografia della Serie Solifera siciliana e considerzoini geologische relative: Memorie descritt. *Carta Geol. D'Italia* 23:1–275.

Olliver, J. G. and J. K. Warren. 1976. Stenhouse Bay gypsum deposit. *South Australian Min. Res. Rev.,* no. 145, pp. 11–23.

Patterson, R. J. and D. J. J. Kinsman. 1981. Hydrologic framework of a sabkha along Arabian Gulf. *AAPG Bulletin* 65:1457–1475.

Playford, P. E. 1980. Devonian "Great Barrier Reef" of Canning Basin, western Australia. *AAPG Bulletin* 64:814–840.

Pray, L. C. and M. Esteban. 1977. Road logs and locality guides. In *Upper Guadalupian Facies.* See Hileman and Mazzullo 1977.

Privett, D. W. 1959. Monthly charts of evaporation from the northern Indian Ocean (including the Red Sea and Persian Gulf). *Quarterly Journal of the Royal Meteorological Society* 85:424–478.

Purser, B. H. and G. Evans. 1973. Regional sedimentation along the Trucial Coast, SE Persian Gulf. In B. H. Purser, ed., *The Persian Gulf,* pp. 211–232. New York: Springer-Verlag.

Sarg, J. F. 1977. Sedimentology of the carbonate-evaporite facies transition of the Seven Rivers Formation (Guadalupian, Permian) in southeast New Mexico. In *Upper Guadalupian Facies,* pp. 451–478. See Hileman and Mazzullo 1977.

Sarg, J. F. 1981. Petrology of the carbonate-evaporite facies transition of the

Seven Rivers Formation (Guadalupian, Permian), Southeast New Mexico. *Journal of Sedimentary Petrology,* 51:73–95.

Schmidt, V. 1977. Inorganic and organic reef growth and subsequent diagenesis in the Permian Capitan Reef Complex, Guadalupe Mountains, Texas, New Mexico. In *Upper Guadalupian Facies,* pp. 93–132. See Hileman and Mazzullo 1977.

Scholle, P. A. and R. B. Halley. 1980. *Upper Paleozoic Depositional and Diagenetic Facies in a Mature Petroleum Province. A Field Guide to the Guadalupe and Sacramento Mountains.* U.S. Geological Survey, Open File Report no. 80-383. Washington, D.C.

Shumard, G. G. 1858. Observations on the geological formations of the country between Rio Pecos and Rio Grande in New Mexico near the line of the 32nd parallel. *St. Louis Academy of Science Transactions* 1:273–289.

Silver, B. A. and R. G. Todd. 1969. Permian cyclic strata, northern Midland and Delaware basins, West Texas and southeastern New Mexico. *AAPG Bulletin* 53:2223–2251.

Tait, D. B., J. L. Ahlen, A. Gordon, G. L. Scott, W. S. Motts, and M. E. Spitler. 1962. Artesia group (Upper Permian) of New Mexico and West Texas. *AAPG Bulletin* 46:504–517.

Todd, R. G. 1976. Oolite-bar progradation, San Andres Formation, Midland Basin, Texas. *AAPG Bulletin* 60:907–925.

Tyrrell, W. W. 1969. Criteria useful in interpreting environments of unlike but time-equivalent carbonate units (Tansill-Capitan-Lamar), Capitan Reef Complex, West Texas and New Mexico. In G. M. Friedman, ed., *Depositional Environments in Carbonate Rocks,* pp. 80–97. SEPM Special Publication no. 14. Tulsa, Okla.

Von der Borch, C., B. Bolton, and J. K. Warren. 1977. Environmental setting and microstructure of subfossil lithified stromatolites associated with evaporites, Marion Lake, South Australia. *Sedimentology* 24:693–708.

Ward, R. F., C. G. St. C. Kendall, and P. M. Harris. 1986. Late Permian (Guadalupian) facies and their association with hydrocarbons. *AAPG Bulletin* 70:239–262.

Warren, J. K. 1982a. The hydrological setting, occurrence, and significance of gypsum in late Quaternary salt lakes in South Australia. *Sedimentology* 29:609–637.

Warren, J. K. 1982b. The hydrological significance of Holocene tepees, stromatolites, and boxwork limestones in coastal salinas in South Australia. *Journal of Sedimentary Petrology* 52:1171–1201.

Warren, J. K. 1988. Sea-marginal setting: Sabkhas and salinas. *AAPG Memoir.* Tulsa, Okla. In press.

Warren, J. K. and C. G. St. C. Kendall. 1985. Comparison of sequences formed in marine sabkha (subaerial) and saline (subaqueous) settings—Modern and ancient. *AAPG Bulletin* 69:1013–1023.

West, I. M. 1965. Macrocell structure and enterolithic veins in British Purbeck gypsum and anhydrite. *Proceedings of Yorkshire Geological Society* 35:47–58.

Wilde, G., S. D. Kerr, T. McClarin, and E. Mears, eds. 1962. *Permian of the Central Guadalupe Mountains, Eddy County, New Mexico: A Guidebook.* West Texas, Roswell, and Hobbs Geological Societies, Publication no. 62-48. Roswell, N. Mex.

Williamson, C. R. 1977. Deep sea channels of the Bell Canyon Formation (Guadalupian), Delaware Basin, Texas–New Mexico. In *Upper Guadalupian Facies,* pp. 409–432. See Hileman and Mazzullo 1977.

Williamson, C. R. 1978. Depositional processes, diagenesis, and reservoir properties of Permian deep sea sandstones, Bell Canyon Formation, Texas–New Mexico. Ph.D. diss., University of Texas, Austin.

Young, A. and J. C. Vaugh. 1957. Addis-Johnson-Foster–South Cowden Field, Ector County, Texas. In *Oil and Gas Fields in West Texas,* pp. 16–22. Bureau of Economic Geology, University of Texas at Austin, Publication no. 57-16. Austin.

Yurewicz, D. A. 1976. Sedimentology, paleoecology, and diagenesis of the massive facies of the lower and middle Capitan Limestone (Permian), Guadalupe Mountains, New Mexico and West Texas. Ph.D. diss., University of Wisconsin, Madison.

Yurewicz, D. A. 1977. Origin of the massive facies of the lower and middle Capitan Limestone (Permian), Guadalupe Mountains, New Mexico and West Texas. In *Upper Guadalupian Facies,* pp. 45–92. See Hileman and Mazzullo 1977.

3

Depositional Interaction of Siliciclastics and Marginal Marine Evaporites

C. ROBERTSON HANDFORD

The chance discovery of anhydrite in an Arabian Gulf coastal salt flat by Curtis and others (1963) marked the beginning of landmark sedimentological studies of the sabkhas near Abu Dhabi, United Arab Emirates, by Shearman, Evans, Kinsman, and Butler during the 1960s (Till 1978). As a result of their studies, the Abu Dhabi sabkhas became the standard model against which all sabkha evaporite facies were compared. But it also became the only model available, for despite the fact that other equally impressive sabkha environments were known, modern sedimentological work was predominantly carbonate-oriented (Till 1978). The bias toward carbonate studies and documentation of sabkhas along the Arabian Gulf coast was undoubtedly boosted by the realization that the vertical succession and the

C. Robertson Handford is an Assistant Professor in the Department of Geology. University of Arkansas, Fayetteville.

association of supratidal dolomite and nodular anhydrite, overlying intertidal algal-laminated carbonates and burrowed subtidal sediments, form a valuable key for interpreting ancient regressive sequences of similar lithologies that make up reservoir and seal facies in certain giant petroleum fields. Thus, studies of dolomitization and its possible relationship to sabkha environments became important to exploration concepts in dolomite-evaporite strata.

Arabian Gulf research yielded immensely important concepts and information that revolutionized our understanding of evaporite deposition; however, the sabkha model of evaporite deposition was, as a result, misused and overextended. We now know that certain ancient evaporites that were initially interpreted to be of sabkha origin may actually be shallow subaqueous deposits. We also know that there are significant variations in modern sabkhas, both in terms of the physical processes that mold them and the sediments that they comprise (Handford 1981). This is most evidently manifested in coastal sabkhas that, unlike those in Abu Dhabi, are dominated by siliciclastic rather than carbonate sediments. Thus, evaporite studies have now reached the point that clastic sedimentology reached over a decade ago; that is, paralleling the recognition that one should not compare all ancient deltaic deposits to the modern Mississippi River delta system, it is now acknowledged that the Abu Dhabi system cannot be a faithful analog for all ancient sabkha evaporites. This article will survey both modern and ancient marginal marine evaporite systems dominated by a clastic matrix.

MIXING OF EVAPORITE AND SILICICLASTIC SEDIMENTS

Marginal marine evaporite systems include sabkhas (evaporitic, tidal, or wind-tidal flats) and salinas (partially to completely isolated, shallow, hypersaline bodies of seawater) whose coastal setting generally dictates that coastal processes such as tidal flooding are paramount in shaping them. But this is not always the case, as both wind and fluvial processes often have significant impact on coastlines, and hence on the style of evaporite deposition. The effects can be recorded by the sediments. Furthermore, siliciclastic sediment derived from marine, fluvial, and eolian environments can be directly introduced into marginal marine evaporite settings.

Three principal scenarios foster the introduction of siliciclastic sediment into marginal marine evaporite settings:

1. Fluvial-dominated, sandy to muddy coastal mud flats and small salinas or evaporite pans can form along the distal margins of fan deltas and coalesced alluvial fans that have prograded seaward from moderate- to high-relief coastal areas (figure 3.1). They may also form along low-relief coastlines, especially where through-flowing streams can deposit their loads as deltas or in embayments (figure 3.2). In each case the delta or fan delta system supplies the sediments from updip; a foundation across which storm

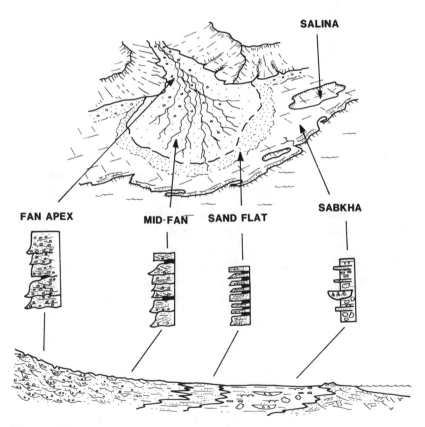

Figure 3.1
Schematic diagram of fan delta and fringing sabkha-salina environments/facies.

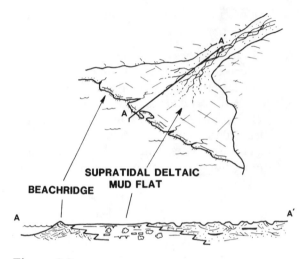

Figure 3.2
Ephemeral stream delta and associated clastic-rich sabkha developing in an embayed low-relief coastal area.

or wind tides can occasionally sweep, thereby reworking sediment and flooding the surface with marine water; and the clastic medium within which evaporites can form.

Modern settings of this type include the fan deltas fringed by narrow coastal sabkhas and salinas in the Red Sea—Gulf of Suez, the Wooramel delta and mud-rich sabkha of Gladstone Embayment in Shark Bay, Western Australia (Davies 1970), and the Ranns of Kutch, an immense sabkha-salina system that has developed near the mouth of the Indus River in India (Glennie and Evans 1976).

2. Although most of the coarse-grained sediment that reaches the sea via rivers is deposited locally, substantial quantities of sand, silt, and clay can be moved alongshore by currents and waves and eventually deposited in strike-oriented coastal environments such as barriers, lagoons and associated tidal flats, or sabkhas and salinas (figure 3.3). Both the landward shores of lagoons and the leeward sides of barriers are chief areas of tidal-flat and sabkha development. Little erosion occurs in back-barrier lagoons because they are protected from oceanic wave attack; thus mud and silt can be deposited by tides and rivers that empty into the la-

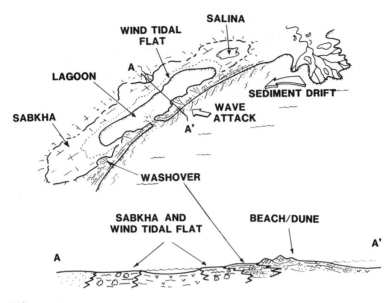

Figure 3.3
Sabkha and salina systems established alongshore from delta system and leeward of barrier complex.

goons to form prograding mud flats. In contrast, the seaward sides of barriers are frequently attacked by storms. As a result, sediment is stripped away from the shoreface, washed over and through breaches in barriers, and deposited as washover fans on the leeward sides of barriers. Tidal and wind-tidal flooding rework the finest sediment, add additional sediment, and eventually form fringing mud flats to the washovers. Laguna Madre in southern Texas (Miller 1975; Brown et al. 1977; Weise and White 1980) is a classic modern case of just such a scenario. Another example of a sabkha and salina system developing along strike from a river source of sediment is the extensive mud flats in the northwestern Gulf of California, Baja California, Mexico (Thompson 1968; Castens-Seidell 1984).

3. Most coastal evaporite settings include eolian environments, often as major components. Eolian dune complexes frequently occupy sabkha and salina margins such that there is a merging of environments and intercalation of eolian and evaporitic sediments (figure 3.4). Sand sources for the eolian systems can be direct

weathering of quartz-bearing (or feldspar) bedrock; deflation or wind-stripping of sand from alluvial plains; and wind erosion of sand-rich coastal shorelines and beaches. Among the best-known localities of sabkha and evaporite-pan deposition in eolian settings are the Jafurah sand sea, Saudi Arabia (Fryberger et al. 1983), and the Umm Said sabkha, southeastern Qatar along the Arabian Gulf (Shinn 1973).

Dip-Fed Marginal Marine Evaporite Systems

Fluvial inflow to coastal environments is through ephemeral or perennial systems, but where evaporites are forming, ephemeral inflow is favored. Inflow to coastal evaporite environments is exactly the same as that described by Hardie et al. (1978) for saline lakes. There, ephemeral streams flow only after storms or perhaps seasonally, as

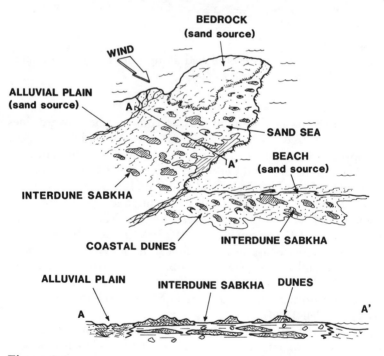

Figure 3.4
Eolian-dominated sabkhas forming in dune systems that form downwind of bedrock, alluvial plain, and coastal dune sand sources.

snow melts in the high mountains and forms significant meltwater runoff. In either case sedimentation is intermittent, occurring only when the streams are charged with runoff, and large quantities of sediment can be deposited over very short intervals of time at the mouths of these streams. In fact, this mechanism can often be the only mechanism available to introduce clastic sediment into a coastal evaporite environment.

As pointed out by numerous workers including Evans (1978), the early stages of continental breakup by rifting to form marine basins are often accompanied by the deposition of continental/marine evaporites that pass landward into thick piles of alluvial-fan and ephemeral-stream terrigenous clastic sediments. Following is a brief review (based largely upon Hardie et al. 1978) of alluvial fan and ephemeral stream environments that often flank marginal marine evaporite systems.

Rifted marine basins formed by ocean spreading are commonly bounded by high-relief block-faulted mountains and fringing alluvial fans. Each fan consists of a thick wedge of boulders, gravel, and sand that form a cone-shaped pile of sediment with a concave radial profile and a convex cross-fan profile (figure 3.1). From its apex, which is often at the mouth of a boulder-strewn floor of a mountain canyon, the fan may slope 4° or more down to its sand toe, 1,000 feet or more below and several miles away. The fan surface is laced with a radial system of braided ephemeral stream channels that emanate from the fan apex. Channels are deep and floored by boulders and coarser debris near the top, but as the sediment grades downslope into finer detritus, the channels become shallower and less distinct. They eventually lose their identity and merge with the sand-flat subenvironment that marks the transition from alluvial fan to mud flat.

Alluvial fan channels in arid regions are inactive and dry during most of the year but following intense, hot-afternoon thunderstorms, the passage of major rain-bearing fronts, or with snow melt in the high mountains, these dry channels can suddenly be transformed into actively eroding or migrating pathways of sediment transport. Although depositional activity on alluvial fans is sporadic and short lived, major depositional units are created across the fan from its apex, through the midfan, and down to the fan toe.

At the fan apex with its incised or deep channels, the sediments are usually of channel-fill, debris-flow and sieve-deposit origin. Channel fills are fining upward sequences of gravel that record deposition in earlier-cut channels. Sieve deposits consist of longitudinal tongues of frame-supported gravel through which floodwaters can sink, carrying with them finer sediment, such that only the gravel and boulders are left at the surfaces. Debris flows are thick lobes of poorly sorted max-trix-supported deposits of mud, sand, and gravel/boulders that were transported as viscous slurries of sediment.

Shallow braided channels and longitudinal bars dominate the mid-fan region. The bars are gravelly and consist of crude horizontal stratification and perhaps some planar cross-stratification at their downstream ends. They are separated by shallow channels floored by coarse sand and finer gravel. Together they form vertical se-quences of crosscutting gravel-sand lenses.

Toward and across the fan toe, braided channels become shallower and subsequently lose their identity. There a sand-flat subenviron-ment is created, with unconfined sheetflood deposits of horizontally laminated sand. Shallow ponds may form from flooding of the sand flat and result in the development of waverippled sand and suspen-sion deposition of mud drapes over sheetflood deposits. Subaerial exposure, however, normally dominates the sand-flat and fan-toe subenvironments, such that the sediment is commonly reworked by the wind. As a result, dune fields may form across these environ-ments, thus helping to create a mixed eolian-fluvial depositional rec-ord, similar to that of the desert wadis described by Glennie (1970).

Tufa or travertine can form diagnostic though mostly locally con-fined deposits around springs that discharge from the toes of alluvial fans where coarse sediment (aquifer) has intersected impermeable muds. As spring waters emerge, travertine and tufa may form mounds or sheetlike deposits around the orifices or along the banks and floors of outflow channels. In some cases, ooids and pisoids may be de-posited locally in the fluvial channels (McGannon 1975) or around the distal margins of the fan as more extensive deposits (Risacher and Eugster 1979).

Gulfs of Suez and Elat. The mountainous desert coastline of the Sinai Peninsula bordering the gulfs of Suez and Elat includes sandy coastal plains crossed by fan deltas and wadis. Sabkhas have formed along their distal fringes (Sneh and Friedman 1984). In the south, where Precambrian rocks make up the Sinai hinterland, quartz-feldspar sand is shed from the highlands and deposited as the dominant sediment in the intertidal-supratidal zones. Carbonate sediments increase in abundance into the subtidal zone as it passes offshore into coral reefs.

Sediments are reworked by longshore currents along the seaward fringes of wadis and fan deltas and result in the development of spits, sand waves, shallow protected lagoons, and tidal flats with channels. The intertidal zone is about 400 to 800 m wide. Flat-topped sand waves are present in the lower intertidal zone, while a rippled sand flat with shallow channels makes up the higher intertidal zone. Sabkhas have formed in the supratidal zone and are characterized by a flat to slightly hummocky surface commonly cemented by a halite crust. Layers of fibrous gypsum, gypsum crystal mush, and hard layers of gypsum are interbedded with sand, mud, peloids, and ooids. In some areas eolian dunes of oolitic carbonate sand migrate across the sabkhas.

Northwest Gulf of California. Another clastic-rich sabkha that has formed at the toe of an alluvial fan system in a rifted marine basin is present in the northwest Gulf of California (Castens-Seidell 1984; Thompson 1968). The Baja California peninsula consists of a mountainous spine 3,000 m high, around which numerous alluvial fans have built an apron of coarse clastic sediments over 20 km wide. These fans pass downslope into the 20-km-wide sabkha and tidal flats near the mouth of the Colorado River (figures 3.5 and 3.6). Although most of the sediment comprised by this sabkha system was derived from the Colorado River, the alluvial fans do constitute an additional source of siliciclastic sediment from updip sources.

Like other alluvial fans, those in Baja California can be subdivided into a fan apex, midfan, and a fan toe. These subenvironments have already been generally discussed, but Castens-Seidell (1984) provides the most detail on the transitional fan-toe and sand-flat subenvironments. Following is a synopsis of her descriptions.

The fan toe is laced by very shallow braided-stream washes. Small eolian dunes, cresote bushes, cacti, and other clumps of vegetation hold sediment together and divert stream flow. The washes are floored by granular to coarse sand. In vertical sequence, fan-toe deposits con-

SALTON SEA

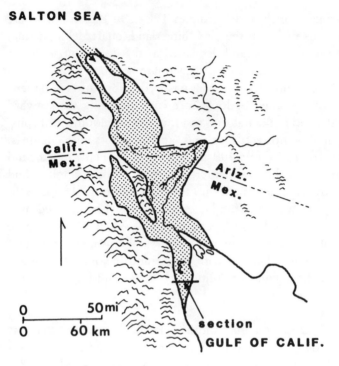

GULF OF CALIF.

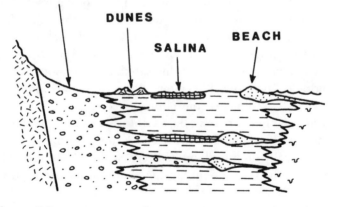

Figure 3.5
Northwest Gulf of California mud flats and schematic cross section
through the alluvial fan and mud-flat systems (modified from
Thompson 1968 and Walker 1967).

Figure 3.6
Aerial view south toward flooded supratidal mud flats and Salina Omotepec (February 1983). Fresh to brackish water covered most of the sabkha and salina for several months and dissolved evaporite deposits.

sist of shallow channel scours filled with cross-stratified pebbly sand. Additionally, there are units of discontinuous horizontally stratified sand and fine gravel, and fining upwards sedimentation units (3–8 cm thick) with pebbly bases that grade upward into flat-laminated coarse sand and thin caps of cross-stratified medium-grained sand. The latter probably represents waning current flood deposits.

Sedimentary structures are disrupted by burrowing insects, root growth, and soil development. When these three are combined, massive, mottled sedimentary units with caliche nodules are formed.

A sand flat marks the transition from fan toe to sabkha. Here the braided washes are lost on the flat sandy surface and flows become unconfined sheetfloods. Sand-flat sediments form a sandy apron and interfinger with evaporitic muds of the sabkha. Castens-Seidell (1984) recognizes proximal and distal sand-flat zones. Fitting with the flat topography, few bed forms are present, and sedimentary packages are dominated by wavy to horizontally laminated lenses of granular and micaceous sand. Some of the features are interpreted as antidune laminae, primarily because many of the laminae dip slightly upstream. In addition to the antidune laminae, there are small scours, cross-laminated sands, fining-upward sedimentation units, thin mud laminae, and lenses of gypsum-needle sand with fenestrae. Bedding can be disrupted by burrowing, root growth, desiccation, and intrasediment growth of gypsum crystals (rosettes).

Castens-Seidell (1984) notes that sediment size decreases from the proximal to distal sand flat. Fine micaceous sand, silt, and mud make up the detrital sediment in the distal parts of the sand flat. Cores and trenches show that small-scale cross-stratification is present, usually over horizontally laminated sand, overlain in turn by mud drapes bearing plant matter. Mud drapes thicken seaward concomitantly with an increase in gypsum lenses and layers. Triplet packages of sand-mud-gypsum appear, and with the first appearance of clastic gypsum sand, the transition to the gypsum-pan subenvironment is recorded. Diagenetic gypsum is also present as discoidal rosettes, overgrowths on clastic gypsum, and a crust of intergrown crystals in sediment.

With the passage of major Pacific storms, the mud flats in the northwestern Gulf of California are flooded by rainfall and runoff. In early 1983 and 1984, the entire sabkha surface was flooded (<1 m deep) and transformed into a fresh to brackish water lake (figure 3.6). Castens-Seidell (1984) observed that under these circumstances, sand-flat sediments can be reworked into strandline deposits. Swash zones of inclined-laminated sand with possibly keystone vugs form along the ephemeral beach, as do current-modified wave ripples. As floodwaters recede, these sediments are once again subjected to wind reworking, desiccation, insect burrowing, plant colonization, and efflorescent salt-crust formation. Continued reworking by the wind has led to the formation of dune fields along fan toes, sand sheet deposits, and surfaces covered by probable adhesion ripples.

Gladstone Embayment, Western Australia. An example of a fluvial-dominated sabkha forming along a low-relief coastal setting is present in Gladstone Embayment of Shark Bay, Western Australia (figures 3.7 and 3.8; see also Davies 1970). The embayment's northern

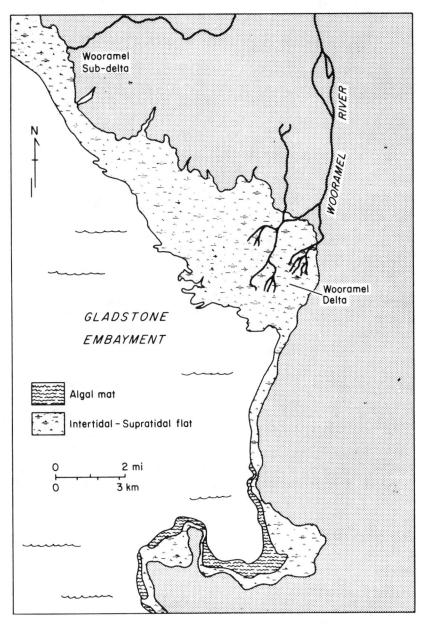

Figure 3.7
Regional setting of Wooramel Delta in Gladstone Embayment, Shark
Bay, Western Australia (modified from Davies 1970).

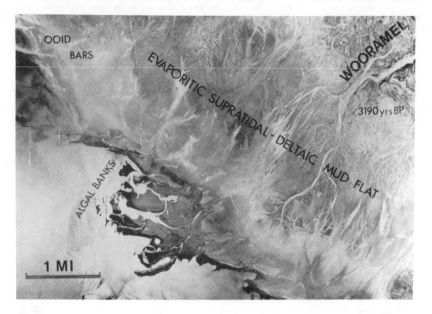

Figure 3.8
Aerial view of northwest margin of Wooramel Delta (compare to figure 3.7) showing interdigitation of braided fluvial channels and evaporitic supratidal flat (photograph courtesy of G. R. Davies).

margin is marked by the Wooramel River, an ephemeral stream that has built a prograding delta (28 sq km) across the intertidal zone. The delta is fed by two distributaries that have broken through a dune barrier that once marked the predelta coastline. Since the Wooramel River is normally dry, deposition on the delta is restricted to times of flooding after heavy rains. Deltaic sediments are red sand, silt, and clay that were deposited as a thin sheet over Recent subtidal to intertidal carbonates.

The intertidal-supratidal zones are nearly 5 km wide and in the south they are dissected by shallow tidal channels. Tidal range is less than 1 m, but when tides are combined with strong northwesterly winds that accompany storms, the sabkha is flooded.

Coastal sediments are predominantly biogenic carbonates (forams, coralline algae). Algal-laminated sediments are prominent in the intertidal zone and consist of tufted, convoluted, and smooth mat types that mechanically bind both carbonate and siliciclastic sediment. The supratidal sediments include skeletal debris washed up by high tides or reworked by the wind. As a result, some eolian dunes have formed in the supratidal zone. The porous, carbonate-rich supratidal flats are rarely encrusted with halite, but the siliciclastic mud flats flanking the delta are commonly covered by a 1- to 3-cm-thick crust of halite. Gypsum is more common than halite, however, and occurs as a mush of small, lenticular and twinned crystals in the upper 10 cm of the carbonate sediment; as large lenticular crystals >3 cm wide in fine-grained aragonitic sediment; and as large, vertically oriented sand crystals.

Cores taken by Davies (1970) in the outer part of the Wooramel delta show a succession of skeletal carbonate sands at the base, over-lain by interlaminated deltaic silt and marine carbonate and a cap of firmly crystalline gypsum mush (figure 3.9). This sequence clearly illustrates the periodic deposition of siliciclastic deltaic sediments from updip sources via flooding over intertidal-supratidal deposits of ma-rine carbonate. With exposure and evaporation of pore waters, gyp-sum is precipitated within the sediment and at the surface.

Ranns of Kutch, India. The Ranns of Kutch bordering the Arabian Sea in western India (Glennie and Evans 1976) encompass the largest marginal marine sabkha system in the world today (figure 3.10). These salt flats, which lie along the southeastern flank of the Indus River delta, extend more than 300 km inland (west to east) and are 80 to 100 km wide. Although this great salt-encrusted area is flooded an-nually by as much as 2 m of seawater (driven into the Ranns by southwest monsoon winds from the Arabian Gulf), its surface ap-pears to be just above normal high tide, thus making it a largely supratidal environment.

The Ranns of Kutch occupy grabens along the coast, and were once marine gulfs; however, the adjacent Indus and Nara rivers prob-ably contributed much of the sediment that subsequently filled the Ranns to form a vast mud flat. In addition to the sediment deposited directly onto the Ranns by fluvial flooding, significant quantities of fine sediment derived from the Indus River are forced up the Ranns by the monsoon winds, and there is also some input to the outer Ranns by longshore transport and tidal currents.

Marine flooding generally occurs only during the monsoon season,

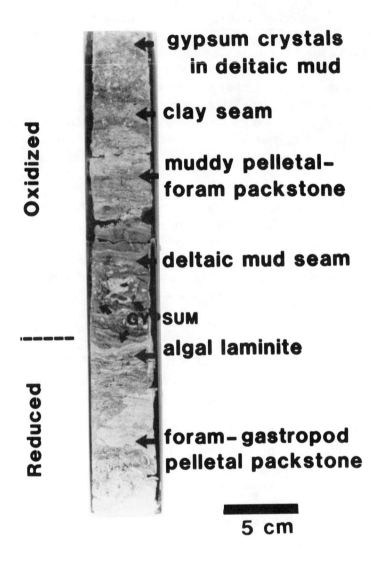

gypsum crystals
in deltaic mud

clay seam

muddy pelletal-
foram packstone

deltaic mud seam

GYPSUM

algal laminite

foram-gastropod
pelletal packstone

Oxidized

Reduced

5 cm

Figure 3.9
Wooramel Delta core with supratidal layers of gypsiferous mud and
pelletal-foram packstone overlying intertidal algal mat and subtidal
packstone (photograph courtesy of G. R. Davies).

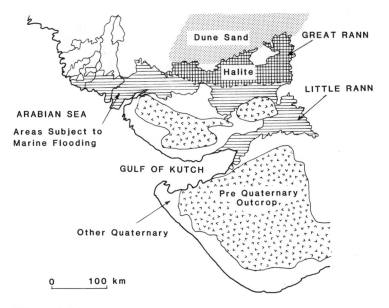

Figure 3.10
Coastal sabkha and associated environments of the Ranns of Kutch, west coast of India (modified from Glennie and Evans 1976).

and as that passes, floodwaters recede and evaporation begins. A halite crust up to 120 cm thick has been reported to form, and once the sabkhas are subaerially exposed, evaporite crystallization continues in the still-moist sediment. Glennie and Evans (1976) suggest that much of the halite crust is probably dissolved each year by flooding marine waters. Gypsum, however, may survive.

Few cores were taken from the Ranns of Kutch. Those that Glennie and Evans (1976) illustrated consist of alternating clay with silt, and sand with sand-sized gypsum crystals that grew from interstitial waters. A regressive record should be present, with subtidal sands and finer sediment overlain by gypsiferous clay and possibly thin halite layers of great lateral extent.

Strike-Fed Marginal Marine Evaporite Systems

It might appear to the casual observer flying over our modern coastlines that the final resting place of clastic sediments carried to the sea by rivers would be around river mouths and deltas. Of course, a closer examination would reveal that this is not the case, as large quantities of sediment are moved offshore and alongshore by waves, wind, and tidal currents. That which is moved alongshore in open shelf seas is principally moved by wind- and wave-driven currents; tidal currents tend to be more important in partially enclosed seas and blind gulfs (Johnson 1978). In both cases, sediment ranging from sand to clay can be added to the shorelines by nearshore processes, causing coastlines to migrate and prograde. As a result, coastal barrier-bar, spit, and lagoon systems or wide tidal flats with or without major beach ridges, chenier complexes, or dissecting tidal channels can form. Providing that arid conditions prevail, evaporites may develop and comprise a major sedimentary component in restricted lagoons, tidal flats, and salinas leeward of the coastal barriers.

Laguna Madre, Texas. Laguna Madre is a highly restricted and very shallow hypersaline lagoon that lies behind, or leeward, of Padre Island off southern Texas (figure 3.11). Brown et al. (1977), Weise and White (1980), and many others have surmised that Padre Island formed from offshore sand shoals that coalesced through spit development and migration during sea-level stabilization several thousand years ago. Sediments that make up the island and former shoal systems were derived from the Rio Grande to the south and the Colorado and Brazos rivers to the north. It is believed that these rivers transported sediment out across the continental shelf during a lower sea-level stand. With sea-level rise, these sediments were reworked by waves and currents to form the sand shoals that eventually became Padre Island. Much of Padre Island is currently situated in a zone of longshore convergence, and if so situated in the past, converging longshore currents from the south and north would have been largely responsible for island development and sediment mixing.

Padre Island is broken by just one man-made pass; thus Laguna Madre is highly restricted and is filled with hypersaline water during most of the year. Though Padre Island is a longshore current- and wave-dominated environment, Laguna Madre is protected. The lagoon is not deep enough, nor is there enough fetch to result in significant wave energy during most of the year. However, with the tropical storms and hurricanes that occasionally batter the coast, Padre

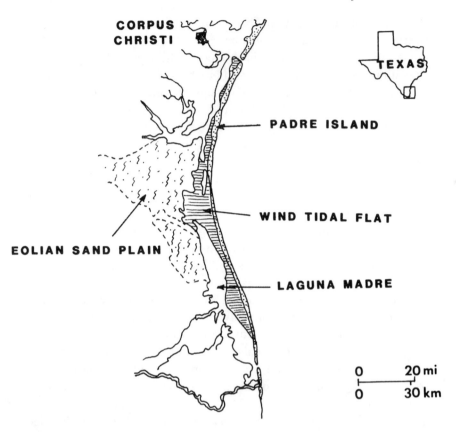

Figure 3.11
Southern Texas coastline, with Padre Island and Laguna Madre mud flats.

Island is modified and serves as an important sediment source for filling the lagoon.

Most of the island is low lying, as the highest points correspond with the fore-island dune ridge. As such, it is subject to breaching by storm surges generated by wind waves and related high tides. Channels are cut across the island by hurricanes, and funnel storm water carrying large volumes of sediment derived from the island and shoreface. Storm-water current velocities decrease once the open lagoon is reached, and consequently, washover fans of storm-deposited sand are created there. Through time and with the periodic storm deposition of sand on the washover fans, the width of the lagoon has narrowed, and in places, it has been filled.

Deposition by wind tides has aided in the lagoon-filling process. Prevailing southeasterly winds and northers give rise to wind tides in Laguna Madre and cause tens of square miles of low-lying areas flanking the lagoon to flood within hours. These wind tides rework sediment originally deposited by storms on washover fans, and deposit mud from suspension across tidal flats and in local depressions on these flats.

Intertidal areas are marked by extensive development of algal mats. Sediment beneath the mats consists of intercalated clay deposited from suspension, sand blown onto the flats from the barrier island, and gypsum precipitated from evaporating interstitial brines during periods of exposure. Areas that are infrequently flooded are made up of sandy flats with eolian dunes and local algal mats.

Several types and modes of gypsum occurrence have been observed by Miller (1975). Sucrosic gypsum occurring as thin layers and pods has formed in association with clay laminae and algal-mat sequences, and nodules of sucrosic gypsum are present as burrow fillings and voids formed by desiccation of algal mats. Both small and large (>1 cm) lenslike crystals of gypsum are present singly, in aggregates, or as rosettes in dry sandy sediments, as well as in clay–algal mat sediments.

Shallow depressions present across the wind-tidal flats are flooded by wind tides and heavy rains. Most are less than 25 cm deep and 100 to 800 m across. Some of these develop into salt pans as the ponded water evaporates, and develop halite crusts about 2 to 10 cm thick (figure 3.12). The crusts are ephemeral and no halite is preserved in the sedimentary record.

Northwest Gulf of California. Another modern example of a clastic-dominated sabkha principally fed along strike is the extensive mud flats bordering the northwest Gulf of California (figure 3.5). Although prior discussions have pointed out that this sabkha receives some sediment from alluvial fans that lie updip, the vast bulk of sediment comprising the system was derived from along strike through the Colorado River delta.

Prior to human tampering, the Colorado River supplied enormous

Figure 3.12
Laguna Madre supratidal flat and halite-encrusted salt pan (photograph courtesy of J. A. Miller).

amounts of very fine sand, silt, and clay to the Gulf of California.
Thompson (1968) showed through textural and mineralogical analyses that the Colorado River was the principal source of mud to the
sabkha and that the sands making up beach ridges in the southern
part of the system were derived from alluvial fans to the south. Gulf
waves transport these sands alongshore to the north and deposit them
along the intertidal margin of the mud flat. Meanwhile, silt and clay
carried in suspension to the Gulf by the Colorado River undergoes
net westward transport by rotary tidal currents during late ebb and
early flood, so that high concentrations of suspended sediment are
pushed toward and maintained along the Gulf's northwest shore. Thus,
with the high tides and minimal wave action, a predominantly muddy
shoreline can prograde.

Both Thompson's (1968) and Castens-Seidell's (1984) work, which
included the compilation of subsurface core data and near-surface

sampling, further indicated a progradational history. The stratigraphic sequence shallows upward from subtidal to intertidal and supratidal sabkha deposits. Castens-Seidell (1984) used Thompson's (1968) work and information she gathered from shallow cores and trenches to construct an idealized sabkha cycle that summarizes the depositional processes operating on the sabkha and the resultant facies that are formed there (figures 3.13 and 3.14). Intertidal and supratidal facies that cap the cycle include laminated, prism-cracked mud overlain by shallow subaqueous pan or salina deposits of layered gypsum and intercalated mud (to be discussed later). Gypsum layers are commonly contorted or even diapir-like, such that adjacent sedimentary layers are deformed. Contorted layers are also commonly truncated, attesting to planation during a former period of subaerial exposure.

Eolian-Dominated Marginal Marine Evaporite Systems

Coastal eolian dune complexes and sand seas owe their origin to, among other factors, the presence of a source and continuous supply of sand upwind from the eolian system. Sand may be supplied from several sources:

1. Bedrock, such as basement and intrusive complexes or terrigenous clastic sedimentary rocks, can directly supply significant amounts of sand when weathered in place.
2. Alluvial plains can supply massive amounts of sand to eolian systems. Although rainfall is sporadic in deserts, nonetheless it helps increase rates of weathering and erosion of bedrock. Sediment is entrained by ephemeral streams and transported downslope to coastlines that may lie nearby or hundreds of miles away. The ephemeral nature of such streams means of course that channels are dry during most of the year, and thus sand is free to be re-

---

Figure 3.13
Core recovered in 1979 at Salina Omotepec is made up of supratidal red mud overlain by layered mud, gypsum, and halite. Halite layer (7 cm below the top) is Shearman's (1970) recent halite rock, which was largely, if not totally, removed by dissolution, probably during the flooding events of early 1983 (see figure 3.6).

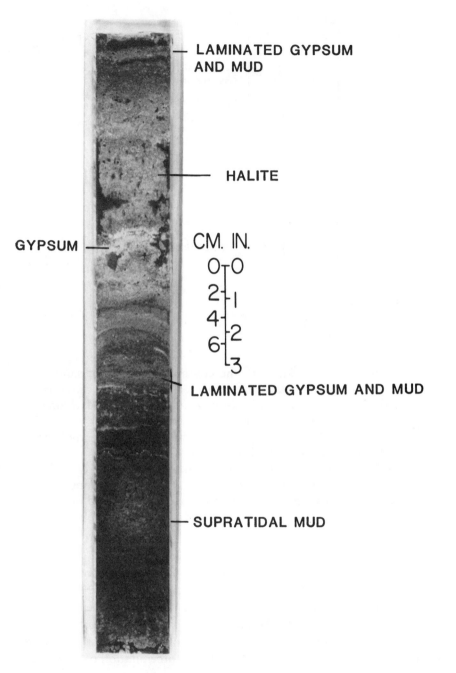

LAMINATED GYPSUM AND MUD

HALITE

GYPSUM

CM. IN.

0 — 0
2 — 1
4 —
6 — 2
   — 3

LAMINATED GYPSUM AND MUD

SUPRATIDAL MUD

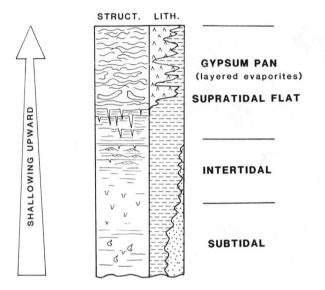

Figure 3.14
Idealized depositional cycle for Gulf of California mud flats (modified from Castens-Seidell 1984).

moved by the wind. With the continuous supply of sand to alluvial plains to replace that removed by the wind, large eolian systems can form downwind of the alluvial source. The Simpson Desert in Australia, the Great Kavir in Iran, and the Algodones-Sonoran sand dunes in California and Mexico own their origin to fluvial sources of sand (Glennie 1970).

3. As discussed previously, large volumes of sand that reach the sea via fluvial input are reworked and transported alongshore by waves and currents to form coastal barriers, beaches, and spits. That which is deposited high on the beach by storms is subject to wind removal. Thus strong onshore winds commonly blow these sands further onshore to create coastal dune complexes.

In each of the above cases, sand may accumulate in coastal areas (figure 3.4) as eolian deposits. Furthermore, eolian dune systems are invariably characterized by the presence of interdune depressions at or near the water table. As such they are subject to nearly constant wind deflation, a process that of course promotes evaporation of water from the depressions and can lead to the precipitation of evaporites.

Gypsum, anhydrite, and halite commonly form in interdune depressions as either precipitants from standing bodies of water or as interstitial water precipitants in the sedimentary matrix.

Very little research has been directed towards interdune evaporite deposits, and in fact, only recently have sedimentologists specifically addressed the origins of interdune deposits. Only Ahlbrandt and Fryberger (1981) and Fryberger et al. (1983) have attempted to provide any details regarding sedimentary features in evaporitic interdunes. Thus we probably understand less about interdune evaporites than about other marginal marine evaporite deposits.

Although the periodicity of climate change, wind regime, and sand supply rates are probably more important than dune migration rates or size of interdune areas in determining the thickness of interdune deposits (Simpson and Loope 1985), additional factors control evaporite deposition in interdune environments. These, of course, include the availability of brines and their geochemistry, the elevation of the groundwater table through the seasons, and evaporation rates.

If the groundwater table remains at or just below the surface, evaporite minerals can precipitate as a surface crust or within the sediment. A higher water table, such that a salina is formed, will of course result in the deposition of layered, subaqueous evaporites. Fluctuating groundwater tables are common (Fryberger et al. 1983) and thus both layered and displacive evaporites may form. Since sand dunes are water-storage devices (Fryberger et al. 1983), the style of deposition and type of evaporites present may be an indication of dune water-storage capacity, storage time, discharge rates, hydraulic head, and so forth.

The thickness of interdune evaporite deposits can be highly variable. Recent work by Simpson and Loope (1985) has questioned the validity of the climbing bed hypothesis to explain stacking of dune and interdune deposits. They have shown that low sand supply rates can result in the preservation of thin, lenticular dune deposits in a succession of mostly thick, amalgamated interdune deposits. The thickness of these interdune deposits is controlled by the number of interdune areas that have passed a particular point. Thus a sequence of interpreted interdune deposits can represent deposition in either single interdune areas or several that have left behind an amalgamated depositional record (Simpson and Loope 1985).

Sedimentary features of sabkhas studied by Fryberger et al. (1983) in the Jafurah sand sea, Saudi Arabia (figure 3.15) include intercalated, contorted layers of evaporites (gypsum, anhydrite, halite) and windblown sand. Algal mat development and formation of salt polygons and ridges lead to sediment deformation and disruption. Large gypsum sand rosettes are common in brine-saturated sediments, as is nodular or layered anhydrite.

The features noted above are identical to those observed in other sabkha-salina environments. However, their presence in eolian-deposited stratigraphic successions serves to distinguish such successions from marine- and fluvial-dominated sequences.

SILICICLASTIC AND EVAPORITE DEPOSITION IN SALINAS

Recent investigations (Castens-Seidell 1984; Handford 1981, 1982; Logan 1982; Lowenstein and Hardie 1985; Warren 1982) have shown that subaqueous deposition of evaporites in shallow salinas is an extremely important process, and that many ancient evaporites are replete with sedimentary features attributable to this type of deposition. Salina deposits are associated with either carbonate or siliciclastic sediments. Following is a brief summary of the principal points that characterize the depositional processes and facies in silicilastic-rich salinas as documented by the authors listed above.

A salina, as defined here, is an evaporitic depression, almost wholly enclosed or isolated, that is periodically flooded by either marine or nonmarine waters to form a shallow and frequently ephemeral standing body of water. However, this lakelike body will shortly shrink or even disappear as floodwaters begin to evaporate. Hardie et al. (1978) showed that solute concentrations rise during evaporation such that a brine is evolved and evaporite minerals are precipitated. At that point the salina is made up of two subenvironments—a shallow subaqueous salt pan, surrounded by a wet saline mud flat. Evaporites that form in the salt pan are subaqueous products, while those forming concomitantly in the flanking saline mud-flats (or sabkha) record intrasedimentary growth. As a result, layered evaporites characterize the salt pan, and muddy clastic sediments (exclusive of the carbonate fringe discussed by Warren 1982), crowded with evaporite crystals, commonly typify the saline mud flat. The boundary separating these

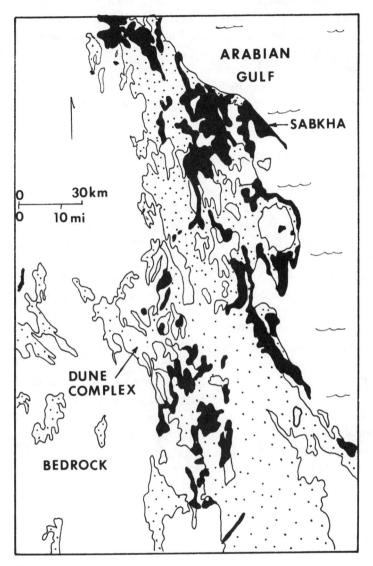

Figure 3.15
Map of northern Jafurah sand sea on east coast of Saudi Arabia.
Note the presence of dunes and interdune sabkhas from the shoreline
to 40 mi inland (modified from Fryberger et al. 1983).

two subenvironments is transitional, complex, and dynamic through time. This is because the salt pan expands and contracts rapidly and frequently, commonly over a large area, during repeated episodes of flooding and evaporation.

Salinas occur in both marine and nonmarine or continental settings. Furthermore, they may be large or small. Tiny salinas may dot the coastline, or in contrast, large salinas, such as Lake MacLeod in western Australia, may be present. Coastal salinas can also form at or below sea level (Lake MacLeod), or even above sea level, perched atop impermeable supratidal muds, as in the case of Salina Omotepec in the northwest Gulf of California tidal flats.

The alternate periods of flooding, evaporative concentration, and desiccation in salinas produce unique sedimentary layers whose features and origins were carefully described by Castens-Seidell (1984) and Lowenstein and Hardie (1985) for siliciclastic-dominant environments, and by Warren (1982) for carbonate-gypsum salinas. Castens-Seidell showed that Salina Omotepec evolves through initial flooding and evaporative concentration stages from a lake to a gypsum pan and eventually to a halite pan. Following is a synopsis of the stages and their depositional products.

During the flooding stage, water from meteoric or marine sources deposits silicilastic sediment across the formerly dry salina as a thin layer of mud. Previously deposited gypsum is commonly reworked across the surface by floodwaters and accumulates as crystal lags and intraclasts in the freshly laid down mud deposit. Flooding also results in erosional truncation of polygonal gypsum ridges, from which most of the clastic gypsum is derived.

Shortly after flooding and deposition of the siliciclastic mud layer, the surface is colonized by an algal mat with a variety of contorted morphologies. It serves as a substrate for later gypsum precipitation, including surface crusts and cements that grow beneath the blistered, wartlike surface. A crinkled, contorted sequence of algal laminae intercalated with and overlain by gypsum is produced.

The gypsum-pan stage is marked by crystallization of gypsum from the brine-air interface, followed by bottom nucleation of vertically oriented, bladed prisms to form a grasslike mat or crust. It is contorted due to inheritance from the contorted underlying algal mat and enhanced by competitive growth of gypsum, which buckles and domes the layers even more.

Continued evaporation of the brine eventually leads to supersaturation with respect to halite and its subsequent precipitation. At this stage, halite initially precipitates as tiny skeletal hopper crystals at the brine surface, and these eventually coalesce to form rafts that sink to the bottom of the salt pan (Lowenstein and Hardie 1985; Shearman 1970). Later, halite precipitates on the sunken rafts to form crystals that grow upwards with a distinctive vertical fabric of chevrons and cornets.

Successive stages of evaporation may be interrupted by new flood events. Sudden input of dilute meteoric or fresh marine waters into a salt pan can introduce additional terrigenous sediment and form dissolution features in the evaporites, some of which may be diagenetically rehealed as floodwaters become evaporatively concentrated (Lowenstein and Hardie 1985; Shearman 1970).

If the salt pan is allowed to completely desiccate, the dry pan surface will evolve into a polygonally disrupted crust (figure 3.16). Puffy efflorescent crusts form in the polygonal cracks and atop the ridges around the cracks as interstitial water is drawn to the surface by evaporative pumping. As the water table drops below the surface, evaporite minerals precipitate in voids and as displacive crystals in the brine-soaked sediment (Handford 1982; Lowenstein and Hardie 1985).

The resulting sedimentary record of salina flooding and desiccation will therefore include interbedded layers of flood-deposited mud, crystalline halite, and gypsum/anhydrite. Crystals of halite or gypsum present in the mud layers are most likely to represent intrasedimentary displacive precipitants, but if the crystals (gypsum) appear worn or abraded, they were most likely reworked and mechanically transported to the depositional site. The siliciclastic mud, which was washed in by floodwaters, subsequently serves as a substrate for algal mat, gypsum, and halite precipitation; a host for intrasedimentary displacive evaporite growth; and a storage facility in which or on top of which brines may accumulate.

Figure 3.16
Halite-filled mudcracks on high supratidal surface, Baja California, Mexico.

ANCIENT MARGINAL MARINE SILICICLASTIC-RICH EVAPORITE SYSTEMS

Thick sequences of siliciclastic-rich marginal marine evaporites underlie the Permian Basin in Texas, Oklahoma, and New Mexico, the Gulf Coastal Plain from Texas to Florida, and various Rocky Mountain basins in Colorado, Wyoming, and Utah. In each case, these strata contain a recurring set of lithofacies associations, sedimentary structures, textures, and fabrics that have analogous counterparts in modern marginal marine evaporite environments.

Three stratigraphic units of sabkha-salina origin have been chosen for comparison with modern systems for the purpose of distinguishing their depositional styles (figure 3.17). Previously published lithofacies maps of these units are presented here in order to show, first, probable source areas of siliciclastics that occur in the evaporites; and second, lithofacies patterns which aid in the interpretation of the most likely modes of siliciclastic input into the evaporite depositional system.

Knowledge of these in combination with sedimentological information obtained from cores would be invaluable for determining depositional style. Moreover, from a resource point of view, they would be extremely important for predicting and understanding sand-body trends, their depositional origins, and hydrocarbon potential. If, as some suggest (Kirkland and Evans 1981), evaporites are a hydrocarbon source, then sand bodies such as dip-fed fluvial sandstones, barrier bar, beach ridge, and washover sandstones, and eolian dune sandstones that are in facies contact with evaporites can be consid-

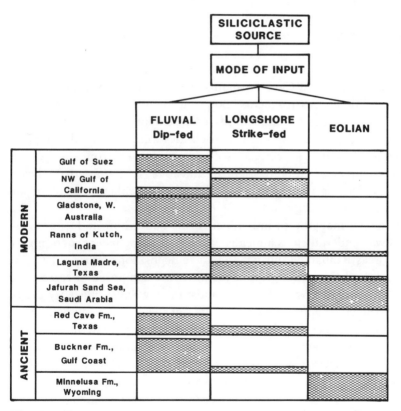

Figure 3.17
Relative importance (shaded) of various modes of siliciclastic sediment input to selected clastic-rich modern and ancient sabkha systems.

ered potential hydrocarbon reservoirs, even in the absence of "conventional" source rocks.

Red Cave Formation (Permian; Texas)

In a regional stratigraphic study, Handford and Fredericks (1980) showed that the Red Cave Formation consists of cyclic red-bed siliciclastic and dolomite-anhydrite facies (figure 3.18) that pass southward across the Texas Panhandle into marine carbonate shelf and shelf-margin facies bordering the deep Midland Basin. Northward, Red Cave evaporites disappear and sandstones (locally gas-bearing) become prevalent. Three clastic members were recognized that sandwich two major evaporite units (figure 3.18). Thus, in vertical succession, the Red Cave is made up of three carbonate-evaporite to siliciclastic sequences.

Although cores are scarce, those present in key areas strongly indicate that the Red Cave Formation was deposited in wadi plain to coastal sabkha environments. Both carbonate- and siliciclastic-rich sabkhas were present but alternated in extent through time. Siliciclastic mud-rich sabkhas formed as clastics were abundantly supplied from updip and transported downdip to sabkha environments. Cores from updip areas and dip-oriented sandstone isolith patterns indicate that sandstones were principally deposited in ephemeral stream or wadi plain systems.

Regional mapping of individual clastic and evaporite facies members was undertaken as a means of reconstructing paleogeography of the Texas Panhandle during Red Cave deposition (figure 3.19). Although Red Cave stratigraphic members are composed of regressive, diachronously deposited facies, boundaries between sedimentary cycles are assumed to approximate time lines at the culmination of each of the underlying cycles. As a result, the paleogeographic maps imply that sabkhas reached a maximum width of approximately 60 mi (98 km). However, that does not mean that tidal flooding reached 60 mi inland. Progradation of sabkha environments probably caused the earliest-formed supratidal surfaces to become gradually isolated from marine flooding, such that through time, large sabkha areas, though possibly constructed by seaward progradation, eventually became continental sabkhas. A similar pattern of sabkha isolation from

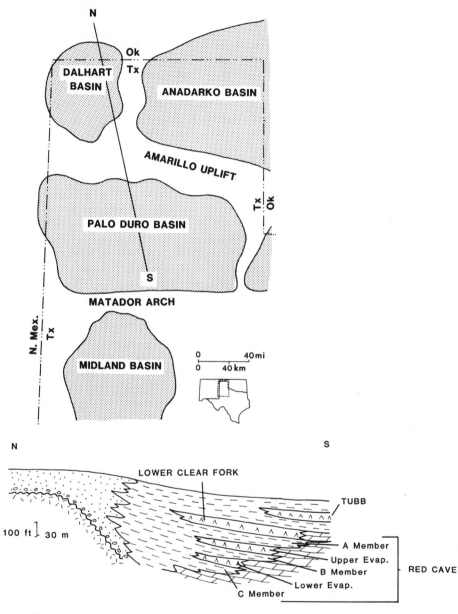

Figure 3.18
Regional geologic setting of Texas Panhandle and schematic north-south cross section through Permian evaporites.

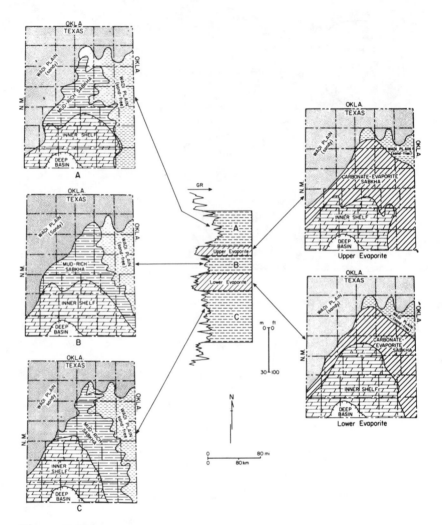

Figure 3.19
Paleogeography of the Red Cave Formation. Cyclic clastic and carbonate-evaporite facies reflect alternating styles of sabkha deposition
that were brought on by the periodic availability and supply of clastics to sabkha environments (Handford and Fredericks 1980).

marine flooding has occurred at Laguna Salada, Baja California, Mexico (Thompson 1968). Currently located 70 mi (115 km) from the Gulf of California, Laguna Salada was once a marine tidal flat, but was subsequently isolated from the Gulf by progradation of the Colorado River delta and associated mud flats.

Buckner Member (Jurassic; Louisiana, Arkansas)

The Buckner Member of the Haynesville Formation is an assemblage of red shales and sandstones with nodular-anhydrite and bedded halite that pinch out northward against Paleozoic rocks and interfinger downdip (seaward) with high-energy ooid grainstones of the oil-producing Smackover and Haynesville formations. The character of the Buckner rocks suggests deposition in marginal marine sabkhas and salinas that lay between an alluvial plain to the north and an ooid shoal system to the south (figure 3.20). Recent work (Harris and Dodman 1982) has shown that porous Smackover grainstones are shingled in cross section and pinch out updip into reservoir-sealing units of lower Buckner evaporites. On the basis of this observation, it is surmised that local salinas and sabkhas were present in low areas between the chenierlike, oolitic beach ridges (figure 3.21). These smaller evaporite complexes probably passed northward into much more extensive salinas and saline mud flats. Facies distribution patterns almost certainly show that siliciclastic sediments making up these marginal marine evaporite systems were supplied by ephemeral streams or an alluvial plain system that drained the Paleozoic Ouachita Mountains north of the Gulf basin. Very little strikewise transport of siliciclastics into the Buckner evaporites occurred, because nearly pure oolitic carbonates comprise the marine side of the coastal system.

Minnelusa Formation (Permian; Wyoming)

Alternating marine and eolian deposits are present in the Minnelusa Formation in northeastern Wyoming and form cyclic depositional sequences. Achauer (1982) delineated two shallowing upward cycles (figure 3.22) in Minnelusa cores that include subtidal dolomite over-

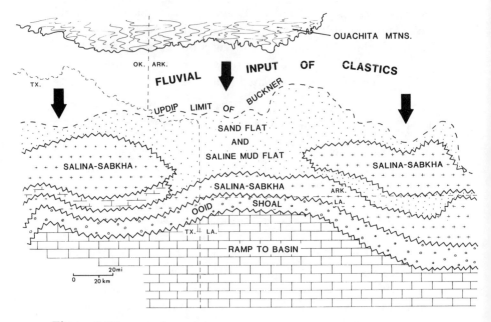

Figure 3.20
Lithofacies patterns and inferred environments of the Buckner Formation, northern Louisiana and southern Arkansas (modified from Anderson 1979).

lain by intertidal, algal-laminated and desiccated dolomicrites and sandstones, and supratidal nodular anhydrite.

Using cores and calibrated gamma-ray logs, Achauer (1982) mapped one of the dolomite units across a ~30 sq mi area (figure 3.23). He discovered that the dolomite is restricted to elongate, narrow corridors separated by eolian sandstone facies. In a slight modification of Achauer's interpretation, it is suggested that these dolomites accumulated in swales, lagoons, and ponds between eolian dunes that prograded into the shallow sea from a more extensive coastal dune system or sand sea. Shinn (1973) and Fryberger et al. (1983) have documented eolian sand dunes that are prograding into the Arabian Gulf, and Achauer (1982) included a map of the Abu Dhabi area showing geomorphic patterns that mimic his mapped dolomite distribution (figure 3.24).

N   **S**

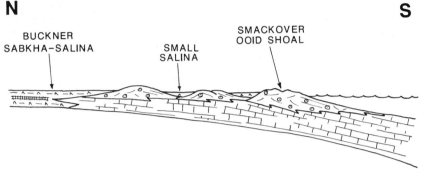

Figure 3.21
Depositional profile through Buckner and upper Smackover ooid-shoal
and sabkha-salina environments (modified from Harris and Dodman
1982).

CONCLUSIONS

Unlike carbonate-dominated marginal marine evaporite systems in
which the carbonate sediments are almost entirely of marine origin,
sands and muds comprising siliciclastic-rich systems are derived from
a number of sources: fluvial systems that directly introduce sediment
into evaporite settings; coastal shoreface systems in which siliciclastics are transported alongshore prior to being swept by storms into
coastal sabkhas and salinas; and eolian and sand seas or coastal sand
dune complexes that are juxtaposed to marginal marine evaporite
environments. In the first two cases, muddy clastics interbedded with
evaporites are likely to result, whereas in the third case dunes can
migrate directly into a sabkha or salina so that interbedded sands
and evaporites may form.

In general, the influx of siliciclastics establishes a sedimentary
foundation across which sabkha evaporites can form and therefore
expand or prograde. Examples include the distal margin of fan deltas,
wadis or ephemeral stream deltas in protected embayments, estuaries, behind coastal barriers in washover fans, landward sides of
lagoons, and interdune depressions. However, if the influx of clastics
is too rapid, a sabkha may be destroyed or forced to shift laterally
(not expand). For example, if a fluvial system begins to deposit an

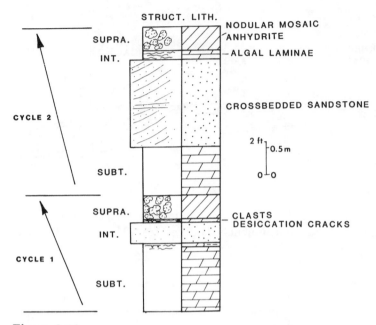

Figure 3.22
Graphic log of two depositional cycles in the upper Minnelusa Formation, South Rozet Field, Wyoming (modified from Achauer 1982).

excessive amount of sediment, the sabkha may be smothered by clastics or diluted by fresh river water such that previously deposited evaporites may dissolve. Rapidly migrating eolian dunes that encroach on a sabkha or salina can also smother evaporite deposition. The sedimentary rock record of such a history may be marked by a succession of extensive sabkha evaporites sharply overlain by interbedded eolian sandstones and thin interdune evaporites.

Salinas can develop in siliciclastic-rich settings as depressions behind coastal barriers, interdune depressions with a high groundwater table, in poorly-drained wadis or tidal channels, distal margins of fan deltas, and as ponds on poorly drained supratidal flats. Local tectonics can exaggerate existing depressions or create new ones for subsequent salina development. For example, Shearman (1970) suspected that the Salina Omotepec halite-filled depression is controlled by faulting. Similar to that experienced by sabkhas, excessive silici-

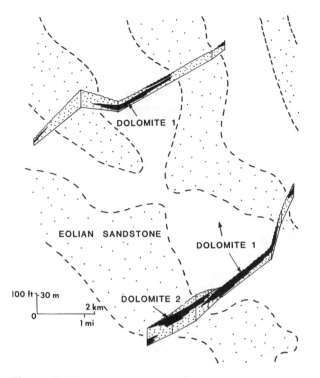

Figure 3.23
Distribution of dolomite, upper Minnelusa Formation, South Rozet Field, Wyoming (modified from Achauer 1982).

clastic influx can also destroy salinas. Flood deposits of wadi and fan-delta sediment and storm washovers from coastal barriers may rapidly fill shallow salinas and lead to evaporite dissolution; migrating eolian dune complexes may bury salinas in a short span of time, though perhaps without the dissolution effects that accompany deposition of water-laid clastic sediments.

So, one should recognize that the depositional interaction of siliciclastics and evaporites is, at the same time, simple and complex. Sediment sources and mechanisms of influx are recognizable and understandable for both modern and ancient evaporites; however, when one considers how the rates and amounts of siliciclastic and evaporite deposition (as well as evaporite preservation) are affected by subsi-

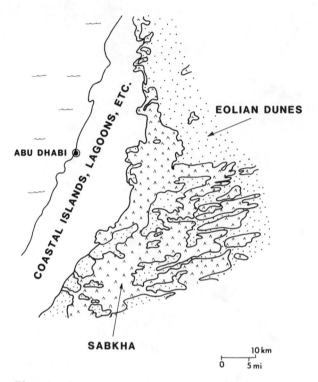

Figure 3.24
Map of coastal dune–sand sea systems and sabkhas near Abu Dhabi, along the Arabian Gulf (modified from Achauer 1982 and Glennie 1970).

dence, local tectonics, relative budget of both fresh water and saline water, climatic patterns through time, and so forth, matters are greatly complicated and more difficult to assess. Our ability to understand the evaporite rock record has increased greatly and quickly, but with each step forward new questions arise.

REFERENCES

Achauer, C. W. 1982. Sabkha anhydrite: The supratidal facies of cyclic deposition in the upper Minnelusa Formation (Permian), Rozet Fields area,

Powder River Basin, Wyoming. In C. R. Handford, R. G. Loucks, and G. R. Davies, eds., *Depositional and Diagenetic Spectra of Evaporites—A Core Workshop,* pp. 193–209. SEPM Core Workshop no. 3. Tulsa, Okla.

Ahlbrandt, T. S. and S. G. Fryberger. 1981. Sedimentary features and significance of interdune deposits. In F. G. Ethridge and R. M. Flores, eds., *Recent and Ancient Nonmarine Depositional Environments: Models for Exploration,* pp. 293–314. SEPM Special Publication no. 31. Tulsa, Okla.

Anderson, E. G. 1979. *Basic Mesozoic Study in Louisiana—The Northern Coastal Region and the Gulf Basin Province.* Louisiana Geol. Survey Folio Series, no. 3. Baton Rouge, La.

Brown, L. F., J. H. McGowen, T. J. Evans, C. G. Groat, and W. L. Fisher. 1977. *Environmental Atlas of the Texas Coastal Zone—Kingsville Area.* Austin, Tex.: Bureau of Economic Geology, University of Texas at Austin.

Castens-Seidell, B. 1984. The Anatomy of a Modern Marine Siliciclastic Sabkha in a Rift Valley Setting: Northwest Gulf of California Tidal Flats, Baja California, Mexico. Ph.D. diss., Johns Hopkins University, Baltimore, Md.

Curtis, R., G. Evans, D. J. J. Kinsman, and D. J. Shearman. 1963. Association of dolomite and anhydrite in the Recent sediments of the Persian Gulf. *Nature* 197:679–680.

Davies, G. R. 1970. Algal-laminated sediments, Gladstone Embayment, Shark Bay, western Australia. In B. W. Logan, G. R. Davies, J. F. Read, and D. E. Cebulski, eds., *Carbonate Sedimentation and Environments, Shark Bay, Western Australia,* pp. 169–205. AAPG Memoir no. 13. Tulsa, Okla.

Evans, R. 1978. Origin and significance of evaporites in basins around Atlantic Margin. *AAPG Bulletin* 62:223–234.

Fryberger, S. G., A. M. Al-Sari, and T. J. Clisham. 1983. Eolian dune, interdune, sand sheet, and siliciclastic sabkha sediments of an offshore prograding sand sea, Dharan area, Saudi Arabia. *AAPG Bulletin* 67:280–312.

Glennie, K. W. 1970. *Desert Sedimentary Environments.* Vol. 14 of *Developments in Sedimentology.* Amsterdam: Elsevier.

Glennie, K. W. and G. Evans. 1976. A reconnaissance of the Recent sediments of the Ranns of Kutch, India. *Sedimentology* 23:625–647.

Handford, C. R. 1981. A process-sedimentary framework for characterizing Recent and ancient sabkhas. *Sedimentary Geology* 30:255–265.

Handford, C. R. 1982. Sedimentology and evaporite genesis in a Holocene continental-sabkha playa basin—Bristol Dry Lake, California. *Sedimentology* 29:239–253.

Handford, C. R. and P. E. Fredericks. 1980. *Facies Patterns and Depositional History of a Permian Sabkha Complex: Red Cave Formation, Texas Panhandle.* Bureau of Economic Geology, University of Texas at Austin, Geology Circular no. 80-9. Austin, Tex.

Hardie, L. A., J. P. Smoot, and H. P. Eugster. 1978. Saline lakes and their deposits: A sedimentological approach. In A. Matter and M. Tucker, eds., *Modern and Ancient Lake Sediments*, pp. 7–41. Int. Assoc. Sedimentologists, Special Publication no. 2. Oxford: Blackwell.

Harris, P. M. and C. A. Dodman. 1982. Jurassic evaporites of the U.S. Gulf Coast: The Smackover-Buckner contact. In C. R. Handford, R. G. Loucks, and G. R. Davies, eds., *Depositional and Diagenetic Spectra of Evaporites—A Core Workshop*, pp. 174–192. SEPM Core Workshop no. 3. Tulsa, Okla.

Johnson, H. D. 1978. Shallow siliciclastic seas. In H. G. Reading, ed., *Sedimentary Environments and Facies*, pp. 207–258. New York: Elsevier.

Kirkland, D. W. and R. Evans. 1981. Source-rock potential of evaporitic environment. *AAPG Bulletin* 65:181–190.

Logan, B. W. 1982. *Evaporite Sedimentation and Basin Evolution, Lake MacLeod, Western Australia*. Field Seminar Handbook, University of Western Australia. Nedlands, Western Australia.

Lowenstein, T. K. and L. A. Hardie. 1985. Criteria for the recognition of salt-pan evaporites. *Sedimentology* 32:627–644.

McGannon, D. E. 1975. Primary fluvial oolites. *Jour. Sed. Petrology* 45:719–727.

Miller, J. A. 1975. Facies characteristics of Laguna Madre wind-tidal flats. In R. N. Ginsburg, ed., *Tidal Deposits: A Casebook of Recent Examples and Fossil Counterparts*, pp. 67–73. New York: Springer-Verlag.

Risacher, F. and H. P. Eugster. 1979. Holocene pisoliths and encrustations associated with spring-fed surface pools, Pastos Grandes, Bolivia. *Sedimentology* 26:253–270.

Shearman, D. J. 1970. Recent halite rock, Baja California, Mexico. *Trans. Inst. Min. Metall. Section B.* 79:155–162.

Shinn, E. A. 1973. Sedimentary accretion along the leeward SE coast of Qatar Peninsula, Persian Gulf. In B. W. Purser, ed., *The Persian Gulf*, pp. 200–209. New York: Springer-Verlag.

Simpson, E. L. and D. B. Loope. 1985. Amalgamated interdune deposits, White Sands, New Mexico. *Jour. Sed. Petrology* 55:361–365.

Sneh, A. and G. M. Friedman. 1984. Spit complexes along the eastern coast of the Gulf of Suez. *Sedimentary Geology* 39:211–226.

Thompson, R. W. 1968. *Tidal Flat Sedimentation on the Colorado River Delta, Northwestern Gulf of California*. Geol. Soc. America Memoir no. 107. Boulder, Colo.

Till, R. 1978. Arid shorelines and evaporites. In H. G. Reading, ed., *Sedimentary Environments and Facies*, pp. 178–206. New York: Elsevier.

Walker, T. R. 1967. Formation of red beds in ancient and modern deserts. *Geol. Soc. America Bull.* 78:353–368.

Warren, J. K. 1982. The hydrological setting, occurrence, and significance of gypsum in late Quaternary salt lakes in South Australia. *Sedimentology* 29:609–637.

Weise, B. R. and W. A. White. 1980. *Padre Island Seashore—A Guide to the Geology, Natural Environments and History of a Texas Barrier Island.* Bureau of Economic Geology, University of Texas at Austin, Guidebook no. 17. Austin, Tex.

4

Subaqueous Evaporite Deposition

B. CHARLOTTE SCHREIBER

Because we have never observed a large, functional hypersaline marine-fed sea in which evaporites are actually forming, we tend to forget that such a water body may receive diverse deposition throughout the basin, due to local variations in topography, evaporation, inflow, and ionic concentration. The only really large modern, hypersaline water bodies presently available for study are nonmarine, largely the product of long periods of aridity in basins with interior drainage. Among these basins there are some with a substantial mineral contribution also coming from groundwaters sourced by older evaporite deposits, as in the case of the Dead Sea or in the saline lakes of Spain. Our actualistic models for study of marine-fed subaqueous evaporites

B. Charlotte Schreiber is a Professor in the Department of Geology, Queens College, City University of New York, and a Visiting Senior Research Scientist, Lamont-Doherty Geological Observatory, Columbia University, Palisades, New York.

are the many solar saltworks (salinas) and saline lakes located all over the world along arid and semiarid seashores. At present all of the known saline water bodies are located in shallow basins, although many ancient deposits seem to have been located in morphologically deep basins. The largest of these modern marginal marine, evaporatic water bodies include Los Roques (Venezuelan Antilles), the Salin du Midi and Salin Giraud (southern France), the Salinas of Cabo di Gata and Santa Pola (Spain), Salina Margherita di Savoia (near Bari, Italy), and some of the coastal lakes of southern and western Australia (see Kendall and Warren, article 2).

Concentrations of seawater components in large shallow-water bodies may vary from area to area, and synchronous accumulations of sulfate may form in one area (or subbasin), while halite (or any other phase) may form in another. Within a morphologically deep basin the depth of water may also vary, depending on the restriction of inflow of water and regional climatic fluctuations, but laterally synchronous mineralogical changes are not very likely (Kendall, article 1). Thus, over time, a single evaporite deposit may be formed under varying water depths as well as from changing water chemistry (see below, "Water Depth"). An interesting model for the stratigraphic interpretation of such changing depths has been presented by Jauzein and Hubert (1984). Water within such a hypersaline sea may be marine, nonmarine, or mixed in origin. Commonly the constituent minerals, their trace elements, their isotopic compositions, and their proportions in the deposit make evident the source or sources of the water, but at times even these differences are equivocal (see Pierre, article 6).

Substantial evaporite deposits (a kilometer or more in thickness) which form over large areas within a very short time period (a few hundred thousands of years), could only be supported by a largely marine influx. Where the depositional period is greater, on the order of millions of years, nonmarine dominance is equally likely. Therefore the single greatest constraint on the accumulation of thick nonmarine deposits is that of time. The 2 + km thickness of the largely marine Messinian evaporites of the Mediterranean took 250,000 to 300,000 years to form (Ryan et al. 1974), while the continental Miocene of the Ebro Basin (0.6 km thickness) apparently required 10 to 15 million years (Birnbaum 1976), even though it is rimmed by considerable amounts of Mesozoic sediments (with evaporites) that acted as a concentrated source area.

CONSIDERATIONS OF IMPORTANCE IN THE STUDY OF EVAPORITES

Water Sources

A very important aspect of evaporite deposition is that the meteoric input into any restricted basin, whether it be rainfall or groundwater, affects both the morphology and the chemistry of the deposits. In fact, it seems almost impossible to develop an evaporite deposit that has no component of nonmarine water. For this reason the chemistry of each basin must be examined with great care, cross-checking all measurable parameters in order to estimate what proportion of the basin's water was truly marine. The evolution of evaporative water bodies, and a discussion of the resultant minerals, is given in Hardie, Smoot, and Eugster (1978) and then, in greater detail, in Hardie (1984).

A number of evaporite basins that are basically within marine settings have some part of their fill originating from nonmarine intervals. The Delaware Basin sequence (Permian), which begins in an open-marine setting, is topped by the Salado (recycled marine, mixed with continental) and the Rustler (nonmarine) formations, as has been demonstrated by Lowenstein (1983) and Lowenstein and Hardie (1985). This sedimentary sequence represents the filling of a marine basin into which marine waters no longer flow. Not all evaporite basins end their depositional history with nonmarine sequences. The Messinian deposits of Sicily begin with marine-sourced gypsum and halite and then pass to potassic salts, composed of recycled marine salts (reworked in part by nonmarine water). However, this sequence is then overlain by halite and then gypsum, which are apparently sourced by marine water. The evaporites are then capped by an open-marine carbonate ooze of Plio-Pleistocene age (Decima and Wezel 1973). Hence, it can be seen that not all basins follow the same depositional patterns, and that variable climatic controls, as well as tectonic restrictions, govern deposition.

Tectonic Settings

Many rift basins, of all ages, are floored by evaporites sitting atop early continental deposits. In some regions these evaporites begin as partly marine-sourced precipitates (Lac Assal, Djibouti, Afar), but in the majority of instances geochemical evidence suggests that initial evaporites are entirely nonmarine (based on mineral assemblages, trace element groupings, and stable isotope values). In many examples the upper portion of the evaporite sequence seems to evidence more and more marine influx, and intercalated beds of marine fauna are present. In those cases where no floral or faunal traces are noted, reliance on geochemistry is necessary, although it is not always clear and definitive.

Pull-apart, back-arc, foreland, and simple fault-controlled basins may also become the locus of evaporite deposition, wherever the climate and water sources are appropriate. The Dead Sea, a strike-slip-controlled, pull-apart basin, contains several phases of evaporite deposition which are largely nonmarine (Neev and Emery 1967); however, marine water influx into comparable settings is easily imagined. The eastern Tyrrhenian Basin, adjacent to western Italy, is a good example of an evaporite-floored marginal basin with a complex back-arc history, in which marine carbonate oozes are followed by terrestrial clastic sequences and are overlain by both continental- and marine-sourced evaporites (1986 Ocean Drilling Program [O.D.P.] coring, during Leg 107). The many Tertiary foreland basins of Spain and the Italian Apennines contain both marine- and continental-fed evaporites with underlying interfingering, and overlying marine deposits, often formed under very diverse conditions. The most elegant instance of one such foreland deposit has been documented by Vai and Ricci-Lucchi (1977); in it, shallow-water gypsum deposits (largely marine) lie atop and are overlain by 1+ km of marine deep-sea turbidites (Pliocene; Ricci-Lucchi 1975).

Biota

LOOK IN DARTON

The range of biota living in a saline pond is very restricted; however, both the total population and accumulated organic production is very high (for a thorough discussion of the biota of saline water bodies see Larsen 1980; Cornee 1983, 1984; Thomas and Geisler 1982; Javor 1985; Por 1985; and Sammy 1985). Examination of the successive faunal and floral groupings in the various ponds of a functioning saltworks presents a reasonable picture of the biota present in a totally natural setting.

The first inflow ponds of a saltworks have a salinity that ranges from near-marine to moderately elevated, 37‰ to 55‰ (Perthuisot 1982; Landry and Jaccard 1984), and contain a fairly broad range of diverse marine flora and fauna, but the dominant species are not the forms most common just outside of the ponds. Additionally, green algae cover many surfaces, and mats of them float at the surface. Thousands of cerithid gastropods browse on the floating algal mats. In the successive ponds, fewer species of plants and animals are present, and cyanobacteria (blue-green algae) take over as photosynthesizers. In the first ponds that contain gypsum (ca. 150‰), the blue-greens (*Microcoleus* and *Lyngbya*) and *Dunaliella viridis* also become the substrate for concentrations of other bacteria (*Aphanotece halophytica*).

In the halite ponds (> 325‰) the earlier blue-greens disappear, and *Halobacterium* and *Duniella salina* are prevalent. It is these bacteria that generate the characteristic pink color of the ponds. They also serve as food for the brine shrimp *Artemis salina,* which also take on the same pink coloration. These algae and bacteria control the pH of the ponds (7.5 to 11) during their life processes, and upon their death they sink to the bottom. An entirely different assemblage of bacteria, primarily anaerobic forms (*Desulfovibrio desulfuricans,* for example), take over in the bottom sediments and produce lower pHs (down to ca. 6). This lower limit is buffered by the presence of *Thiobacillus* and by the available calcium carbonate within the bottom muds; however, there is some suggestion that in noncarbonate settings, the pH may even become lower (Stuart Birnbaum, 1982, personal communication). In this way the biota control the solubilities and concentrations of various ions in the brines (Birnbaum and Wireman 1984).

Water Depth

Another significant problem exists in any very restricted basin—that of the position of the level of the water. In a morphologically deep basin, the water level may be basin full; however, where the rate of water inflow and circulation is sufficiently limited as to permit concentration and formation of salts, the basin is also likely to have sporadic sea-level fluctuation. This lowering in water level within a basin has been termed "drawdown," and can range from minor fluctuation in level to complete desiccation (Maiklem 1971). When this variation in water level takes place, it may occur within a fairly short time span, even though the evaporation of water is presumably slowed by the increasing salt content (Hsü 1972).

What actually happens during drawdown in a basin is fairly complex. First, evaporation rates rise rapidly due to adiabatically driven atmospheric circulation above the lowering water surface (Manspeizer 1981). The increasing salinity of the water during concentration, however, slows down the rate of evaporation due to the associated rise in surface tension, but insolative heating of the water also takes place, because the increasing water density and the red coloration caused by bacteria refract incoming light and heat back into the water. The heating of the water then has the effect of accelerating evaporation, offsetting the lowered rate of evaporation from increased salinity. The heated water can then continue to evaporate, and the saline water continues to become more and more concentrated.

Another component enters into our modeling of water depths within evaporite basins, that of surface area versus volume (Jauzein and Hubert 1984). Because most basins have a tapered shape, i.e., the walls slope inward and are not vertical, the available surface area decreases as the sea level falls during evaporation. If the salinity is rising during this period, there will be a point at which the amount of surface area is too small to permit further drawdown (at the new elevated salinity and temperature). In fact the water in the basin may remain at a stable but reduced depth, if salinity remains fairly constant; or the water level may actually rise as the salinity increases, because a greater surface area is required to achieve evaporation. This can produce a slight transgression, as seen in the sediment record, with an increase in salinity.

ENVIRONMENTS OF SUBAQUEOUS EVAPORITE DEPOSITION

It has been just within the past twenty-five years that intensive studies of *in situ* evaporite deposits have been carried out and the working models formulated for the origin of both supratidal and subaqueous deposits. While these studies have focused on deposition along supratidal margins of normal marine basins (Kendall and Warren, article 2, and Handford, article 3 above), examination of evidence from the geologic record has long suggested that many evaporite bodies were also deposited within existing marine basins. These deeper marine basins became constricted and developed hypersaline conditions within the water which they enclosed, but the evaporites overlie normal, open-marine deposits. The question then is whether the evaporite deposits form in deep hypersaline water, or only when the basin begins to dry out, suffering drawdown and even desiccation.

Data obtained from the Messinian (Upper Miocene) sediments of the Mediterranean Basin and from many modern salinas (saltworks) have shown that specific subaqueous evaporitic deposits do indeed form, as well as the better known supratidal ones, and their particular subfacies are clear and demonstrable. The primary composition, texture, and forms that comprise these subaqueous deposits are, however, only partly recognized. Incomplete understanding stems from two basic problems. First, analogous sedimentary settings are relatively rare; i.e., no large marine basin that is presently evaporitic is available for study. Second, evaporites are exceedingly prone to alteration, both while on the surface and after burial. Because of this ease of alteration, the ancient record left to us has proven largely misleading (see "Diagenesis," below).

The range of environments under which evaporites can form and accumulate is as diverse as that for carbonates, and, in a general way, for every type of carbonate and siliciclastic sediment, there is an equivalent evaporite. The energy regimes of the formative water bodies are recorded in both carbonate and evaporite sediments as current-controlled bedding and other structures. Additionally, mineral species and crystal morphology are governed not only by ionic compositions and concentrations, but also by temperature, evaporation rates, biological processes, and impurities such as organic inclusions and clay.

The settings of the various environments of evaporite deposition are shown schematically in figure 4.1. The physical parameters indicated in the figure are deliberately without numerical values, but general limits can be set. For example, currents generated by wave action are usually restricted to water depths of about 3 to 5 m or less, but this is not to say that currents do not flow at greater depths. However, we know that the association of current-deposited structures with oscillation ripple marks, desiccation features, and edgewise conglomerates (so-called rip-up breccias) characterize deposition in shallow water (see "Clastic Evaporites" below).

The terms "tidal," "intertidal," and "subtidal" are commonly employed in description of shallow-water evaporites, for want of better terms; however tides are surely limited in most such restricted basins, and what is called the "tidal range" is probably on the order of 15 cm

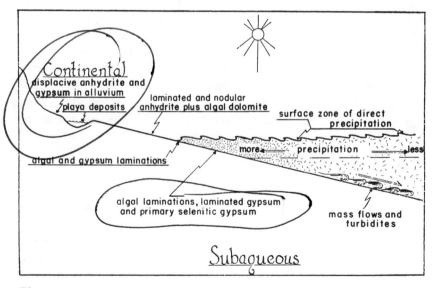

Figure 4.1
Environments of deposition for evaporite sediments (Schreiber, Catalano, and Schreiber 1977).

to 1 m (as in the modern Mediterranean). "Depth of wave base" might be better than the term "tidal" for the zone of shallow-water fluctuations. The association of clastic gypsum, anhydrite, and even clastic halite with the development of mass flows, olistoliths, and turbidites suggests that basinal relief exists in parts of many evaporite basins. Such a designation again sidesteps absolute numbers, but the initiation and deposition of nonevaporite turbidites, studied in modern settings, appears to require water depths of at least 20 to 30 m or more. The absolute thickness of any sedimentary sequence (bottom to top), minus the appropriate loading subsidence, can also aid in making closer approximations of original basin depth on a case-by-case basis (see "Clastic Evaporites" discussion below).

A similar problem of definition exists in the use of the term "photic zone." In the clearest water known in modern seas, blue-green algae, which produce $CaCO_3$ precipitate, have been found living at depths of 100 to 120 m (about 1% light penetration), but 50 m is the more usual limit. Below these depths there are apparently no blue-green algae that can form the *in situ* calcified mats, crusts, oncolites, and other structures that are commonly associated with evaporites. Hypersaline water tends to be poorly oxygenated (Kinsman, Boardman, and Borcsik 1974); hence, it is reasonable to suppose that the resultant preservation of organic residues in suspension severely reduces the depth of light penetration. Therefore, "photic zone" can signify a range from a depth of 120 m to 10 to 20 m or even less, depending upon the organic input and its preservation within the water column.

PRIMARY SUBAQUEOUS DEPOSITS

The accumulation of subaqueous evaporites may take place as carbonate, gypsum, halite, carnallite, and many other phases. While anhydrite is the most common sulfate in the subsurface, it has been found to form only in the vadose zone of the supratidal, and in the hot brine deposits on the floor of the Red Sea. There appears to be a kinetic barrier to nucleation of anhydrite within water, and it has not been observed forming in nature or in the laboratory at temperatures resembling natural conditions (Hardie 1984). However, rising temperatures and salinites cause ready transformation from gypsum to anhydrite. The most rapid sedimentation of evaporites occurs in shallow water, and is thus prone to reworking during storms and also through the agency of slumping and mass flow. This section reviews the primary depositional microfacies thus far observed within evaporative carbonate, gypsum, and halite formed in subaqueous settings.

Evaporative Carbonate

Carbonates, formed within evaporative hypersaline water, are usually composed of aragonite, calcite, and some small amount of strontium carbonate. The carbonate is characteristically devoid of easily recognizable fossils and has always been considered to be abiotic, the product of direct inorganic precipitation ("whitenings"). In actual fact these saline environments teem with life, and it appears that most of the carbonate is associated with various life processes. The calcium carbonate (aragonite and calcite) is largely algal and/or bacterial, and even the strontianite is largely organic in origin (Javor 1985). In the modern marine salina deposits, much of the $CaCO_3$ is in the form of aragonite. Marginal portions of the Messinian (Upper Miocene) of Sicily are made up of such massive carbonate deposits and in many areas also contain the casts and molds of intercalated halite crystals and halite beds (figure 4.2; Decima, McKenzie, and Schreiber 1988). In the subsurface, the halite is still present. Similar evaporitic carbonates are known in many areas, for example in the Upper Silurian of the Michigan Basin.

In the Messinian deposits of the Mediterranean, where only slight diagenesis has taken place, much of the original aragonite still remains. In these, and to a lesser degree in the altered calcite, there are the remains of algal structures (charae), cyanobacterial filaments and structures, coccoid, bacterial clusters, and brine shrimp (and other) pellets (figure 4.3). The organic content of many of these carbonates may be appreciable (up to 20%), and the ready bacterial breakdown of these organic components results in the early generation of methane and other organic gases.

Gypsum

The growth of gypsum shows a surprisingly variable morphology, ranging from long slim prismatic needles, elongate swallow-tailed twins,

Figure 4.2
Evaporitic carbonate limestone with halite pseudomorphs and voids. Taken from carbonates (Calcare di Base) of the Messinian (Upper Miocene), near Favara, Sicily.

spatulate forms with one curved face, lenticular forms comprised of two curved faces, to stubby tabulate forms. The classical tabulate form shown in most text books is actually the least common morphology. The reasons for the growth of the various crystal faces are only partially understood. However, observations in modern gypsum-precipitating environments have permitted some understanding concerning growth patterns to be developed (e.g., Cody and Hull 1980; Cody and Cody 1988).

Crystals of primary gypsum grow in aqueous hypersaline environments. Free-growing gypsum may form within saturated water as single prismatic crystals, as twinned crystals, clusters, and crusts, and as beds or pavements of coalesced crystals. These beds may range from 1 to 2 cm in thickness to massive banks of gypsum crystals, growing in orderly rows up to 7 m high. The morphology of these crystals is quite diverse, depending on the impurities present in the

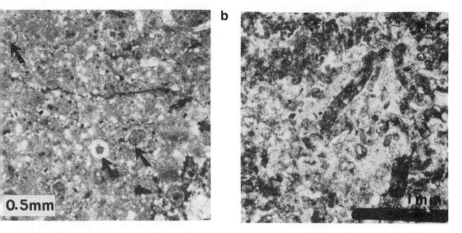

Figure 4.3
Photomicrographs of evaporitic carbonate thin sections, showing (a) charae, pellets, and fine coccoidal clusters: and (b) algal filaments and brine shrimp pellets in recrystallized calcite spar. From Calcare di Base, Messinian; Favara, Sicily.

water (Orti-Cabo and Shearman 1977). Other gypsum beds, made up of great thicknesses (>40 m) of small crystals (2–3 cm), are also found in many areas, with the successive rows of crystals now sitting one upon the other, with and/or without interbeds.

In large brine ponds, growth of gypsum is initiated in the shallowest marginal parts of the water body, where evaporation is highest per volume of water. Although deposits then build basinward very rapidly, comparatively small amounts of gypsum are observed forming synchronously in the deeper parts of saline water bodies. Here the precipitation of gypsum does not take place within the deeper anoxic water, but near the surface, and settles to the bottom (a type of "decantation"). The average growth increment of gypsum crystals or crusts, as studied in a number of solar ponds around the Mediterranean and along the coast of California, is about 1.25 cm per year (Schreiber and Kinsman 1975). However, the fastest rate of growth observed to date by the writer is within conduits and on waterwheels in a solar saltworks (Salinas Viejas) at Cabo di Gata, near Almeria, Spain. It is through these conduits that brine is moved from one crystallizing pond to another, and the conduits become encrusted with as much as 6 cm of gypsum in one year. Accumulation rates are probably lower in nature, due to less-than-optimal concentration conditions and also due to sporadic periods of dissolution.

Primary gypsum is comparatively rare in the subsurface and is best seen in Tertiary deposits that have not been deeply buried. Many of these crystalline deposits were originally considered to be metasomatic because of the huge size of the crystals (up to 7 m). One of the first, well-described deposits of massive crystalline (primary) gypsum was from the Eocene (Ludian) of the Paris Basin (the original quarries for "plaster of Paris"). It was studied initially by Hageau (Hippolyte) in 1837, and later LaCroix (1897) wrote an extensive description of the sequence. Much argument as to the origin of this deposit ensued, and finally Fontes, Gonfiantini, and Tongiorgi (1963) showed, on the basis of isotopic studies of the sulfate, that this particular deposit was largely formed from nonmarine waters. This interpretation has now been further substantiated by paleontological evidence.

The gypsum-bearing Messinian (Upper Miocene) deposits of Sicily were first studied by Mottura in 1871–1872; unfortunately this work appeared in a special volume that initially was not widely read. It was not until the works of Ogniben (1955, 1957), Hardie and Eugster (1971), Parea and Ricci-Lucchi (1972), and the publication of the results of the Deep Sea Drilling Project (D.S.D.P.), Leg 13 (Ryan, Hsü et al. 1973) that these large, well-preserved Messinian deposits became widely known. Much of the basic work on subaqueous gypsum microfacies comes from these deposits (Decima and Wezel 1973; Schreiber 1973; Schreiber et al. 1976) and only in the past ten years have the modern analogs been studied in detail (Busson 1982; Orti-Cabo and Busson 1984). The large areal extent and the thickness, as well as the essentially unaltered nature of these Miocene deposits, renders them of the utmost importance in unraveling evaporite facies. Other well-preserved deposits of primary gypsum include the Langhian (Middle Miocene) deposits of the Paratethys of Poland (Kwiatkowski 1972) and the Quaternary to Recent deposits at Marion Lake, Australia (Von der Borch, Bolton, and Warren 1977; Warren 1982, 1985).

Morphologies. Primary gypsum may grow as single prismatic crystals (figure 4.4), as simple twins (figure 4.5), and in clusters or in

Figure 4.4
Prismatic gypsum crystals (some twinned) forming on the floor of a saline lake (shallow water). (Sample taken from Lac Assal, Djibouti, by Monique Seyler, Lamont-Doherty Geological Observatory.)

massive beds (figure 4.6). These massive beds of gypsum are commonly made up of orderly rows of vertically standing crystals. The longest and most perfectly developed examples of crystals seen to date are found in Cyprus, where "twinned" crystals stand about 7 m tall and extend from the bottom to the top of an entire bed. These crystals are only 15 to 20 cm across and are regular and uniform throughout their entire length. The modern analog for this crystal morphology is found forming in Lac Assal (figure 4.7a), a marine-fed lake in Djibouti (Afar, Ethiopia). Such twinned forms, in fact, have an apparent variable "twin" angle that is governed by incorporated impurities (see figure 4.7b and figure 4.8) and also by interference in the growth of the crystals by halite (seasonal growth). In the latter case, during phases of rapid evaporation, salinities rise and halite nucleates on the edges and corners of the gypsum crystals (seasonal growth). Upon return to the gypsum phase, the renewed growth of gypsum is disrupted by the residual halite, causing aber-

Figure 4.5
Simple gypsum twin, Messinian (Upper Miocene), Eraclea Minoa, Sicily.

rations in crystal form, crystal splits, and discontinuities in crystal growth.

Whereas the composition of the very large crystals from Cyprus is exceedingly pure, the gypsum deposits of Poland contain large quantities of clay and some organic residues. In the pure gypsum of Cyprus, the crystals are fairly regular and closely resemble ideal twin structure. In the less pure deposits, the crystals are split and exhibit palmate or fanlike structures that may attain diameters of a meter or more (figure 4.9). Associated with these fanlike forms are skeletal crystal growths along crystal margins, with large open spaces and split structures that are infilled with clay, clastic gypsum, and *in situ* displacive lenticular gypsum. Similar palmate beds are found in other Tertiary deposits, but none are so grand as those seen near Kraków, Poland (figure 9b).

Other beds of gypsum are comprised of layer upon layer of smaller (1–2 cm), less orderly crystals. In some areas the morphology of the small crystals seems to be controlled by crystals in the underlying gypsum layers (nucleating surfaces), while in other cases they are dis-

a

b

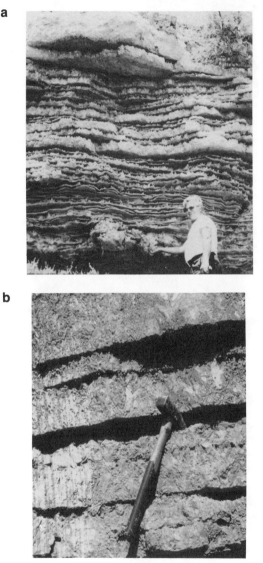

Figure 4.6
Gypsum crystal morphologies seen in the Messinian (Upper Miocene), Eraclea Minoa, Sicily: (a) clustered groups of twinned gypsum;
and (b) massive beds of twinned gypsum.

a

1 cm

b

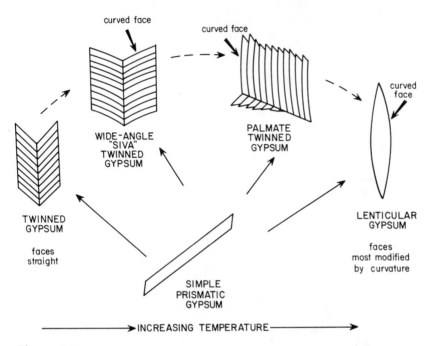

curved face

curved face

curved face

WIDE-ANGLE
"SIVA"
TWINNED
GYPSUM

PALMATE
TWINNED
GYPSUM

curved
face

TWINNED
GYPSUM

faces
straight

LENTICULAR
GYPSUM

faces
most modified
by curvature

SIMPLE
PRISMATIC
GYPSUM

INCREASING TEMPERATURE

Figure 4.8
Review diagram of primary gypsum crystal morphology. Evolution
of characteristic twin forms is due to growth modification in specific
(sensitive) twin faces, which become more and more curved due to
changes in temperature, pH, and the presence of organic impurities
(Orti-Cabo and Shearman 1977; Cody and Cody 1988).

Figure 4.7
Gypsum twins with diverse morphologies: (a) ideal gypsum twin form,
Modern; Lac Assal (collected by M. Seyler); and (b) "wide-angle"
twins, Messinian (Upper Miocene), southeastern Spain.

a

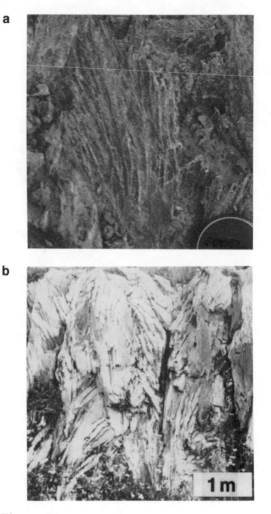

b

Figure 4.9
Variants on wide-angle twin morphology: (a) palmate twinning
(Messinian; Salemi, Sicily); and (b) extreme forms of palmate and
wide-angle twinning in gypsum (Middle Miocene; Gartowice, Poland).

Figure 4.10
Small upright rows of primary gypsum (Upper Miocene; Vita, Sicily):
(a) successive rows formed in continuity; and (b) successive rows
formed as discrete nucleating surfaces.

a

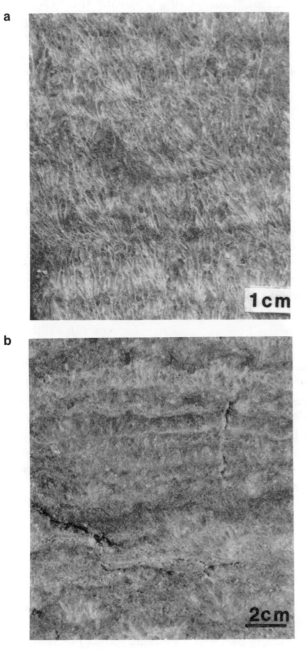

b

crete and wholly unrelated (figure 4.10). Originally these sections formed as halite-gypsum alternations, but the halite has since been dissolved (Schreiber and Schreiber 1977). These beds are common in the Messinian of the Apennines, Sicily, Spain, Crete, and Cyprus, and make up a large proportion of the gypsum that has formed.

Minute quantities of impurities are often incorporated as zones within gypsum crystals. These zones, 1 to 2 mm wide, lie parallel to the crystal faces and evidently record the shapes of the crystals at various stages in their growth. Observations from modern salinas suggest that each zone is the product of a new, dilute water influx. In some crystals, dissolution and renewed growth may occur both on a fine incremental level and also on a massive scale, as long as the dissolution surface is not sealed by the dissolution residue. Such breaks, caused by dissolution, appear to extend across all of the crystals in the same bed. In those beds where the dissolution residues were thick and impervious, a new generation of crystals is present above the break. In other instances the impurities are algal laminae and stromatolites, still in growth position, entombed in the gypsum.

The development of clusters or stellate groupings of gypsum is common in crystals where nucleation takes place on a soft substrate such as gypsiferous marls or algal peats. In this situation the crystals sink into the underlying substrate under their own weight as they grow. A view of the base of one of these beds shows the pattern illustrated in figure 4.11. In some instances, the crystals may sink very deeply into the substrate, causing the crystals to cant. If the crystals are not totally buried, growth may continue, but at a somewhat displaced angle, so that the crystals grow to resemble Paleozoic horn corals, which sometimes have grotesque curvatures from continued growth adjustments.

Nucleation can also take place on the surface of algal mats. Often the blistered and irregular growth forms of the algae are preserved by the first growth of the encrusting gypsum, which renders the irregularities rigid. Early infilling of the hollows under the blisters by growth of interlocking gypsum crystals is common and appears to occur at the same time as the overlying crust is forming. The ultimate appearance, along a section cut perpendicular to the bedding surface, is a hummocky sequence of drumlinlike structures, the hummocks being 15 to 20 cm in length and 4 to 6 cm high. Larger blister struc-

6 cm

Figure 4.11
Base of twinned gypsum cluster that sank into the substrate during growth, forming a typical depression cone. Messinian; Eraclea Minoa, Sicily.

tures seem to become suppressed, possibly due to loading by successive layers.

Increasing amounts of organic impurities in formative water also cause rapidly growing clusters of gypsum crystals to fan out and develop into large-scale domelike structures. When seen in vertical section, the beds resemble cuts through cabbages (called "cavoli structures"). These fanned crystal clusters, as seen forming in Marion Lake, Australia, grow almost like reef structures, extending up through the water, standing proud above the lake floor. These gypsum mounds cease growth upon reaching the water surface. The same domed structures in the Messinian gypsum of Sicily may also display planar upper surfaces along many of the beds (figure 4.12). This unusual domed structure, together with their truncated upper surfaces (formed in very shallow water), is readily preserved upon burial and through

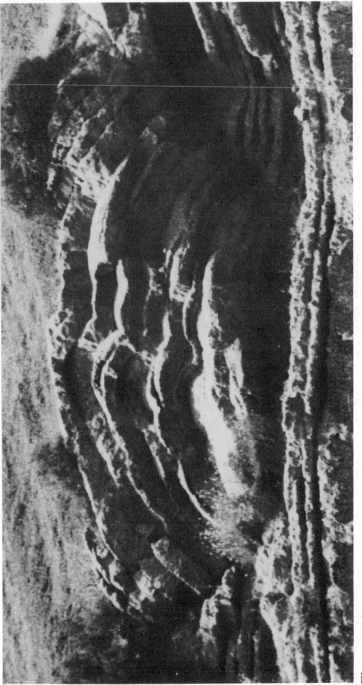

Figure 4.12
Domed gypsum bed ("cavoli structure") showing flat upper surface due to growth limitation controlled by water surface. Dome is 10 m in height. Messinian; Eraclea Minoa, Sicily.

several phases of diagenesis (see "Diagenesis"), and can serve as a guide to interpretation of ancient deposits even when all of the original crystal forms are obliterated. Such domed deposits commonly alternate with algally dominated dolomitic marls and carbonates in which ripples, cross-bedding, desiccation cracks, and salt casts are common.

Gypsum may also grow interstitially within the saturated sediments of both the phreatic portions of sabkhas, playas, and high intertidal zone algal mats (see Kendall, article 1; West et al. 1985), and within the phreatic subaqueous environment. The cemented and/or displacively formed crystals within sediment are commonly lenticular or partially lenticular in habit and may be poikilotopic, incorporating the host sediment in various amounts depending upon the mechanical properties of the sediment and the growth rate of the gypsum (figure 4.13; Shearman 1981).

Halite

Halite may form under a great number of subenvironments, and is only weakly affected by the pH or Eh of the water (Gornitz 1965; Southgate 1982). For this reason, halite may form in water of any depth or with nearly any impurities as long as it is above saturation. A few organic impurities (urea and humic acid, for example) do alter the preferred crystal form of halite, but no compound seems to totally inhibit crystallization. It must also be noted that rapid precipitation of halite permits the development of large numbers of fluid inclusions within the halite, as well as the incorporation of various solid impurities, such as sand, silt, and clay (when available). The fluids and/or particulate impurities are systematically incorporated along the growth faces of the crystals; hence their position in crystals is usually regular.

Morphologies. The earliest halite in a sequence commonly forms at the air-water interface as floating hoppers and rafts of laterally linked hoppers (Dellwig 1955). The halite begins to precipitate as tabular crystals at the air-water interface in direct response to concentration due to evaporation. As the crystals continue to grow, they develop into inverted pyramid forms, hanging at the surface (Shearman 1978;

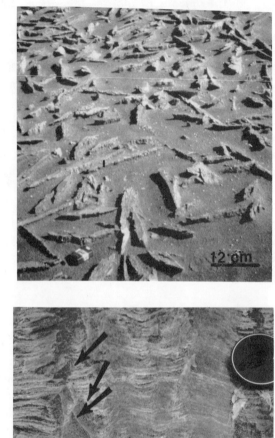

Figure 4.13
Displacive to through-growing *lenticular* gypsum forming within an evaporitic carbonate mud: (a) Modern deposit at Sebkha el Melah de Zarzis (courtesy of J. P. Perthuisot); and (b) through-growing lenticular gypsum (visible at arrows) in a laminar algal bed (Upper Miocene; near Bari, Italy).

figure 4.14). When the pyramids and rafts of pyramids sink, they then become the nuclei of halite cubes, but may retain both their hoppered structure (internally) and/or their raftlike linkage (figure 4.15). Southgate (1982) pointed out that traces of certain organic components may serve to inhibit the formation of these "floaters."

Bottom nucleation in halite may develop a number of microfacies patterns. The first general study of such salt deposition was by Valyashko (1952), but it was not until Shearman's (1970) study of halite formation in the salt flats of Southern California that the morphology of large-scale salt deposition was put into a geological context. Shearman (1970, 1978) shows that bottom-nucleated halite is deposited in flooded salt pans and that repeated desiccation and flooding can develop a characteristic crystal form marked by chevron-shaped fluid inclusions. These crystals are arranged in beds separated by truncated surfaces that represent periods of desiccation and then flooding by new water. This process results in a distinct and recognizable rock morphology (figure 4.16). Repeated phases of flooding and desiccation produce cycles of salina-pan muds, overlain by halite layers with truncated upper surfaces, and finally form broken surface crusts, polygonal in plan. Reflooding causes partial solution and truncation of the upper salt surface and may bring in salt-pan muds (carbonates, siliciclastics, and/or anhydrite).

Other variations of bottom-nucleated halite textures, including halite cornets (hollow cones or teeth of halite), were first described by Valyashko (1952) and are illustrated by Arthurton (1973). Very unusual halite deposits, composed of elegant skeletal, pagoda, and reticulate salt crystals, are also shown to form at the bottom of saline water bodies during the process of desiccation (Southgate 1982). Pseudomorphs, imprints, and molds of these unusual skeletal forms are not uncommon in many evaporite-related deposits, although they are readily overlooked because they are present on bedding planes and have little appreciable thickness (figure 4.17).

Massive beds of halite, as well as those which are finely laminar, composed of tiny salt cubes (1–3 mm), are common in many deposits. These tiny halite crystals seem to form very rapidly within the water column, settle to the bottom, and are called "cumulates" (Lowenstein 1982). The cumulate halite may form sporadic cross-bedding and ripples; however, they are usually either in massive, even

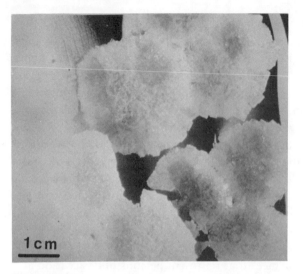

Figure 4.14
Inverted pyramid halite hoppers, floating attached to water surface in modern saline lake. (Courtesy of Peter Scholle, Southern Methodist University.)

Figure 4.15
Halite rafts, made up of halite hoppers; seen in slabbed core. Bresse Basin, Oligocene. (Sample courtesy of Robert Moretto, University of Lyon.)

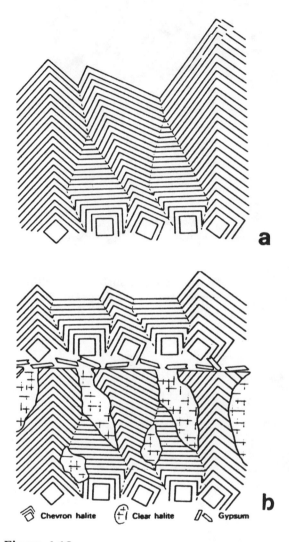

Figure 4.16
Halite growth in a salt pan with sporadic desiccation and exposure:
(a) initial growth of halite with chevron growth pattern marked by
fluid inclusions; and (b) second layer of growth, after renewed flooding, truncation of upper surface, and void filling (clear halite) in lower
initial layer (Shearman 1970).

Figure 4.17
Imprint of skeletal halite found on a bedding surface of dolomitic mudstone in Salinan (Upper Silurian) beds near Gypsum, Indiana. (Photo courtesy of Charles F. Kahle.)

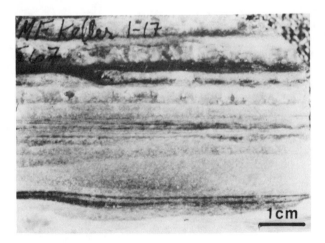

Figure 4.18
Finely laminar halite with low-angle cross-bedding from interreef facies in the A-1 Evaporites of the Michigan Basin (Upper Silurian). (Sample courtesy of Shell Oil Company.)

beds, or in the form of laminae or "varves" (figure 4.18). It is difficult to be sure whether the laminar forms represent deep-water deposition, or whether they are merely formed in shallow, density-stratified water bodies.

Halite commonly grows displacively within the mud of exposed salt flats (Gornitz and Schreiber 1981; Handford 1982). Displacive cubes, incorporating mud from the surrounding matrix, are known from many locations and may be found in both siliciclastic and carbonate settings. In many instances the cubes are skeletal- to pagoda-shaped (with extended corners and edges; see figure 4.19). These crystals can become very large (20–25 cm being reported; see Handford 1982). Because exposed salt flats are prone to flooding by undersaturated water, in many instances the salt cubes undergo partial solution (synsedimentary), and corroded or irregular forms are found in the rock record (Shearman 1978). Where the solution of halite occurs well after burial, then the original cube shapes are preserved, but the early solution permits the matrix to infill the voids, forming halite pseudomorphs. A review of halite morphology is given in figure 4.20.

Figure 4.19
Pagoda-shaped halite cube. Kirkham mudstones, Upper Triassic, England. (I.G.S. photo no. 8345; courtesy of A. A. Wilson.)

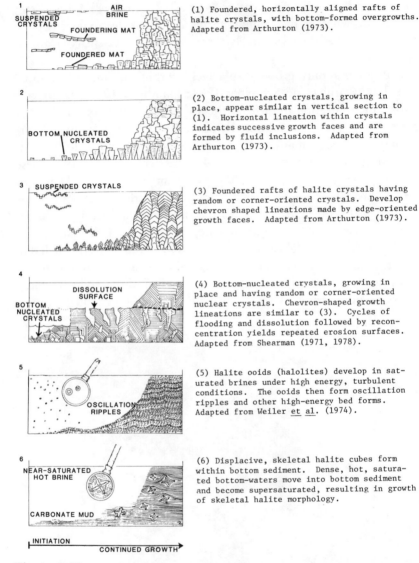

(1) Foundered, horizontally aligned rafts of halite crystals, with bottom-formed overgrowths. Adapted from Arthurton (1973).

(2) Bottom-nucleated crystals, growing in place, appear similar in vertical section to (1). Horizontal lineation within crystals indicates successive growth faces and are formed by fluid inclusions. Adapted from Arthurton (1973).

(3) Foundered rafts of halite crystals having random or corner-oriented crystals. Develop chevron shaped lineations made by edge-oriented growth faces. Adapted from Arthurton (1973).

(4) Bottom-nucleated crystals, growing in place and having random or corner-oriented nuclear crystals. Chevron-shaped growth lineations are similar to (3). Cycles of flooding and dissolution followed by recon-centration yields repeated erosion surfaces. Adapted from Shearman (1971, 1978).

(5) Halite ooids (halolites) develop in sat-urated brines under high energy, turbulent conditions. The ooids then form oscillation ripples and other high-energy bed forms. Adapted from Weiler et al. (1974).

(6) Displacive, skeletal halite cubes form within bottom sediment. Dense, hot, satura-ted bottom-waters move into bottom sediment and become supersaturated, resulting in growth of skeletal halite morphology.

Figure 4.20
Facies produced by various modes of halite growth and emplace-ment. (Composite diagram from Gornitz and Schreiber 1981, based on Arthurton 1973, figure 2; Shearman 1970, 1978; and Weiler et al. 1974)

Potassic Salts

There are few well-documented textural studies of potassic salt deposits (potash evaporites). In their monumental study of the potash deposits of New Mexico and Texas, Schaller and Henderson (1932) were the first students of this sediment type, and little was done with actual sediments for the next thirty years, except for chemical analysis and basin modeling. Holser (1966) was the next person to study potassic sediments, with his analysis of sedimentation in Baja California and his discovery of modern polyhalite precipitates ($Ca_2MgK_2(SO_4)_4 \cdot 2H_2O$). Further study of this site by Pierre (1985) has given us a clearer understanding of the complex chemistry and the synsedimentary mode of replacive formation. Lowenstein (1982, 1983) and Lowenstein and Hardie (1985) have made a very clear point concerning potassic salts—they are largely syndiagenetic or diagenetic in origin, with formation being related to the replacement and alteration of earlier deposits of minerals such as gypsum and halite.

Hardie (1984) suggests the phrase "modified primary anhydrite" for these quasi-primary potash deposits, emphasizing the early nature of many of the alterations in mineral type. Lowenstein (1982) and Lowenstein and Hardie (1985) feel that most potassic deposits form in very shallow to desiccated settings, governed by ephemeral changes in water composition (figure 4.21).

Clastic Evaporites

Clastic Gypsum. Resedimented gypsum in almost all grain sizes commonly occurs in many large evaporite complexes. Mechanical remobilization takes place under conditions similar to those that affect other solid materials, and the sedimentary products can be interpreted in much the same manner. All evaporite environments shown in figure 4.1, ranging from continental to basinal marine, can be considered potential source areas and also sites of clastic deposition.

Reworking of evaporitic sediments within a hypersaline basin should prove to be a common event, for the following reasons:

THE SALINE PAN CYCLE

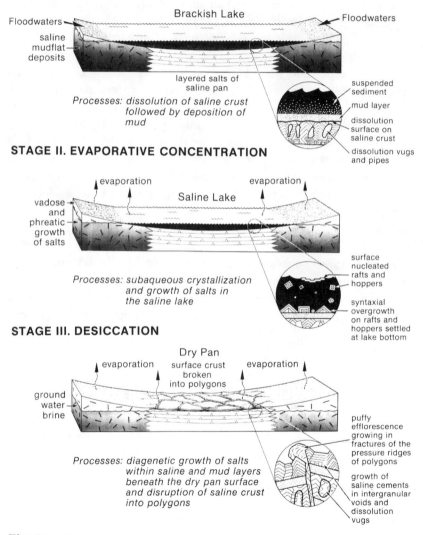

Figure 4.21
A summary of the basic elements of the saline salt-pan cycle (Lowenstein and Hardie 1985).

1. The extremely high rates of sedimentation in shallow portions of the water body result in rapid progradation basinward. This could lead to slope instability in the evaporites, and between the evaporites and the immediately underlying nonevaporites.

2. Sea-level fluctuations are probably common in most evaporite basins, because any marine basin that has become markedly hypersaline has done so due to a restricted supply of water. Thus, any cut-off in marine inflow would result in a large net deficit of water. Rapid drawdown of surface level in the basin would result, and evaporites formed around the margins of the basins would become exposed and left to the mercy of gravity, wind, and meteoric water. This mechanism is a very significant source of reworked sediment.

3. Sporadic changes in the composition of the water body due to variation in inflow from both marine and nonmarine sources could dissolve some of the more soluble, earlier-formed salts. "Sapping" of these soluble materials could result in slope instability and mechanical release of the less soluble components. Changing salinity may also cause syneresis (shrinkage) in the intercalated clays between evaporite beds. This variation in the volume of the clay might also render the gypsum beds unstable (particularly on a slope). Both mechanisms (solution and shrinkage) could provide large quantities of material for reworking.

Massive, unsorted gypsrudite and gypsarenites can be developed as slumps, slides, talus accumulations, and collapse breccias (which result from *in situ* removal of soluble salts). The Messinian deposits of the Apennines are replete with examples (Parea and Ricci-Lucchi 1972; Ricci-Lucchi 1973; Vai and Ricci-Lucchi 1977). Similar deposits also occur in Sicily and in Cyprus (Bommarito and Catalano 1973; Schreiber et al. 1976). Simple drawdown of the water surface level of only a few meters, as has occurred in Marion Lake, Australia, results in very coarse lag deposits of gypsum mixed with associated algal carbonates along the new shoreline (Dr. Marjorie Muir, 1976, personal communication). A comparable deposit to that of Marion Lake is present in the Apennines (Vai and Ricci Lucchi 1977).

Laminar gypsum. Laminar gypsum and anhydrite, either pure or in couplets with carbonate, is common in many evaporitic sediments (figure 4.22a). The laminae are generally thin, between 1 and 8 mm

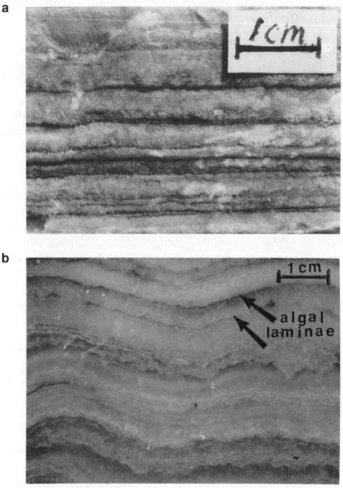

Figure 4.22
Laminar gypsum alternating with thin interbeds of micritic carbonate
(from Upper Miocene; Sicily): (a) even beds composed of gypsum silt
particles; and (b) crenulated gypsum laminae alternating with fine
algal interbeds (showing algal filaments in thin section).

in thickness, and are commonly even and parallel. These laminar beds may also be somewhat uneven and crenulate, or may exhibit cross-stratification, climbing ripples, and oscillation ripples (see "Clastic Evaporites"). The crenulate gypsum laminae are commonly interbedded with micritic dolomite or calcite, which displays algal structures and *in situ* algal filaments (figure 4.22b). Tiny lenticular gypsum crystals (1–3 mm) are present within the carbonate interlaminae and appear to have formed displacively within the carbonate. Cornée (1983, 1984) has suggested that formation of these algal-related gypsum crystals is controlled by the microenvironment established within algal-bacterial mats and that the organic chemistry affects solution and precipitation, lit-par-lit.

It is clear in many instances that the laminae are the product of clastic reworking deposition in shallow water (Hardie and Eugster 1971). The presence of alternating layers of algal mats and algal stromatolites within some laminar gypsum layers represent the other type of hypersaline sediments known in similar modern environments. Interbeds of very even, regular laminae among the reworked cross-bedded sequences can represent ephemeral deepening of the water or of quiet periods, perhaps associated with establishment of density stratification within the water body. The thick beds of laminar gypsum (or anhydrite upon later diagenesis), without evident algal or current structures, appear to represent deposition below wave base or below well-established density stratification. This may come about in one of several different ways:

1. Precipitation of fine-grained sulfate occurs at the air-water interface, with the sulfate particles settling to the bottom as a fine crystal "rain" (for discussion see Dean and Anderson 1978). This process is also called "decantation" in both the Italian and French literature.
2. Deposition occurs below the halocline, with seasonal freshening and precipitation taking place at the bottom (Warren 1985).
3. Sulfate silts and sands from the shallow portions of the basin are reworked by density currents.

These laminar gypsum rocks have diverse and somewhat confusing names in the literature and these names tend to be used without reference to their mode of origin. This is particularly the case for young unaltered evaporites, seen at the surface, which are called "balatino" in Italy but "marmara" in Cyprus; both of these names have found their way into the general literature. The significance of the various names is lost because it becomes unclear whether the writer is employing them genetically, as an indication of supposed origin, or simply as a descriptive term.

Observations of laminar gypsum, made in modern salinas (solar salt pans), suggest that the laminae are initially composed of silt- to sand-sized gypsum crystals, either single or twinned, or of cleavage fragments that lie with their maximum dimension parallel to bedding (figure 4.23). These small crystals have been observed to grow in two ways, either as thin crusts on the floors of brine pans, with the long axes of the crystals oriented perpendicular to the bedding, or as free-floating, delicate acicular crystals at the air-water interface where nucleation is very rapid. The crystals then sink and/or are reworked along the bottom, where they accumulate and form laminae in the same way as other clastic materials. Those crystals that grow as crusts along the bottom are readily broken off and can be reworked in a similar manner. Where the crystals of gypsum remain in open contact with the water, they develop overgrowths and continue to grow until they no longer fall into the category of easily moved sand- or silt-sized grains. However, a very strong disturbance, such as a storm,

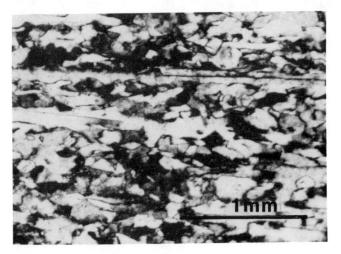

Figure 4.23
Laminar gypsum seen in thin section (seen with crossed polars).

may break even these large crystals into fragments. While primary anhydrite has never been observed to form in a subaqueous setting except at considerably elevated temperatures, it is possible that gypsum-inhibiting organic components may permit formation of anhydrite under some circumstances, and therefore anhydrite may form similar clastic laminites. If anhydrite cannot be nucleated in water, then anhydrite merely follows gypsum diagenetically.

In some laminar gypsum beds, not only is there evidence of shallow-water current action, but also there are features produced by desiccation, such as edge-wise conglomerates (rip-up breccias; see figure 4.24). Other features indicative of shallow water depths include fossil brine shrimp and bird footprints. Various organic remains including leaves and pine needles and cones have been found within and between clastic gypsum beds, but these could have been reworked into water of any depth.

In some deposits where gypsum laminae are associated with delicate alternations of carbonate, both calcite and dolomite, there are indications which suggest that the carbonates were a product of the activity of cyanobacteria and green algae. Although there are algae that live far below the photic zone, the forms and types of algal carbonates found in this association (at least within the occurrences in

Figure 4.24
Rip-up or edgewise conglomerate of laminar gypsum clasts.

the Recent, the Messinian of the Mediterranean, the saline lakes of Australia and Baja, and the Tertiary of Poland) are limited to the photic zone (Von der Haar 1976; Monty 1965, 1972, 1976; Von der Borch, Bolton, and Warren 1977; Cornée 1984). Planktonic algae may sink to the bottom and form a part of the bottom sediments, but discrete intercalations having true algal structure are unlikely to develop in this fashion. In the Messinian, tiny delicate Cerithid gastropods (air-breathing, algae eaters) are sometimes found on the surface of the carbonate interlaminae. These, together with bacterial and algal structures, serve to strengthen the argument in favor of carbonate formation *in situ* for these deposits, rather than redeposition by settling through a column of water.

There are at least two basins in the Messinian of Sicily, the Gibellina and the Ciminna, in which laminar, organic-rich beds of gypsum mixed with carbonate are associated with gypsum mass flows and turbidites. The gypsum laminae, made up of tiny gypsum needles and fragments, show sporadic current features such as cross-bedding and plastic deformation structures, but other features normally associated with shallow-water environments are absent. It seems likely, therefore, that these small gypsum crystals formed at or near the water surface or are fragments broken from larger crystals, and that they have moved downslope or sunk through the water column into a basin of some considerable depth and finally come to rest in an environment far different from that in which they originated.

Coarse Clastic Gypsum. If dissolution proceeds to an extreme, many of the gypsum crystals are moved out of their original growth positions and form a residual breccia. Gypsarenites and gypsrudites sometimes form graded beds, or they may show all manner of induced current structures (Schreiber et al. 1976; Schlager and Bolz 1977; Ciaranfi et al. 1978). They may be pure, mixed with carbonate, or with any other type of clastic material. Gypsum may be found as a major component of channel deposits, beach bars, delta structures, or as simple intercalations between beds of laminar and massive gypsum. Reworking may carry "clastic" gypsum into water depths sufficient to generate true turbidite structures. A gypsum turbidite bed, one of several hundred in a sequence found in the tectonically controlled Gibellina Basin in Sicily, is illustrated in figure 4.25. In

Figure 4.25
Gypsum turbidite bed. Messinian; Gibellina, Sicily.

this example, the turbidites are underlain and in a few places inter-calated with both undifferentiated and graded mass flows. The total thickness of the turbidite sequence is about 120 m. The turbidites are overlain by laminar gypsum and a fairly thick sequence of *in situ* beds of shallow-water selenite. The latter contain regular laminae of algal filaments, brine shrimp eggs, and fecal pellets, and they are of shallow-water origin.

In order to interpret this succession, it must be realized that although the basin was a tectonically controlled depression, its absolute depth was not known, nor the actual depth of water at any one time. The uneven thickness of the observed turbidite succession suggests that the basin was possibly not very deep or very large. However, the *in situ* selenite beds overlying the turbidites do contain brine shrimp pellets and algal structures, fixing that water depth as quite shallow. This combination suggests that although the water was comparatively deep during deposition of the turbidites, it had shallowed considerably by the time growth of selenite commenced, either by filling or because of an absolute change in water depth.

Microscopic observation and discussion. Gypsum oolites (or "gyp-solites") were noted by Ciaranfi et al. (1978) among gypsarenite particles in subaqueous shoreline deposits from the southern Apennines. These gypsum oolites, and similar ones from Cyprus, are found in dunelike cross-bedded structures (figure 4.26a) in which all of the grains have become worn to some degree as a result of mechanical abrasion, cleavage-controlled breakage, and dissolution of edges and

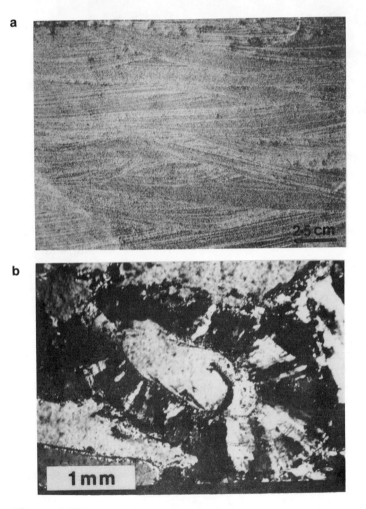

Figure 4.26
High-energy festooned cross-beds in gypsum sands: (a) festoon cross-beds; and (b) photomicrograph of a thin section of sample from (a), showing clastic gypsum and "gypsolites." Messinian; near Bari, Italy.

corners. These oolites also exhibit radial overgrowths of fine gypsum crystals forming a rim around the original particle (figure 4.26b). The overgrowth rim often contains several layers of growth, each marked by a row of minute carbonate inclusions, usually also arranged in concentric fashion. It is therefore suggested that gypsum ooids or rim overgrowths may be formed under high-energy, hypersaline conditions, perhaps with alternate quiet periods, in the range of gypsum precipitation.

Halite. An analogy may be drawn between clastic gypsum and halite. While most halite is cemented into place almost immediately due to the rapid rate of halite precipitation, windstorm deposits form and may rework halite into cross-beds and ripples within very short time intervals. The clastic halite deposits of the Dead Sea (Weiler, Sass, and Zak 1974), in which the halite precipitates and their oolitic overgrowths form in the shallows under conditions of highly agitated water and shoreward winds, are a reasonable parallel to the clastic gypsum oolites discussed above, both as to form and site of occurrence.

Downslope reworking of halite together with other upslope clastics into an evaporative basin is also possible. In the Messinian of Sicily, reworked clastic components mixed with fragmental halite has been observed in a number of salt mines (Arvedo Decima, 1972, personal communication). In those instances where the interfingering beds of nonevaporite clastics are substantial, formational water moves through the clastic beds and gradually dissolves the marginal evaporites. Basin-center halite contains only limited amounts of clastic components, except windblown sand and silt (carbonate and siliciclastic) and clay (Hsü, Cita, and Ryan 1973). The clays move out from the margins, across saline basins, following the interface surfaces of stratified brines (Sonnenfeld 1985). Despite the high pH values of the saline water, many layers do indeed have residues of rounded windborne deposits, perhaps preserved because once the water becomes saturated with respect to silica, no further solution takes place.

DIAGENESIS

Evaporites suffer the effects of diagenesis more readily than any other sediment. Alteration begins with deposition; the earliest change is cementation, and most, but not all, evaporites are cemented as they are formed. Even the simplest changes in climate may cause substantial variation in both the mineralogy and the morphology of the deposits. This ease of transformation, both during formation and more particularly after burial, has led some students of the problem to consider much of evaporite diagenesis as a sort of "progressive metamorphism" (Borchert and Muir 1964). This treatment has received much criticism, but consideration of evaporite diagenesis as a type of metamorphism serves to illustrate the conceptual problems of evaporite diagenesis. There are, incidently, true metamorphosed evaporites, and anhydrite remains a stable mineral up to 1,000°C.

A simple overview of evaporite diagenesis may be made by dividing diagenesis into five groupings: changes due to influx variation (salinity and influx concentration); variation in basin level; changes upon burial; exhumation with rehydration, calcitization, and solution; and deformation. Considerable effort has been spent in reaching an understanding of the many aspects of the burial diagenesis of evaporites, and for a review of these concepts the reader is directed toward the works of West (1964, 1965, 1979), Holliday (1970, 1973, 1978) and Holliday and Lake (1978). In particular, it must be appreciated that syndiagenetic changes within all evaporites are the norm. (See also Kendall and Warren, article 2; Handford, article 3).

Influx Variation

For a water body to become sufficiently saline for the formation of evaporites, there must be a high rate of evaporation and a marked restriction in the total water budget. This restriction must be sufficient to permit evaporation of 80% to 90% of the inflow. If part of the inflowing water is nonmarine, then the total evaporation must be even greater if evaporites are to form. The anticipated sequence for direct chemical precipitation (as in a chemical beaker of marine water) is $CaCO_3 \rightarrow CaSO_4 \rightarrow NaCl \rightarrow$ "high salts" (Usiglio 1849). In nature there are a number of problems in achieving this direct sequence of precipitates. First, most calcium carbonate is removed by biogenetic processes through the agencies of algae and bacteria. Second, additional ions enter the system from the air (particularly CO_2) and through surface runoff from adjacent land (Ca^{+2}, Mg^{+2}, and

HCO_3^-). Third, some of the early precipitates react with concentrated residual fluids to form a somewhat different mineral assemblage than is ideal (Harvie et al. 1980). Also, each batch of seawater entering the system must evaporate further than the previous batch before it can mix with previously concentrated water (which is stratified below it), because the earlier batch of residual waters is extremely concentrated and may be much more dense (Holser 1966; Kühn 1968).

Changes in salinity and in ionic concentration relative to the minerals on the floor of a brine basin take place very readily. In fact, a simple rainfall or even a period of more humid weather may remove or corrode previously formed minerals (Kinsman 1976). The engineers who operate the Salin de Giraud in southern France have modeled each of these changes simply to maximize their salt crop (Boudet, Capdequi, and Marchand 1985). In China, the managers of a saltworks have evolved a cover system for their salinas to prevent dissolution in the face of unwanted new water (Liu et al. 1985).

Solutional surfaces develop on a seasonal basis in any natural brine pond. Etched crystal surfaces and layers of crystals are the norm, and solutional dismemberment of individual beds or entire deposits are commonly found in many places. Solutional sapping, with the removal of more soluble beds—for example, in the case of halite-gypsum alternations—results in broken crystals, beds, and zones, in an otherwise continuous section. This may occur on a synsedimentary basis or at any later period in the burial and exhumation history of the region. Removal of halite from a carbonate-halite deposit, for example, may produce a totally dismembered breccia (figure 4.27). The Broken Beds of Dorset, England, are not only disrupted by exhumational solution, but they are further altered by the calcitization of the sulfates (West 1975).

Variation in Sea Level

Because an evaporitic basin is so restricted as to inflow, it may readily suffer either flooding or drawdown (Maiklem 1971). Rehydration of sediments formed in supratidal settings by both rising seas or by continental runoff results in early change from anhydrite to gypsum as well as the deposition of additional gypsum cements (Anas Gunatilika,

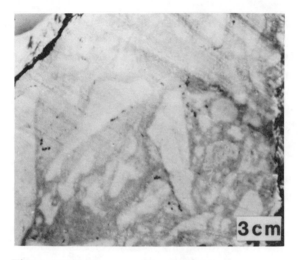

Figure 4.27
Evaporitic carbonate breccia due to solution of halite. Near Favara, Sicily.

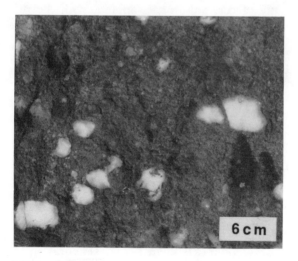

Figure 4.28
Massive gypsum breccia due to slumping. Messinian; from the Apennines.

1984, personal communication). The reverse is also true, in that subaqueous evaporites (crystals or clastic layers) may be subjected to the sabkha environment, and gypsum become converted to anhydrite, halite dissolved, and carbonates dolomitized. Great fluctuations in basin level leave early deposits stranded upslope, and these are prone to slumping, mass flow, and mechanical and chemical reworking by meteoric water. Such slumps and reworking are probably the cause of the massive gypsum beds (figure 4.28) depicted by Vai and Ricci-Lucchi (1977).

Modest drawdown also permits karst formation, canyon cutting, and the development of salt and sulfate "swells." The drawdown, when it occurs very soon after deposition, results in exposed surfaces with soil development. In many deposits we find a residual terra rosa (soil), marking the exposure (for example, Site 132, Leg 13 of the D.S.D.P. study of the Mediterranean, Ryan, Hsü et al. 1973; and Site 653, O.D.P. Leg 107, Kastens, Mascle, Auroux et al. 1987). Karst fill, commonly but not always formed in a subaerial setting, may be the site of animal and plant concentrates, as is the case in any cave setting. Salt and sulfate residues can result in local relief, which later represent neat stratigraphic traps.

Burial Diagenesis

The usual effects of burial upon a sedimentary section are compounded in the case of evaporites. Compaction and mechanical dewatering are not unlike those processes in a carbonate sequence, except that the migrating fluids may also be very effective as agents of alteration in the course of their movement. On the other hand, the dehydration of gypsum and other hydrated evaporites represents a formidable difference from the simple dewatering of most other sediments. From study of many anhydrite deposits, which were originally formed as gypsum, it has become obvious that the loss of water may take place a little at a time unless fluid migration is prevented by a seal, such as shale. Shearman (1985) pointed out that some anhydrites are nodular while others are not, and the nodular varieties stand apart from the rest in their appearance and in their crystal fabrics. Many of the nodular ones formed from syndepositional alteration of gypsum. The non-nodular ones, in discrete, even beds, may well have formed from the dehydration of gypsum (Schreiber et al. 1976; Davies 1977).

Rouchy (1980) was the first to explain the step-by-step conversion of crystalline gypsum to anhydrite, although Davies (1977) first demonstrated that the internal morphology of crystalline gypsum beds may carry over into anhydrite (figure 4.29). Many anhydrite beds are, in fact, made up of vertically oriented, elongate nodules of anhydrite (a form not usually developed in the sabkha), which were originally gypsum beds (Schreiber, Roth, and Helman 1982; Loucks and Longman 1982; see figure 4.30). Very pure deposits of gypsum, whether laminar or crystalline, have little internal structure, and the conversion results in massive, featureless layers. In the case where fluid escape is deferred due to sealing and overpressure, the escape structures may possibly be preserved, and a 38% volume change is involved.

Fluid migration from other evaporites is even more involved. Carnallite ($KMgCl_3 \cdot 7H_2O$), for example, may go to sylvite (KCl) plus

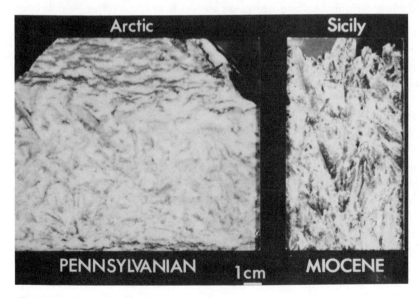

Figure 4.29
Anhydrite pseudomorphs (left) of primary gypsum (right). (Davies 1977.)

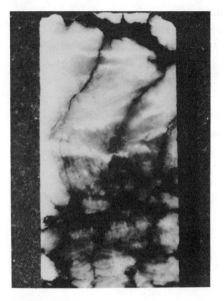

Figure 4.30
Anhydrite pseudomorphs of palmate gypsum. Cretaceous; Maverick
Basin.

bishofite ($MgCl_2 \cdot 6H_2O$), with only a modest water loss (at 40°–50°C).
The bishofite, however, is very soluble, and it may in turn dissolve
in the freed water of crystallization, and totally leave the system in
solution. Such brines (and others) have a secondary effect, causing
the dehydration of associated gypsum, as observed near the Real-
monte Mine (Upper Miocene; Sicily), and tend to be related to local
structural deformation.

Rehydration, Solution, and Calcitization

Exhumation permits rehydration, solution, and calcitization. Rehydration, particularly of anhydrite, requires a considerable volume expansion. Rehydration just at the surface results in spalling and cracking in the new gypsum. Near-surface rehydration does not present a
great problem, because the volume change simply results in slight
uplift. A distance below the surface, hydration is far slower and commonly takes place along fractures and faults. In areas where there

are strain buildups, as in earthquake-prone regions, water builds up along fractures or in aquifers and is forced into anhydrite. The coarsest gypsum forms along the porous and permeable zones in intercalated clastics and carbonates, and the water slowly moves farther in along crystal boundaries and small fractures, forming the typical interlocking mosaic of alabastrine gypsum (figure 4.31). Fractures, forced open by overpressured water, may become the sites of satin spar formation (Mossop and Shearman 1973).

Calcitization of sulfate and of associated dolomite beds commonly takes place in the zone of penetration by meteoric water. Bacteria that utilize sulphate in their biological processes (*Desulfovibrio*) commonly operate in this realm in the presence of meteoric waters, releasing H_2S and converting the sulfate into calcite. The presence of sulfate-saturated water in contact with dolomite also seems to affect the calcitization of dolomite (Clark 1980). Both processes appear to operate in association with rehydrating sulfates.

Deformation

Recent work in the Italian Alps (Permian beds) and in the Apennines (Triassic beds) have shown that on the fine scale, the morphology of

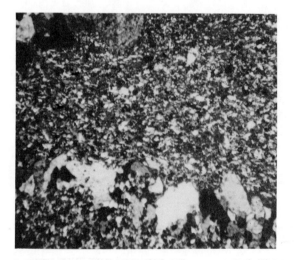

Figure 4.31
Alabastrine gypsum seen in thin section (seen with crossed polars).

Figure 4.32
Ptygmatic folding due to structural deformation. Permian; Italian Dolomites.

deformed sulfates (gypsum and anhydrite) is not unlike that of syn-sedimentary deformation. Enteroliths and soft sediment diapirs appear to be very similar to ptygmatic folds formed by tectonic activity (figure 4.32). Only on the scale of whole beds and sedimentary cycles can we recognize the differences. Evaporitic sulfates are commonly interbedded with dolomitic layers, and this contrast in strength of materials, with plastic versus brittle deformation, provides the clues necessary for the differentiation between the two (figure 4.33).

In the initial steps of response to stress, the carbonates show little more than spaced cleavage and vertical stylolites, while the sulfate, as anhydrite, has flowed within the beds and displays an internally deformed structure (isoclinal folding; figure 4.34). Anhydrite also provides a suitable décollement surface for thrusting, as does halite, due to their low strength, and both are commonly involved in nappe emplacement (Müller and Hsü 1980).

Deformation of halite and other salts is quite complex and may take place through the agency of pressure solution and the formation of spaced cleavage, or by folding and diapirism. Simple deformation in salt, particularly when several different salts are present, creates

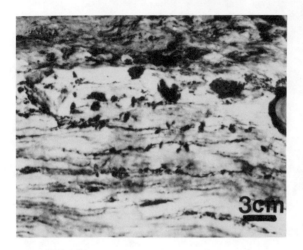

Figure 4.33
Plastic deformation in gypsum (white) with contrasting brittle deformation in dolomitic layers (dark grey). Note that dolomite fragments are rotated and broken. Permian; Italian Dolomites.

Figure 4.34
Isoclinally folded mix of gypsum (after anhydrite) and thin-bedded dolomite. Permian; Italian Dolomites.

Figure 4.35
Plastic deformation in a mix of several salts (halite, carnallite, and keiserite). Messinian; Sicily.

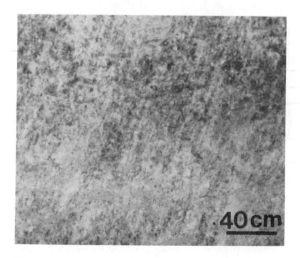

Figure 4.36
Nearly vertical-spaced cleavage, apparent as parallel solution residues in a halite bed (Keuper Salt, Triassic).

dramatic illustration of fold structures (figure 4.35). Spaced cleavage has just recently been recognized in salt, and really only becomes visible in "dirty" or clayey layers (figure 4.36). Intercalated beds of clear halite show no structures, because pressure solution proceeds without a residue to trace the volume loss. Such cleavage structures have long been ascribed to syneresis and/or to desiccation, but they are recognizable as structurally controlled because they are a response to regional deformation, and are aligned through all of the beds where they are visible because of it (evaporitic or not). Simple plastic deformation in salt is the most common feature, although faults are quite commonly observed.

EVAPORITES AND BASIN CONFIGURATION

Subaqueous evaporite deposits represent a significant portion of the evaporitic sedimentary record. Identifiable facies components can be found once core and/or outcrop is examined, as demonstrated in the previous section. The continuum of subaqueous sedimentary facies that is formed may be overprinted by changing sedimentary conditions (syndiagenesis) and later postdepositional diagenesis. The critical features that identify original sedimentation are the most significant pieces of the puzzle that we can find in this study. Relics of crystalline beds, of clastic reworking, of depositional cyclicity, of truncation surfaces, and of soil profiles become the clues that we may use in evaporite facies analysis. Judicious application of trace element studies and isotope analysis helps us to interpret the results of visual examination (see for example McKenzie 1985; Pierre, Ortlieb, and Person 1984).

The relationship of "saline giants" to oil deposits is unquestioned. There are two extreme models in vogue which have been applied to evaporitic basins—the brine-filled basin model, and the desiccated basin model. As with most controversies concerning geological models, the appropriate answer is somewhere in between, and depends upon the basin of choice. Here are presented briefly the sediments in four basins: the Permian of the Delaware and the Zechstein, the Upper Silurian of the Michigan, and the Upper Miocene of the Mediterranean. Without good outcrop and core studies, either model may be employed with impunity, but with the rocks in hand, proper constraints may be applied.

The Messinian of the Mediterranean

The Messinian (Upper Miocene) of the Mediterranean is in a most unusual setting, partially on oceanic crust, and partially on attenuated continental crust. Within this complex setting, the deposit is made up of largely shallow-water and some "desiccated" sequences atop and overlain by open-marine sediments, but intermediate and some deep-water evaporites are also recognized (Schreiber et al. 1976; Vai and Ricci Lucchi 1977). Even the tectonic history of this basin is a maze of contradictions; in some models, phenomenal subsidence rates are invoked throughout the basin to explain its history. Originally these Messinian evaporites were identified from the presence of characteristic reflectors on seismic profiles, but were thought to be Mesozoic in age, as the Mediterranean is rimmed by Mesozoic evaporites. In 1970, the Deep Sea Drilling Project sampled these reflectors and demonstrated that they were Upper Miocene in age (Ryan, Hsü et al. 1973). Subsequent drilling studies, in 1975 (Hsü, Montadert et al. 1978) and recently (Kastens, Mascle, Auroux et al. 1987, O.D.P. Leg 107, drilling during January and February 1986), have extended these observations.

The Miocene evaporites of the Mediterranean range from 0.2 to 2 km in true thickness, and range up to 10 km in regions of great compression (deformed zones in the eastern Mediterranean). In uplifted, exposed portions of these deposits, as in the case of Sicily, the underlying open-marine Tortonian and lower Messinian sediments are overlain by some 800 m of Upper Messinian evaporites. In vertical section, the stratigraphic sequence begins with gypsum, passes into halite and potash, and goes back to gypsum. These evaporites are in turn overlain by deep-water open-marine oozes of Pliocene age (figure 4.37; Decima and Wezel 1973). In the Apennines, the Messinian evaporites (without halite) are sandwiched between sequences of kilometer-thick open-marine turbidites (Parea and Ricci-Lucchi 1972; Vai and Ricci-Lucchi 1977). Seismic sections across the Mediterranean, tied into judicious coring in several areas in the western

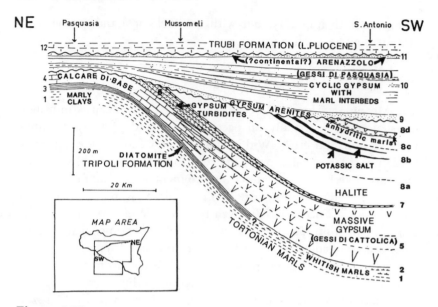

Figure 4.37
Stratigraphic cross section of the Messinian of Sicily (Decima and Wezel 1973).

Mediterranean, show the same stratigraphic sandwich, deep-marine marls passing to evaporites (with a thick halite section) and back into a deep-marine sequence. Clearly the basin became hypersaline in the midst of an open, deep-marine sequence.

In the Tyrrhenian Sea the story is more complex, due to tectonic activity, with back-arc spreading and attenuated continental margins, passing laterally into true continental crust. Rapid basinal subsidence is evident in many sections. For these reasons, basal sedimentation within the Tyrrhenian begins atop oceanic basement in some areas with continental deposits or with shallow marine sediments, and passes upwards into evaporites and thence into deep-marine sedimentation (Kastens, Mascle, Auroux et al. 1987; Malinverno and Ryan 1986). Due to compression and thrusting in Italy itself, more or less from west to east, the growing Apennine mountain chain resulted in crustal loading on parts of the continental crust. Attendant subsidence caused the development of a series of elongate north-south foreland basins, in which turbidite-evaporite-turbidite sequences were developed in the Upper Miocene–Pliocene.

It is all well and good to model much of the Mediterranean basin floor as either shallow or deep at the onset of evaporite formation, but the rocks tell their own sedimentologic story as to conditions during their deposition. In Sicily, lying above deep-marine marls, the lowermost portion of the basinal evaporite section begins with massive layers of crystalline gypsum, parts of which display considerable lateral continuity (figure 4.38). Both the gypsum and the intercalated carbonates contain algal structures representing growth in the photic zone (Schreiber and Friedman 1976). At the basin margins there are massive evaporitic carbonates, bearing halite molds and casts or sporadically interbedded laminar gypsum (Decima, McKenzie, and Schreiber 1988). These marginal carbonates are cut by channels and by desiccation and truncation surfaces, and contain well-developed algal and bacterial structures. Some of the gypsum interbeds in the marginal areas are subaqueous deposits, but others are clearly supratidal.

The middle portion of the Messinian of Sicily is largely made up of halite and potassic salts. Some of these sections are composed of laminar halite and potash, which display lateral continuity clearly traced layer by layer, over 5 to 6 km, and are certainly formed within

Figure 4.38
Photograph of laterally continuous, massive gypsum beds that comprise the lower gypsum cycle of the Messinian deposits, Monte Grande, near Raffadali, Sicily.

some sort of stratified water body. The overlying salts, however, formed in shallow-water to dried basins, with characteristic halite crusts and desiccation polygons. A similarly variable assemblage of halite facies has been drilled on the Mediterranean floor (Ryan, Hsü et al. 1973).

The upper portion of the evaporite section is composed of seven or eight cycles composed of facies formed in restricted water and which range up to evaporite deposition. Each cycle begins with a marl that contains a depauperate fauna and/or a brackish-water diatomite. Each cycle passes upwards into shallow-water gypsum. In some areas the upper part of these cycles are truncated and topped by sabkha caps. These upper cycles are the most widespread of the evaporite zones, and overstep the underlying evaporites. In almost all areas of the Mediterranean these shallow-water evaporites are overlain by deep-water pelagic carbonates of the lower Pliocene. The sea returned to normal salinities and basin-full depths. Whether the reentry of water into the Mediterranean, in the form of a Gibraltar waterfall, was necessary to end the basin restriction is an unanswered question (Hsü, Cita, and Ryan 1973).

The Salina of the Michigan Basin

During the deposition of Upper Silurian sediments, the Michigan Basin was an interior, epeiric basin, with little clastic source area (figure 4.39). We now have enough data to study these sediments with a fair degree of stratigraphic control. The sequence begins with shallow-marine carbonates in the Niagaran, across which a true morphologic basin is gradually developed, deepening in the center (see figure 4.40); reefs and skeletal shoals flank the margins, and pinnacle reefs grow along the slopes. This open-marine reef sequence apparently had a few brief moments of restricted water, as indicated by pauses in sedimentation noted within the reefs, and by brief exposure surfaces and possible thin evaporite pulses amid the marginal carbonates. When the period of true open-water sedimentation drew to a close, capping carbonate facies formed under very restricted conditions, marking the changing environment. This restricted facies is followed by a complex evaporite sequence composed of two major and discrete phases of concentration with associated freshening cycles (in figure 4.40, A-1 Evap. to A-1 Carb., and A-2 Evap. to A-2 Carb.).

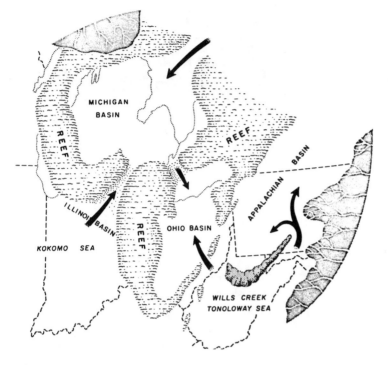

Figure 4.39
Michigan Basin and related seas during the Upper Silurian (Kahle
1978).

These are followed by a great thickness (610+ m, or 2,000 ft) of very-shallow-water evaporites as a final basin fill (B Salt and overlying evaporite section).

These evaporite facies, when described only on the basis of their chemistry, have been most difficult to understand. Even correlation diagrams are awkward, because the very rapid deposition of evaporites along a migrating shoreline has the appearance of synchronous sedimentation, when it is not. A simple test of this concept may be posed. If the basin (which was fairly deep in the center) remained full (deep water) during evaporite deposition, then the preevaporite carbonate section should show no erosion or emergence features in the shelf and pinnacle reef areas, and the evaporite deposits should contain only those features that are appropriate to deep subaqueous

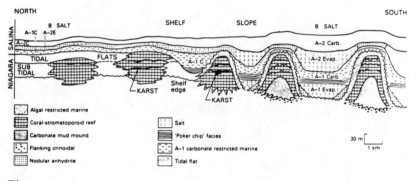

Figure 4.40
Stratigraphic and facies reconstruction of the Upper Silurian of the northern portion of the Michigan Basin, section north to south (adapted from Sears and Lucia 1979, 1980).

deposition. If, however, the basin experienced drawdown, then the shelf and pinnacle carbonates should evidence exposure features and karsts, and the evaporite deposits should be shallow-water to supratidal, with numerous trucation surfaces, etc.

Careful observation of the Michigan Basin shows that the shelf and upper slope deposits contain many exposed tidal-flat and truncated surfaces; severely and multiply altered and dolomitized reefs, grainstones, and shallow-water carbonate muds (the "over-dolomitized carbonates" of Sears and Lucia 1979); and beds of travertinelike cements, and karst and fresh-water alteration zones at the top of and within the reefs (Sears and Lucia 1980). The evaporites and evaporitic carbonates in the lower A-1 Evaporites and most basinal A-1 Carbonates are suggestive of modest water depths (some tens of meters, at least), but the middle and upper A-1 Evaporite and Carbonate, the A-2 Evaporite and Carbonate, and the B Salt all have sedimentary features indicating shallow-water to subaerial deposition (Nurmi and Friedman 1977; see figure 4.41).

The Permian of the Zechstein

The Zechstein (Permian) evaporite deposits of Europe were deposited in a broad basin extending from the British Isles eastward across the North Sea, the Netherlands, Denmark, and Germany, and into eastern Poland and Lithuania; this basin is subdivided by several structural highs (figure 4.42). The sequence is made up of five major depositional cycles (in figure 4.43, Z1 through Z5) that sit atop the Yellow Sands (subaerial desert deposit; see Krinsley and Smith 1981). The Z-1 is a shallow-marine carbonate sequence (Lower and Middle Magnesian Limestone in the British Isles; see Smith 1980a, 1980b, 1981), overlain by a restricted stromatolite biostrome, and finally capped by a shallow-water oolite sequence. Basinward and overlying the Magnesian Limestone is the massive Hartlepool Anhydrite. The succeeding four cycles (Z-2 to Z-5) are largely evaporitic, and the whole sequence is overlain by the Bunter Sand (Triassic, subaerial assemblage).

The evaporite facies present in the Zechstein are diverse, including the laminar, laterally extensive halites and anhydrites (Richter-Bernburg 1955), which represent deposition in a stratified basin which is at least some tens of meters in depth (although absolute depth cannot be determined). In the Hartz Mountains of Germany and in the Netherlands, sulfate turbidites have been recognized, supporting the presence of at least moderate water depths for some period of time (Schlager and Bolz 1977; Clark 1980). Elsewhere, shallow-water anhydrite-

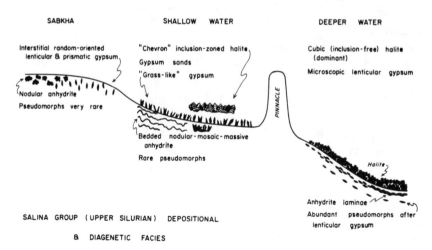

Figure 4.41
Summary diagram of sedimentary elements (facies) developed in the Upper Silurian of the Michigan Basin (Nurmi and Friedman 1977).

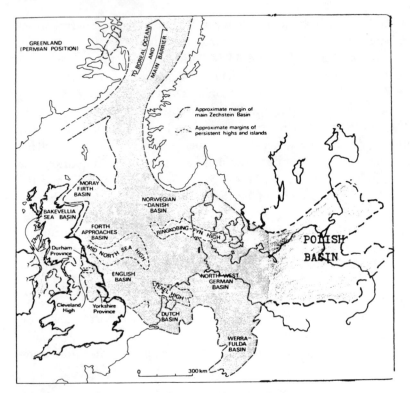

Figure 4.42
Generalized sketch map of the Zechstein Basin showing the main structural units (drawn from Smith 1980b).

after-gypsum—the thick landward portions of the Harlepool and Hayton (Werra) Anhydrites—is clearly in evidence, as is shallow-water to desiccated halite. Severely altered and unusual diagenetic facies are beautifully exposed in the Concretionary Limestones of Durham, England, as are intraformational truncated surfaces and synsedimentary solutional breccias. Thus we have evidence, from the sediments themselves, for changing water depths, cyclicity, and diagenesis. Models alone do not yield a clear picture of the basin without these rock studies.

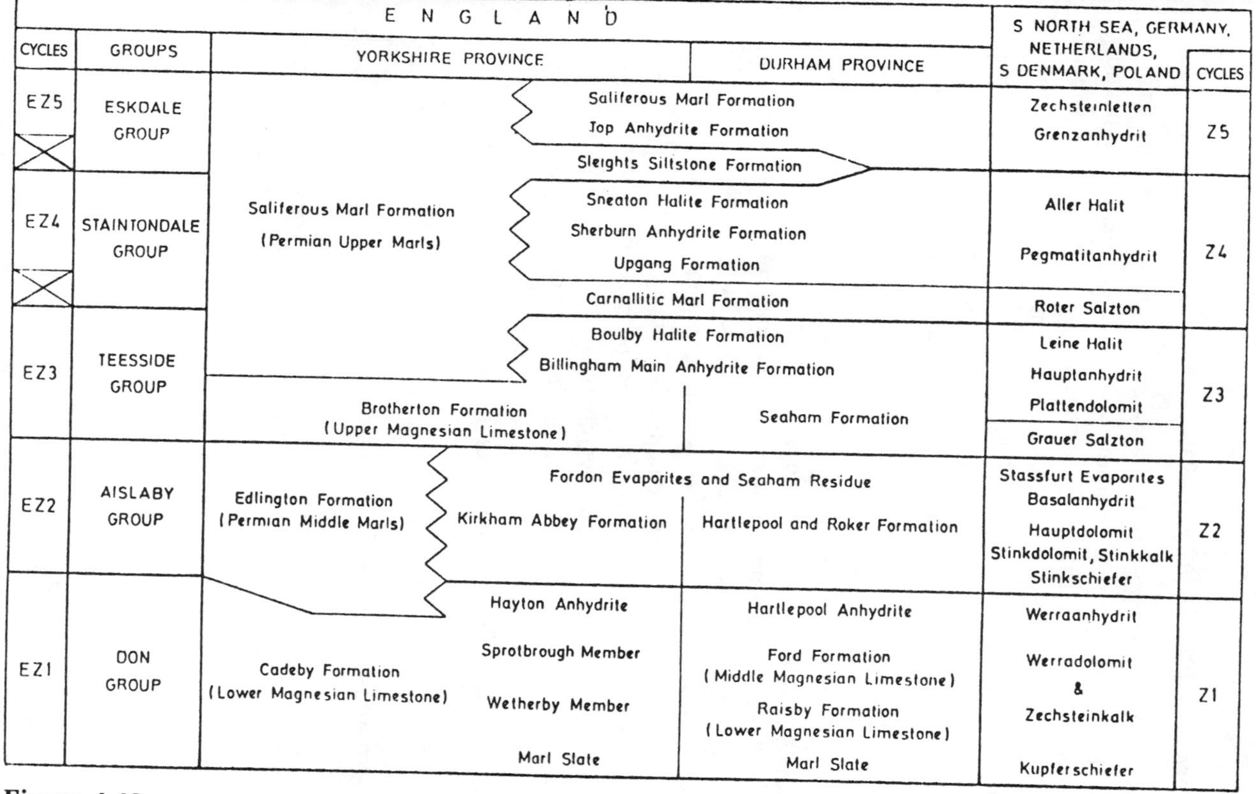

CYCLES	GROUPS	YORKSHIRE PROVINCE	DURHAM PROVINCE	S NORTH SEA, GERMANY, NETHERLANDS, S DENMARK, POLAND	CYCLES
EZ5	ESKDALE GROUP	Saliferous Marl Formation / Top Anhydrite Formation / Sleights Siltstone Formation	Saliferous Marl Formation / Top Anhydrite Formation / Sleights Siltstone Formation	Zechsteinletten / Grenzanhydrit	Z5
EZ4	STAINTONDALE GROUP	Saliferous Marl Formation (Permian Upper Marls)	Sneaton Halite Formation / Sherburn Anhydrite Formation / Upgang Formation	Aller Halit / Pegmatitanhydrit	Z4
EZ3	TEESSIDE GROUP	Brotherton Formation (Upper Magnesian Limestone)	Carnallitic Marl Formation / Boulby Halite Formation / Billingham Main Anhydrite Formation / Seaham Formation	Roter Salzton / Leine Halit / Hauptanhydrit / Plattendolomit / Grauer Salzton	Z3
EZ2	AISLABY GROUP	Edlington Formation (Permian Middle Marls) / Kirkham Abbey Formation	Fordon Evaporites and Seaham Residue / Hartlepool and Roker Formation	Stassfurt Evaporites / Basalanhydrit / Hauptdolomit / Stinkdolomit, Stinkkalk / Slinkschiefer	Z2
EZ1	DON GROUP	Hayton Anhydrite / Sprotbrough Member / Wetherby Member / Cadeby Formation (Lower Magnesian Limestone) / Marl Slate	Hartlepool Anhydrite / Ford Formation (Middle Magnesian Limestone) / Raisby Formation (Lower Magnesian Limestone) / Marl Slate	Werraanhydrit / Werradolomit & Zechsteinkalk / Kupferschiefer	Z1

Figure 4.43

Stratigraphic nomenclature and correlation of the Zechstein. Names formerly in common use in brackets. (From Taylor 1984, after Smith 1980b).

The Permian of the Delaware Basin

The lithologic and stratigraphic characteristics of the Delaware Basin (figure 4.44) during the deposition of the upper Guadalupian and Ochoan sections suggest the presence of a deep basin with deep-water sedimentation (see figure 4.45, and Anderson and Kirkland 1966; Anderson et al. 1972; Anderson 1982). This is topped by desiccated basin-fill, which was deposited in a mix of marine and continental waters and was affected by considerable solution and reprecipitation Lowenstein 1982, 1983). Synchronous shallow-water to subaerial deposition took place behind the basin barrier (Ward et al. 1986).

The evidence for this scenario may be taken from the rock record. During the Guadalupian, turbidites were deposited in the basin center, while reefs grew along its margins, and shallow-water evaporitic carbonates, intercalated with clastics, formed landward of the reefs. By the time of Bell Canyon deposition, the water in the Delaware Basin began to become restricted, and the upper surfaces of Bell Canyon turbidites bear tiny celestite rosettes, perhaps a reflection of the increasingly saline water. The beginning of the Ochoan Group (the Castile Formation) is marked by a restriction of inflow, and the basin went into an evaporitic phase and reefs ceased growing. The thick Castile sequence (550 m, or 1,800 ft) is composed of four deposi-

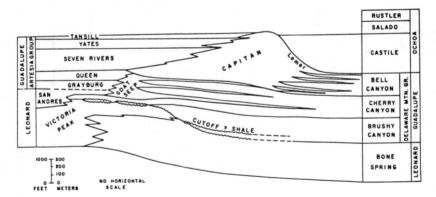

Figure 4.44
Stratigraphical and physical relationships of Permian rocks of the Delaware Basin, Texas–New Mexico (Williamson 1977).

Figure 4.45
Permian tectonic elements on location map of the Delaware Basin,
West Texas–southeastern New Mexico (Williamson 1979).

tional cycles in which much of the section is composed of laminated
alternations of white sulphate and dark-colored carbonate together
with some halite beds. Long sections of these thin laminated beds
can be correlated for a distance of over 100 km, also in keeping with
deposition in a deep-water basin. Near the top of the Castile, the even
laminar character of the sulfate changes, and pseudomorphs of shallow-water gypsum are noted (figures 4.46). The Salado-Rustler sequence, overlying the Castile, represents partial and then total cut-
off from marine water, with nonmarine reworking of earlier marine
sediments as well as direct continental deposition (Lowenstein 1982,
1983; Lowenstein and Hardie 1985).

CONCLUSIONS

Evaporites form in a vast diversity of environments, and as long as
the water of formation is saturated with respect to a given compound, the components remain and are preserved in the sedimentary
record. Dense brines may sink down into underlying sediments, ce-
menting then and otherwise altering them greatly; however, in many
instances overlying brines appear to have little or no direct effect on

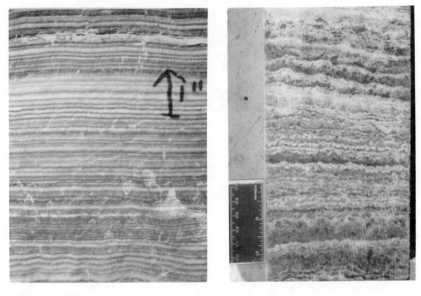

a

Figure 4.46
Thin-bedded anhydrite of the Castile Formation: (a) laminar, even alternations of anhydrite and organic-rich carbonate typical of lower two-thirds of Castile section; and (b) thin-bedded anhydrite pseudomorphs of original gypsum crusts common in the upper portion of the Castile.

underlying sediments. This is probably the result of the rapidity of sedimentation during evaporitic periods, which seal the underlying sediments from incursion by the later-formed brines. Upon burial, compaction moves both pore water and water of crystallization out of the evaporites and into updip or overlying sediments, causing dolomitization, replacement, and plugging. In some regions, such fluids apparently cause calcitization of already dolomitized rocks. While most of these fluids are expected to move updip or up-section, local over-pressure may force water downward or laterally, permeating the more porous beds.

Short-term sea-level variations, on the order of tens of thousands of years, are to be expected in evaporitic water bodies, but total cut-off from the world's oceans seems likely to occur only for shorter periods of time. In these periods the mineralogy of the precipitates becomes expectedly diverse, and the deposits depend in part on waters entering the area from nonmarine sources. Within a basin, sea-level fluctuation (drawdown) will impose new facies conditions upon older layers of evaporitic sediment, and the subaqueous facies may lose their characteristic features by overprinting (in a subaerial setting). Sea-level rise or dilution of waters causes the solution or rehydration of some existing evaporites, and further complicates the petrologic features of the deposit. A significant lithologic and stratigraphic feature of these deposits is the short time interval represented by the evaporites, and, even more significant, the marked internal diagenetic changes within them.

Careful coring, logging, and seismic analysis of these deposits followed by reasonable application of this knowledge is necessary to basin analysis (see Schreiber, Introduction, and Kendall, article 1). Evaporites represent a unique group of facies; they are the key to the finding of many undiscovered bodies of hydrocarbons (see Evans and Kirkland, article 5), and without an understanding of their real meaning, regional stratigraphy cannot help but be wrong.

REFERENCES

Arthurton, R. S. 1973. Experimentally produced halite compared with Triassic layered halite-rock from Cheshire, England. *Sedimentology* 20:145–160.

Anderson, R. Y. 1982. A long geoclimatic record from the Permian. *Jour. Geophysical Research* 87:7285–7294.

Anderson, R. Y. and D. W. Kirkland. 1966. Intrabasin varve correlation. *Bull. Geol. Soc. Amer.* 77:241–256.

Anderson, R. Y., W. E. Dean, Jr., D. W. Kirkland, and H. I. Snider. 1972. Permian Castile varved evaporite sequence, West Texas and New Mexico. *Bull. Geol. Soc. Amer.* 83:59–86.

Birnbaum, S. 1976. Non-marine evaporite and carbonate deposition, Ebro Basin, Spain. Ph.D. diss., University of Cambridge.

Birnbaum, S. J. and J. W. Wireman. 1984. Bacterial sulfate reduction and pH: Implications for early diagenesis. *Chem. Geology* 43:143–149.

Bommarito, S. and R. Catalano. 1973. Facies analysis of an evaporitic Messinian sequence near Ciminna (Palermo, Sicily). In: C. W. Drooger, ed., *Messinian Events in the Mediterranean,* pp. 172–177. Amsterdam and London: North-Holland.

Borchert, H. and R. O. Muir. 1964. *Salt Deposits: The Origin, Metamorphism, and Deformation.* London: Van Nostrand.

248   B. C. SCHREIBER

Boudet, G., P.-A. Capdequi, and P. Marchand. 1985. Software for a saltworks yearly production cycle. In *Sixth Symposium on Salt*, 2:313–322.
See Schrieber and Harner 1985.

Busson, G. ed. 1982. *Géologie Méditerranéenne (Nature et Genèse des Facies Confinés)*. *Annales de l'université de provence*. 9(4):303–591.

Ciaranfi, N., L. Dazzaro, P. Pieri, L. Rapisardi, and A. Sardella. 1978. Preliminary desription of some Messinian evaporite facies along the Abruzzi-
Molese Boundary. In R. Catalano, G. Ruggieri, and R. Sprovieri, eds.,
Messinian Evaporites in the Mediterranean. Memoir Soc. Geologica Ital-
iana 16:251–260.

Clark, D. 1980. The sedimentology of the Zechstein carbonate formation
of Eastern Drenthe, the Netherlands. In *The Zechstein Basin*, pp. 131–
166. See Füchtbauer and Peryt 1980.

Cody, R. D. and A. B. Cody. 1988. Gypsum nucleation and crystal morphology in analog saline terrestrial environments. *Jour. Sedim. Petrol.*
58:000–000.

Cody, R. D. and A. M. Hull. 1980. Experimental growth and primary anhydrite at low temperatures and water salinities. *Geology* 8:505–509.

Cornée, A. 1983. Sur les bactéries des saumures et des sédiments de marais
salants méditerranéens: Importance et rôle sédimentologique. *Documents
du GRECO*, vol. 52, no. 3. Paris: Laboratoire de géologie du Museum.

Cornée, A. 1984. Étude préliminaire des bactéries des saumures et des
sédiments des salins de Santa Pola (Espagne). Comparaison avec les marais
salants de Salin-de-Giraud (Sud de la France). In *Introduction to the Sedimentology of the Coastal Salinas of Santa Pola (Alicante, Spain)*, pp. 109–
122. See Orti-Cabo and Busson 1984.

Davies, G. 1977. Carbonate-anhydrite facies relation in Otto Fiord Formation (Mississippian-Pennsylvanian), Canadian Arctic Archipelago. *Bull.
Amer. Assoc. Petrol. Geol.* 61:1929–1949.

Dean, W. E. and R. Y. Anderson. 1978. Salinity cycles: Evidence for subaqueous deposition of Castile Formation and the lower part of Salado
Formation, Delaware Basin, Texas and New Mexico, In G. S. Austin, ed.,
*Geology and Mineral Deposits of Ochoan Rocks in the Delaware Basin
and Adjacent Areas*, pp. 15–20. New Mexico Bureau of Mines and Mineral Resources, Circular no. 159. Socorro, N. Mex.

Decima, A. and F. Wezel. 1973. Late Miocene evaporites of the Central
Sicilian Basin, Italy. In *Initial Reports of the Deep Sea Drilling Project,
Leg 13*, pp. 1234–1240. See Ryan, Hsü et al. 1973.

Decima, A., J. A. McKenzie, and B. C. Schreiber. 1988. The origin of "evaporative" limestone: An example from the Messinian of Sicily (Italy). *Jour.
Sedim. Petrol.* 58:256–272.

Dellwig, L. 1955. Origin of the Salina Salt of Michigan. *Jour. Sedim. Petrol.*
25:83–100.

Duxbury, A. C. 1971. *The Earth and Its Oceans*. Reading, Mass.: Addison-
Wesley.

Fontes, J. C., R. Gonfiantini, and E. Tongiorgi. 1963. Composition isotopique et origine de la série évaporitique du bassin de Paris (extrait). C. R. Sommares de *Séances de la Soc. Geol. Fr.* 3:92–93.

Füchtbauer, H. and T. Peryt, eds. 1980. *The Zechstein Basin.* Vol. 9 of *Contr. Sedimentology.* Stuttgart: E. Schweizerbart'sche Verlagsbuchhanlung.

Garrison, R., B. C. Schreiber, D. Bernouli, F. Fabricius, R. Kidd, and F. Melieres. 1978. Sedimentary petrology and structures of Messinian evaporitic sediments in the Mediterranean Sea. In K. J. Hsü, L. L. Montadert et al., *Initial Reports of the Deep Sea Drilling Project, Leg 42A,* pp. 571–612. Washington, D.C.: U.S. Government Printing Office.

Gornitz, V. M. 1965. A study of halite from the Dead Sea. Master's thesis, Columbia University, New York.

Gornitz, V. M. and B. C. Schreiber. 1981. A study of halite morphologies as formed within hypersaline carbonate sediments. *Jour. Sedim. Petrol.* 50:787–794.

Hageau, Hippolyte. 1837. *Rapport sur les carrières de Montmartre.* Paris. In 4e.

Handford, C. R. 1982. Sedimentology and evaporite diagenesis in a Holocene continental sabkha: Bristol Dry Lake, California. *Sedimentology* 29:239–253.

Handford, C. R., R. G. Loucks, and G. R. Davies, eds. 1982. *Depositional and Diagenetic Spectra of Evaporites—A Core Workshop.* SEPM Core Workshop no. 3. Tulsa, Okla.

Hardie, L. A. 1984. Evaporites: Marine or non-marine? *Amer. Jour. Sci.* 284:193–240.

Hardie, L. A. and H. Eugster. 1971. The depositional environment of marine evaporites: A case for shallow clastic accumulation. *Sedimentology* 16:187–220.

Hardie, L. A., T. K. Lowenstein, and R. J. Spencer. The problem of distinguishing between primary and secondary features in evaporites. In *Sixth Symposium on Salt,* 1:11–40. See Schreiber and Harner 1985.

Hardie, L. A., J. P. Smoot, and H. P. Eugster. 1978. Saline lakes and their deposits: A sedimentological approach. In A. Matter and M. E. Tucker, eds., *Modern and Ancient Lake Sediments,* pp. 7–42. I. A. S. Special Publication no. 2. Oxford: Blackwell.

Harvie, C. E., J. H. Weare, L. A. Hardie, and H. Eugster. 1980. Evaporation of sea water: Calculated mineral sequences. *Science* 208:498–500.

Holliday, D. 1970. The petrology of secondary gypsum rocks: A review. *Jour. Sedim. Petrol.* 40:734–744.

Holliday, D. 1973. Early diageneses in nodular anhydrite rocks. *Transactions, Inst. Mining and Metallurgy* 82B:81–84.

Holliday, D. 1978. Petrology of Basal Purbeck (Upper Jurassic) gypsum rocks of the Broadock Borehole, Sussex. In R. D. Lake and D. W. Holliday, eds., *Broadock Borehole Report,* pp. 13–28. British Inst. Geol. Sciences, no. 78/3. London: H. M. Stationary Office.

Holliday, D. and R. D. Lake. 1978. Purbeck Beds of the Broadock Borehole, Sussex. In R. D. Lake and D. W. Holliday, eds., *Broadock Borehole Report*, pp. 13–28. British Inst. Geol. Sciences, no. 78/3. London: H. M. Stationary Office.

Holser, W. T. 1966. Diagenetic polyhalite in Recent salt from Baja California. *Am. Miner.* 51:99–109.

Hsü, K. J. 1972. Origin of saline giants: A critical review after the discovery of the Mediterranean evaporite. *Earth Sci. Rev.* 8:371–396.

Hsü, K. J., M. B. Cita, and W. B. F. Ryan. 1973. The origin of the Mediterranean evaporites. In *Initial Reports of the Deep Sea Drilling Project, Leg 13*, pp. 1203–1231. See Ryan, Hsü et al. 1973.

Hsü, K. J., L. Montadert et al. 1978. *Initial Reports of the Deep Sea Drilling Project, Leg 42*, vol. 1. Washington, D.C.: U.S. Government Printing Office.

Jauzein, A. and P. Hubert. 1984. Les bassins oscillants- Un modèle de genèse des Séries salines. *Sci. Géol. Bull. Strasbourg* 37:267–282.

Javor, B. J. 1985. Nutrients and ecology of the Western Salt and Exportadora de Sal saltern brines. In *Sixth Symposium on Salt*, 1:195–206. See Schrieber and Harner 1985.

Kahle, C. F. 1978. Patch reef development and effects of repeated subaerial exposure in Silurian shelf carbonates, Maumee, Ohio. In *Geol. Soc. Amer. 1978 Field Excursion Guide*, pp. 63–115. Ann Arbor: University of Michigan.

Kastens, K. A., J. Mascle, C. Auroux et al. 1987. *Proceedings Initial Report of the Ocean Drilling Program, Leg 107, Part B*. Washington, D.C.: U.S. Government Printing Office.

Kinsman, D. J. J. 1976. Evaporites: Relative humidity control of primary mineral facies. *Jour. Sedim. Petrol.* 46:273–279.

Kinsman, D. J. J., M. Boardman, and M. Borcsik. 1974. An experimental determination of the solubility of oxygen in marine brines. In A. Coogan, ed., *Fourth Symposium on Salt*, 1:343–348. Cleveland: Northern Ohio Geol. Soc.

Krinsley, D. and D. B. Smith. 1981. Selective SEM study of grains from the Permian Yellow Sands of N-E England. *Proc. Geol. Assoc.* 92(3):189–196.

Kühn, R. 1968. Geochemistry of the German potash deposits. In R. B. Mattox, ed., *Saline Deposits*, pp. 427–504. Geol. Soc. Amer. Special Paper no. 88. Boulder, Colo.

Kwiatkowski, S. 1972. Sedymentacja gibsów mioceńshich potudniowej Polski (Sedimentation of Miocene gypsum in southern Poland). *Pd. Alad. Nauk. Muz. Ziemi, Pr.*, no. 19, pp. 3–93.

LaCroix, A. 1897. *Le gypse de paris et le minereaux qui l'accompagnent*. Nouvelles Archives du Museum, 3d series, 9:201–296. Paris: Masson.

Landry, J. C. and J. Jaccard. 1984. Chimie des eaux libres dans le marais salant de Santa-Pola, salina de Bras del Port. In *Introduction to the Sed-*

imentology of the Coastal Salinas of Santa Pola (Alicante, Spain), pp. 37–54. See Orti-Cabo and Busson 1984.

Larsen, H. 1980. Ecology of hypersaline environments. In A. Nissenbaum, ed., *Hypersaline Brines and Evaporitic Environments*, pp. 23–40. Amsterdam: Elsevier.

Liu, D., M. Cai, S. Wang, and X. Wu. 1985. A new techniques of plastic film covering of solar salt production in a rainy climate. In *Sixth Symposium on Salt*, 2:323–330. See Schreiber and Harner 1985.

Loucks, R. G. and M. Longman. 1982. Lower Cretaceous Ferry Lake anhydrite, Fairway Field, East Texas: Product of shallow-subtidal deposition. In *Depositional and Diagenetic Spectra of Evaporites*, pp. 130–173. See Handford, Loucks, and Davies 1982.

Lowenstein, T. K. 1982. Primary features in a potash evaporite deposit: The Permian Salado Formation of West Texas and New Mexico. In *Depositional and Diagenetic Spectra of Evaporites*, pp. 276–304. See Handford, Loucks, and Davies 1982.

Lowenstein, T. K. 1983. Deposits and alteration of an ancient potash deposit: The Permian Salado Formation of New Mexico and West Texas. Ph.D. diss. Johns Hopkins University, Baltimore.

Lowenstein, T. K. and L. A. Hardie. 1985. Criteria for the recognition of salt-pan evaporites. *Sedimentology* 32:627–644.

McKenzie, J. A. 1985. Stable isotope mapping in Messinian evaporitic carbonates of central Sicily. *Geology* 13:851–854.

Maiklem, W. R. 1971. Evaporitic drawdown—a mechanism for water-level lowering and diagenesis in the Elk Point Basin. *Bull. Can. Petrol. Geol.* 19:487–503.

Malinverno, A. and W. B. F. Ryan. 1986. Extension in the Tyrhennian Sea and shortening in the Apennines as a result of arc migration driven by sinking of the lithosphere. *Tectonics* 5:227–245.

Manspeizer, W. 1981. Early Mesozoic basins of the Central Atlantic passive margins. In A. W. Bally, ed., *Geology of Passive Continental Margins*, pp. 4-1 to 4-60. AAPG Education Course Note Series no. 19. Tulsa, Okla.

Monty, C. 1965. Recent algal stromatolites in the windward lagoon, Andros Island, Bahamas. *Ann. Soc. Géol. Belge* 88:269–276.

Monty, C. 1972. Recent algal stromatolite deposits, Andros Island, Bahamas: Preliminary report. *Geol. Rundschau* 61:742–783.

Monty, C. 1976. The origin and developments of cryptalgal fabrics. In M. K. Walter, ed., *Stromatolites*, pp. 193–259. Vol. 20 of *Developments in Sedimentology*. Amsterdam: Elsevier.

Mossop, G. and D. J. Shearman. 1973. Origins of secondary gypsum rocks. *Transactions, Inst. Mining and Metallurgy* 82B:147–154.

Mottura, S. 1871/1872. Sulla formazione terziaria nella zona zolfifera della Sicilia. *Memorie del Ricerca Comitato Geologico d'Italia* 1:50–140.

Müller, W. H. and K. J. Hsü. 1980. Stress distribution in overthrusting slabs and mechanics of Jura deformation. *Rock Mechanics, Suppl.* 9:219–232.

Neev, D. and K. O. Emery. 1967. *The Dead Sea: Depositional Processes and Environments of Evaporites.* Israel Geol. Survey, Bulletin no. 41. Jerusalem.

Nurmi, R. and G. M. Friedman. 1977. Sedimentology and depositional environments of basin-center evaporites, Lower Salina Group. In J. H. Fisher, ed., *Reefs and Evaporites—Concepts and Depositional Models,* pp. 23–52. AAPG Studies in Geology no. 5. Tulsa, Okla.

Ogniben, L. 1955. Inverse graded bedding in primary gypsum of chemical deposition. *Jour. Sedim. Petrol.* 25: 273–281.

Ogniben, L. 1957. Petrografia della Serie Solfifera siciliana e considerazioni geologiche relative. *Mem. Descr. Carta. Geol. Italia* 33:1–275.

Ogniben, L. 1963. Sediment halitico-calitici a structtura grumosa nel calcare di base Messiniano in Sicilia. *Giornale di Geologia,* series 2, 31:509–542.

Orti-Cabo, F. and G. Busson, eds. 1984. *Introduction to the Sedimentology of the Coastal Salinas of Santa Pola (Alicante, Spain).* Revista d'Investigacions Geologiques 38/39:1–235.

Orti-Cabo, F. and D. J. Shearman. 1977. Estructuras y fábricas deposicionales en las evaporitas del miocena superior (Messinense) de san Miguel de Salinas (Alicante, España). *Instit. Invest. Geolog. Diput. Prov. Univ. Barcelona* 32:5–54.

Parea, G. C. and F. Ricci-Lucchi. 1972. Resedimented evaporites in the Periadriatic trough (Upper Miocene, Italy). *Israel Journal Earth Sciences* 21:125–141.

Perthuisot, J. P. 1982. Introduction générale à l'étude des marais salants de Salin-de-Giraud (Sud de la France). Le cadre géographique et le milieu. In *Géologie Méditerranéene,* pp. 309–328. See Busson 1982.

Pierre, C. 1985. Polyhalite replacement after gypsum at Ojo de Liebre Lagoon (Baja California, Mexico): An early diagenesis by mixing of marine brines and continental waters. In *Sixth Symposium on Salt,* 1:257–265. See Schrieber and Harner 1985.

Pierre, C., L. Ortlieb, and A. Person. 1984. Supratidal evaporitic dolomite at Ojo de Liebre Lagoon: Mineralogical and isotopic arguments for primary crystallization. *Jour. Sedim. Petrol.* 54:1049–1061.

Por, F. D. 1985. Anchialine pools: Comparative hydrobiology. In G. M. Friedman and W. E. Krumbien, eds., *Hypersaline Ecosystems,* pp. 136–144. New York: Springer-Verlag.

Ricci-Lucci, F. 1973. Resedimented evaporites: Indicators of slope instability and deep basin conditions in Periadriatic Messinian. In C. W. Drooger, ed., *Messinian Events in the Mediterranean,* pp. 142–149. Amsterdam: North-Holland.

Ricci-Lucci, F. 1975. Miocene paleogeography and basin analysis in the Periadriatic Apennines. In C. Squyres, ed., *Geology of Italy,* pp. 5–111. Tripoli: Petroleum Exploration Soc. of Libya.

Richter-Bernburg, G. 1955. Stratigraphische Gliederung des Deutschen Zechsteins. *Z. Deutsch. Geol., Ges.* 105:843–854.

Rouchy, J. M. 1980. The evaporitic sequences of the terminal Miocene of Sicily and of southern Spain. In G. Busson, ed., *Evaporite Deposits,* pp. 33–39 and 160–173. Houston: Gulf Publishing.

Ryan, W. B. F., M. B. Cita, M. D. Rawson, L. H. Burckle, and T. Saito. 1974. A paleomagnetic assignment of Neogene stage boundaries and the development of isochronous datum planes between the Mediterranean, the Pacific, and Indian oceans in order to investigate the response of the world ocean to the Mediterranean "Salinity Crisis." *Revista Italiano Paleontologia* 80:631–688.

Ryan, W. B. F., K. J. Hsü et al. 1973. *Initial Reports of the Deep Sea Drilling Projects, Leg 13.* Washington, D.C.: U.S. Government Printing Office.

Sammy, N. 1985. Biological systems in northwestern Australian salt fields. In *Sixth Symposium on Salt.* 1:207–216. See Schrieber and Harner 1985.

Schaller, W. T. and E. P. Henderson. 1932. Mineralogy of drill cores from the Potash Field of New Mexico and Texas. United States Department of the Interior, Geological Survey, Bull. no. 833. Washington, D.C.: U.S. Government Printing Office.

Schlager, S. and H. Bolz. 1977. Clastic accumulation of sulphate evaporites in deep water. *Jour. Sedim. Petrol.* 47:600–609.

Schreiber, B. C. 1973. Survey of the physical features of the Messinian chemical sediments. In C. W. Drooger, ed., *Messinian Events in the Mediterranean,* pp. 101–110. Amsterdam: North-Holland.

Schreiber, B. C., R. Catalano, and E. Schreiber. 1977. An evaporite lithofacies continuum: Latest Miocene (Messinian) deposits of Salemi Basin (Sicily) and modern analog. In J. H. Fisher, ed., *Reefs and Evaporites— Concepts and Models,* pp. 169–180. AAPG Studies in Geology no. 5. Tulsa, Okla.

Schreiber, B. C. and G. M. Friedman. 1976. Depositional environments of Upper Mioceme (Messinian) evaporites of Sicily as determined from analysis of intercalated carbonates. *Sedimentology* 23:255–270.

Schreiber, B. C., G. M. Friedman, A. Decima, and E. Schreiber. 1976. Depositional environments of Upper Miocene (Messinian) evaporite deposits of the Sicilian Basin. *Sedimentology* 23:729–760.

Schreiber, B. C. and H. L. Harner, eds. 1985. *Sixth Symposium on Salt.* Alexandria, Va.: Salt Institute.

Schreiber, B. C., and D. J. J. Kinsman. 1975. New observations on the Pleistocene evaporites of Montallegro, Sicily and a modern analog. *Jour. Sedim. Petrol.* 45:469–479.

Schreiber, B. C., M. Roth, and M. L. Helman. 1982. Recognition of primary facies characteristics of evaporites and the differentiation of these forms from diagenetic overprints. In *Depositional and Diagenetic Spectra and Evaporites,* pp. 1–32. See Handford, Loucks, and Davies 1982.

Schreiber, B. C. and E. Schreiber. 1977. The salt that was. *Geology* 5:527–528.

Sears, S. O. and F. J. Lucia. 1979. Reef-growth model for Silurian pinnacle reefs, northern Michigan reef trend. *Geology* 3:299–302.

Sears, S. O. and F. J. Lucia. 1980. Dolomitization of Northern Michigan Niagara Reefs by Brine Refluxion and Fresh Water/Sea Water Mixing. Soc. Econ. Paleont. Miner. Special Publication no. 28, pp. 215–235. Tulsa, Okla: SEPM.

Shearman, D. J. 1970. Recent halite rock, Baja California, Mexico. *Trans. Inst. Min. Metall.* 79:B155-162.

Shearman, D. J. 1978. Evaporites of coastal sabkhas. In W. E. Dean and B. C. Schreiber, eds., *Marine Evaporites,* pp. 6–42. Soc. Econ. Paleont. Miner. Short Course no. 4. Tulsa, Okla.: SEPM.

Shearman, D. J. 1981. Displacement of sand grains in sandy gypsum crystals. *Geological Mag.* 118:303–306.

Shearman, D. J. 1985. Syndepositional and late diagenetic alteration of primary gypsum. In *Sixth Symposium on Salt* 1:41–55. See Schreiber and Harner 1985.

Smith, D. B. 1980a. The shelf-edge reef of the middle Magnesian Limestone (English Zechstein Cycle 1) of northeastern England—A summary. In *The Zechstein Basin,* pp. 3–5. See Füchtbauer and Peryt 1980.

Smith, D. B. 1980b. The evolution of the English Zechstein Basin. In *The Zechstein Basin,* pp. 7–34. See Füchtbauer and Peryt 1980.

Smith, D. B. 1981. The evolution of the English Zechstein Basin. In *Proceedings, International Symposium Central European Permian* (Jablonna, 1978), pp. 9–47. Warsawa: Wydawnictwa Geologiczne.

Sonnenfeld, P. 1985. Evaporites as oil and gas source rocks. *Jour. Petrol. Geol.* 8:253–271.

Southgate, P. N. 1982. Cambrian skeletal halite crystals and experimental analogues. *Sedimentology* 29:391–407.

Taylor, J. 1980. Origin of the Werraanhydrit in the U.K. southern North Sea—a reappraisal. In *The Zechstein Basin,* pp. 91–113. See Füchtbauer and Peryt 1980.

Taylor, J. C. M. 1984. Late Permian-Zechstein. In K. W. Glennie, ed., *Introduction to the Petroleum Geology of the North Sea,* pp. 61–83. Oxford: Blackwell.

Thomas, M. and D. Geisler. 1982. Peuplements benthiques à cyanophycées des marais salants de Salin-de-Giraud (Sud de la France). In *Géologie Méditerranéenne,* pp. 391–412. See Busson 1982.

Usiglio, 1849. Analyse de l'eau de la Méditerranée sur les côtes de France. *Annalen Chimie* 27:92–107, 172–191.

Vai, G.-B. and F. Ricci Lucchi. 1977. Algal crusts autochthonous and clastic gypsum in a cannibalistic evaporite basin: A case history from the Messinian of the Northern Apennines. *Sedimentology* 24:211–244.

Valyashko, M. G. 1952. Halite: Its principal varieties found in salt lakes and its structural features. In *Trudy Vses. Nauchoissled. Inst. Galurgii,* vol. 23.

Von der Borch, D., B. Bolton, and J. K. Warren. 1977. Environmental setting and micro-structure of subfossil lithified stromatolites associated with evaporites, Marion Lake, South Australia. *Sedimentology* 24:693–708.

Von der Haar, S. P. 1976. Evaporites and algal mats at Laguna Mormona, Pacific Coast, Baja California, Mexico. Ph.D. diss., University of Southern California, Los Angeles.

Ward, R. F., C. G. St. C. Kendall, and P. M. Harris. 1986. Upper Permian (Guadalupian) facies and their association with hydrocarbons—Permian Basin, West Texas and New Mexico. *Bull. Am. Assoc. Petrol. Geol.* 70:239–262.

Warren, J. K. 1982. The hydrological setting, occurrence, and significance of gypsum in late Quaternary salt lakes in South Australia. *Sedimentology* 29:609–638.

Warren, J. K. 1985. On the significance of evaporite lamination. In *Sixth Symposium on Salt*, 1:161–171. See Schreiber and Harner 1985.

Weiler, Y., E. Sass, and I. Zak. 1974. Halite oolites and ripples in the Dead Sea, Israel. *Sedimentology* 21:623–632.

West, I. 1964. Evaporite diagenesis in the lower Purbeck Beds of Dorset. *Proc. Yorks. Geol. Soc.* 34:315–326.

West, I. 1965. Macrocell structure and enterolithic veins in British Purbeck gypsum and anhydrite. *Proc. Yorks. Geol. Soc.* 35:47–58.

West, I. 1975. Evaporites and associated sediments of the basal Purbeck formation (upper Jurassic) of Dorset. *Proc. Geol. Assoc.* 86:205–255.

West, I. 1979. Review of evaporite diagenesis in the Purbeck Formation of southern England. *Symposium Sedimentation Jurassique W. Européen*, pp. 407–416. Assoc. Sedimé. Français Publication speciale no. 1, March.

West, I., Y. A. Ali, and M. E. Hilmy, 1985. Facies associated with primary gypsum nodules of northern Egyptian sabkhas. In *Sixth Symposium on Salt*, 1:171–181. See Schreiber and Harner 1985.

Williamson, C. R. 1977. Deep-sea channels of the Bell Canyon formation (Guadalupian) Delaware Basin, Texas–New Mexico. In: M. E. Hileman and S. J. Mazzullo, eds., *Upper Guadalupian Facies, Permian Reef Complex, Guadalupe Mountains, New Mexico and West Texas*, pp. 409–432. Permian Basin Section SEPM, Publication no. 77-16. Midland, Tex: Permian Basin Section SEPM.

Williamson, C. R. 1979. Deep-sea sedimentation and stratigraphic traps, Bell Canyon Formation (Permian), Delaware Basin. In N. M. Sullivan, ed., *Guadalupian Delaware Mountain Group*, pp. 39–74. Permian Basin Section SEPM Publication no. 79-18. Midland, Tex: Permian Basin Section SEPM.

5

Evaporitic Environments as a Source of Petroleum

ROBERT EVANS
DOUGLAS W. KIRKLAND

Until about fifteen years ago, it would have been heresy for some-
one to propose in open court that any evaporitic environment could
generate enough organic matter, deposited with sediment and subsequently matured, to make a viable source bed for hydrocarbons.
These environments were considered to be inimical to life, too harsh
and forbidding. Larger vertebrates shunned such places, fish died from
osmotic desiccation, and what invertebrates there were, washed in
by accident, remained pathetically stunted, if they survived at all. How
hostile the environment was can be gauged from the description in
Grabau (1920) of the shores near the inlet into the Gulf of Kara-

Robert Evans is a Senior Research Associate with Mobil Research and Development
Corporation, Dallas, Texas.

Douglas W. Kirkland is a Research Scientist with Mobil Research and Development
Corporation, Dallas, Texas.

Bugaz, on the eastern side of the Caspian Sea. The inflowing current carries thousands of fish into the hypersaline waters of the Gulf, where they die, to float on the surface or wash to the shore, there to provide an almost limitless bounty for flocks of seabirds. So great is the accumulation of carcasses that the birds at certain times of the year eat only the eyes of the fish, and then only the eye on the upper surface, deeming it unnecessary to turn the fish over to take the other! As Lotze asserted in a major text published in 1957: "Nicht das Leben, sondern der Tod beherrscht die Salzbildungsstätten" (Not life, but death rules the locales of salt deposition). Without life, there could not be significant accumulations of organic matter, and without much organic matter, there could not be formation of a source bed.

In science, as in religion, the faithful do not venture easily into heresy, and they take that bold step only when contradictory data or conviction seem to allow no other course. In the matter of source beds for petroleum, "heretical" thinking started when geologists began to have nagging doubts about assertions such as Lotze's. If the evaporitic, hypersaline environment were essentially sterile, why did it appear that some evaporites were related spatially and temporally to known source rocks in some petroleum provinces (Woolnough 1937; Hedberg 1967; Peterson and Hite 1969)? And what was it about evaporites that made them appear so consistently on the list of characteristics of major petroleum-producing provinces (Moody 1959; Weeks 1961; Perrodon 1968; Halbouty et al. 1970)?

There was no geological Luther to nail ninety-five theses to the door of a church in Wittenberg or anywhere else, but in the sixties a few tentative suggestions about evaporitic source beds appeared in the literature. They dealt mainly with the capability of the evaporitic environment to *preserve* organic matter (Hedberg 1967; Peterson and Hite 1969) and did not seriously address the possibility that the environment might favor the *production* of organic matter. Apparently, there was still a reluctance to believe that such a hostile environment could support life, or perhaps there was an even greater reluctance to overturn dogma. Whatever the reason, geological investigations of the productivity of the hypersaline environment lagged behind biological ones, for by this time it was clear to biologists that salt lakes, saline lagoons, and solar ponds were not so sterile as popular thought would have us believe (Gibor 1956; Carpelan 1957).

Yesterday's heresies became today's truths, which themselves may well become the dogma of tomorrow, to be overturned by future heresies. Once the first hesitant excursions into this new region of geological thought had been made, there developed in the next fifteen years a body of geological literature that dealt seriously with the possibility that evaporitic environments could both produce and preserve significant quantities of organic matter (Dembicki et al. 1976; Oehler et al. 1979; Malek-Aslani 1980; Kirkland and Evans 1981; Palacas 1983; Jones 1984; Hite et al. 1984; Eugster 1985; Sonnenfeld 1985; Ten Haven et al. 1985; Warren 1986). The organic matter would be deposited with evaporitic carbonates and was of a type that would generate oil (Tissot 1981; Sassen et al. 1986; Richardson et al. 1986). Fifty percent of the world's oil may well have originated this way, and heresy or not, the idea can no longer be ignored or dismissed. This essay is intended as a summary of what has been learned about these source rocks.

Formation of a Source Rock

Not every environment is capable of producing and preserving the right combination of organic matter and sediment to form an effective source rock for oil. Three requirements seem to be essential: the environment should produce sufficient organic matter of a type that upon maturation yields oil; the organic matter should not be diluted excessively by sediment during deposition; and, during sedimentation and early diagenesis, the organic matter should not suffer complete oxidation. A fourth requirement—that there should be diagenetic transformation of the organic matter into oil—is less a function of the original than of the subsequent environment, and therefore is not considered in this discussion. In this paper we will iterate that evaporitic environments can satisfy the three requirements and can produce great quantities of rich, oil-prone source rock. We emphasize the deep-water mesosaline setting because it has the greatest potential for ultimate preservation in the record, but we recognize that microbial mats and sediments deposited in shallow-water or supratidal evaporitic settings can, under some circumstances, produce a source rock.

PRODUCTION OF ORGANIC MATTER IN EVAPORITIC ENVIRONMENTS

Life is not sparse in hypersaline brines. On the contrary, the biomass therein is greater than that for a comparable volume of normal seawater, no matter whether the concentrated brine originates in salt lakes, in solar ponds, or in hypersaline lagoons. Lotze (1957) may have been deceived by the paucity or absence of *vertebrates* in these settings, or he may have been accepting, without further investigation, the traditional view of sterile desert wastes or glaring, white salt flats dotted with hot, hostile pools of brine. Whatever the reason, he was seriously in error, for hypersaline environments teem with life, albeit of a very special kind. In the following review, we consider four evaporitic environments from the point of view of their productivity of organic matter: saline lakes; solar ponds; hypersaline lagoons; and sedimentary sabkhas within coastal or inland sedimentary plains.

Saline Lakes

In an earlier review of productivity of evaporitic environments (Kirkland and Evans 1981), we drew attention to the significance of flamingoes. They are filter feeders, and most species depend for their food upon mud that is rich with organic material. Inhabiting only hypersaline environments, the birds achieve spectacular concentrations in situations that some would have had us believe were inimical to life. In Lake Nakuru, Kenya, more than a million flamingoes of two species, *Phoenicopterus ruber roseus* and *P. minor,* consume an estimated 200 tons of food per day from a shallow, saline body of water only 6.5 by 10 km. The waters are colored green by prolific growth of *Arthospira platenis,* a blue-green alga (Hutchinson, 1967). We can deduce that for such a large flock of birds to subsist and persist in that environment, all that they consume must be replaced by primary production of phytoplankton and bacterioplankton within the brine. The lake must be highly productive.

The same conclusion has been drawn for many other evaporitic hypersaline lakes all over the world. With salinities that approach three times that of seawater, the Tanatar lakes of the Kulunda Steppe, USSR, "teem with life" (Isachenko, quoted in Strakhov 1970), both phytoplankton and zooplankton, and so support many species of water birds. Mono Lake in California, with approximately the same salinity, has been described as having "unlimited epilimnic phytoplankton productivity" (Mason 1968). Great Salt Lake is similarly productive, even though the salinity is much greater, more than five times that of seawater. Green algae occur in such numbers in the brine that the waters are colored red in the north arm of the lake, where *Dunaliella salina* dominates, and green in the south arm, where *D. viridis* is in bloom (Post 1981). The contribution to the biomass made by bacteria, principally *Halobacterium* and *Halococcus*, is one order of magnitude greater than the contribution made by the algae (Post 1981), and the bacteria may be the cause of the red water in the north arm (Brock 1979). The brine shrimp, *Artemia salina*, attains a density of more than 5,000 individuals per cu m (Quinn 1966), and the fecal pellets of this small crustacean form masses of many thousands of tons (Eardley 1938). Brine flies (*Ephydra gracilis* and *E. hians*) are present in such numbers that they darken the shores of the lake, and their pupae used to provide an abundant source of sustenance for the formerly indigenous Paiute Indians (Zahl 1967).

Biological activity within the Dead Sea belies its name. This deep hypersaline lake, now oversaturated with respect to halite (Beyth 1980), supports a population of algae (*Dunaliella viridis*); chemotrophic bacteria (*Halobacterium, Halococcus,* and others); other bacteria including sulfate-reducers and cyanobacteria; and amoebae and other protozoa (Larsen 1980). *D. viridis* is present in sufficient numbers in the near-surface waters of the Dead Sea to correspond to a biomass equal to that of *D. salina* in the northern basin of the Great Salt Lake, although the contribution of bacteria to the biomass in the Dead Sea is lower (Larsen 1980).

Recent summaries and surveys of saline lakes, which occur on all the continents, reveal that these evaporitic environments are highly productive biologically. The soda lakes in tropical Africa are recognized as being among the world's most productive ecosystems (Melack 1981), far more productive than seawater, even though the biota in the lakes include few species and may be dominated by only one (Melack 1981). Where measurements of primary production have been made, the values for saline lakes range up to 19,000 mg of carbon assimilated per sq m per day (Lake Aranguadi, Ethiopia; Hammer 1981), whereas the highest rates for the oceans rarely exceed 2,000, these from areas of coastal upwelling on the western side of South America and Africa (Tissot and Welte 1978).

Further demonstration of the productivity of hypersaline environments can be obtained from small bodies of water known as heliothermic lakes, in which a sun-heated layer of warm water exists beneath a layer of cooler water. Heliothermic lakes are also stratified, with a stagnant and more saline water mass beneath a less saline upper water mass that undergoes periodic circulation (Kirkland et al. 1983). Algae and bacteria grow within these hypersaline water bodies, sometimes forming a bacterial plate at the interface between the two water masses, and on other occasions so clouding the water that the effect of insolation in raising the temperature of the lower water mass is diminished (Kirkland et al. 1983).

Solar Ponds

It is now accepted that biological activity within the brines of artificial salt ponds can have an effect upon the yield and quality of salt extracted (Davis 1973; Javor 1983); it may even be indispensable (Baas-Becking 1931). In a study of the Alviso salt ponds in San Francisco Bay, Carpelan (1957) noted that their productivity was 20 to 30 times that of normal seawater. Where the salinity was in the range 4.5‰ to 9.4‰, the single-celled green alga *Strichococcus bacillans* is predominant throughout the year, but where the salinity exceeds 10‰, *Dunaliella salina* and *Stephanoptera gracilis* are conspicuous. *D. salina* blooms at a particular salinity and occurs in such numbers that the brine has been described as having the color and consistency of tomato soup, and the appearance of the species is used as an indicator of brine concentration. On the coast of Bulgaria, similar salt ponds achieve a maximum biomass at salinities between 7‰ and 9‰ (Caspers 1957). For two salt ponds on the Pacific coast, Javor (1983) described the blooms of algae and photosynthetic bacteria that occur at salinities of approximately 10‰ to 18‰, and itemized the biological changes that took place as the brines changed concentration in their passage from pond to successive pond through the evaporative system. Implicit in her descriptions is the notion that, far from being biologically sterile, the salt ponds function best where there is adequate biological productivity.

In order to determine their sedimentological and diagenetic importance in hypersaline environments, Cornée (1983, 1984) studied bacterial populations in a salina (Salin-de-Giraud) on the south coast of France, and another (Santa Pola) on the east coast of Spain. Her studies, and those of Thomas and Geisler (1982) and Thomas (1982), demonstrate unequivocally that life is abundant in these artificial evaporitic settings. Cyanobacteria and photosynthetic bacteria are the dominant elements in microbial benthic populations, forming mats and clumps. In the free brines, halophilic heterotrophic bacteria are abundant, increasing in number as salinity increases. In brines where the salinity exceeds 25‰ to 30‰, extreme halophiles dominate and form a very significant biomass.

Saline Lagoons

Hypersaline lagoons, permanently or intermittently connected to the sea, are also highly productive. Shark Bay, on the west coast of Australia, has become famous for its algal structures, which colonize the littoral environment in two hypersaline basins barred by a sill. Evaporation there exceeds precipitation by a factor of ten, so that with the restricted circulation imposed by the morphometry of the basins, salinity is up to twice that of normal seawater. Primary production by the microbial mats is on the order of 2,700 mg C per sq m per day (Bauld 1984), a rate that exceeds that of the most productive areas of the oceans. In the zones of maximum algal growth, a small pelecypod, *Fragum hamelini*, is present in vast numbers (Logan and Cebulski 1970).

A similar hypersaline lagoon on the south coast of Australia, Spencer Gulf, has also been studied. Mats have developed on high intertidal areas of a prograding northeastern shoreline of a basin where salinities in the open waters exceed those of the open ocean by about 30%. Primary production is greater than 10,000 mg C per sq m per day (Bauld 1984).

On the Pacific coast of Baja California, Scammon Lagoon (Laguna Ojo de Liebre) displays high productivity in its more saline reaches, where salinity may be up to 1.5 times that of seawater. The high phytoplankton productivity at the base of the food chain (Phleger and Ewing 1962) probably provides the food to support thousands of California Gray whales (*Eschrichtius robustus*) that use the lagoon as mating and calving grounds during the winter.

Laguna Madre, Texas, in which the hydrobiology has been studied in detail, is the only large hypersaline lagoon in the United States, and it provides a clear demonstration of the exceptional biological activity that occurs as normal ocean water is concentrated by evaporation. The lagoon formed behind Padre Island, a coastal barrier island (figure 5.1) and includes Baffin Bay, a drowned river valley separated from the main body of water by a shallow, submerged sill made up of masses of empty tubes of serpulid worms (Hedgpeth 1957). Within the lagoon, salinity rises to about twice that of the adjacent Gulf of Mexico, but prior to the construction of the Intercoastal Waterway in 1946, the maximum salinity recorded had been almost 120‰, or more than three times that of the ocean.

Laguna Madre is characterized by extremely high phytoplankton productivity, the base of the food chain. Blooms of red-pigmented green algae are common in the areas of higher salinity in the lagoon and in Baffin Bay, and they form the so-called red water (Simmons 1957). The population of zooplankton, primarily copepods, is extremely high, and Baffin Bay harbors four species of shelled invertebrates that, except for some reefs of oysters, constitute the largest assemblage of living molluscs along the northern Gulf Coast (Parker 1959). The small pelecypod, *Anomalocardia cuneimeris*, reaches a density of 15,000 individuals per sq m in some parts of the northern lagoon, at a salinity of 45‰ (Simmons 1957). To Hedgpeth and to Simmons, both writing in the fifties, Laguna Madre was one of the most biologically productive areas in the world, and within the limits of their knowledge, it was.

There is a similar lagoon on the east coast of the Crimea peninsula in the USSR. This hypersaline body of water, 113 km long and only 1 to 3 m deep, is separated from the Sea of Azov by a narrow spit, but is open at its northern end via a narrow and shallow inlet only 122 m wide and 0.9 m deep (figure 5.2). Salinity within the Sivash, as it is called, ranges from 18‰ at its opening to 166‰ at its southern end (Zenkevitch 1963).

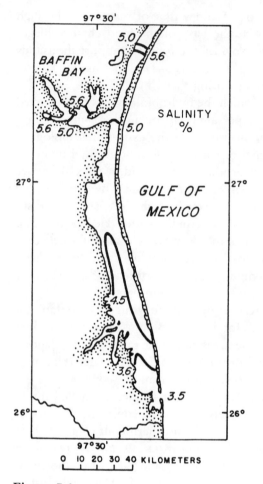

Figure 5.1
Salinity of surface water in Laguna Madre, Texas, July 1957 (re-drawn from Gross 1972:314).

The lagoon is highly productive because of a dense growth of *Cladophora*, a green alga, especially at the southern end, where the salinity is highest (Hedgpeth 1957; Zenkevitch 1963). In the central and southern Sivash, to a minimum salinity of 75‰, the brine shrimp, *Artemia salina*, develops in "enormous numbers," and in the northern area, commercial fish production exceeds that of the Sea of Azov,

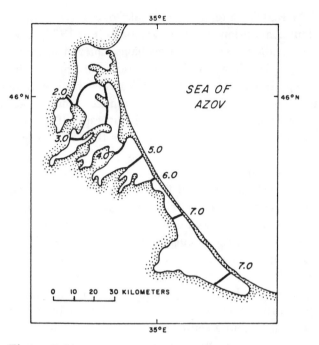

Figure 5.2
Chlorinity of surface water in the Sivash, Ukrainian SSR (redrawn
from Zenkevitch 1963:529). Salinity = 0.03 + (1.805 × chlorinity).

which is itself considered to be the most productive sea in the world.

There is a category of small hypersaline lagoons that do not have
an obvious surface connection with the sea. The term "anchialine
pool" means literally "near the sea," but has taken on the broader
meaning of "pools with no surface connection (with the sea), containing salt or brackish water, which fluctuates with the tide" (Por
1985). Their most important feature is that they have a permanent
subterranean connection with the open sea, but no permanent surface
connection, and they are generally several tens to several hundreds
of meters in diameter, far smaller than the lagoons described above.
They are found on emerged coral shores or shores made of lava flows
that contain a network of crevices, and cannot exist where well-developed fluvial systems debouch into the lagoon. They have been described in locations ranging from islands in the Indian and Pacific

oceans to around the Red Sea and Sinai peninsula, one or two islands in the tropical Atlantic, and islands in the Caribbean (Por 1985). Several of the more hypersaline anchialine pools have achieved scientific notoriety, including the Pekelmeer on Bonaire and Solar Lake on the Sinai peninsula. Hypersaline or not, they are marked by luxuriate biological activity, characterized by the predominance of cyanobacterial mats and photosynthetic bacteria, a rich diversity of aquatic and amphibious coleopterans, the presence of brine flies, and the appearance of *Artemia,* the brine shrimp (Por 1985).

At the start of this discussion of the productivity of hypersaline ecosystems, we mentioned the ability of Lake Nakuru to support an enormous population of vertebrates, specifically flamingoes, the very survival of which belies any notion of the infertility of the hypersaline environment. There is another rather curious example of a vertebrate completely dependent upon the productivity of a hypersaline lagoon: the Laysan teal, *Anas laysanensis,* regarded as one of the rarest ducks in the world (Caspers 1981). It lives only on Laysan Atoll, 4.5 sq km of barrenness at the northwestern end of the Hawaiian chain, 1,000 km northwest of Kauai. Within a central depression on the island is a lagoon in which the salinity normally exceeds 50‰, but can reach 130‰, presumably when storm flooding diminishes, thereby reducing the resupply of water to the lagoon. The simple food chain in and around the lagoon is dependent upon the nutrients in the water, whose composition is modified by the addition of nitrate and phosphate from the guano of seabirds. In the brine is a dense growth of *Dunaliella,* which supports *Artemia salina* and a brine fly, *Neoscatella sexnotata* (Ephydridae), whose larvae also feed on mats of a blue-green alga—*Schizothrix* sp.—on the littoral edge of the lagoon. The flies develop in massive populations and are the source of food for the flock of *Anas laysanensis.* It was this dependence upon flies and not vegetation that enabled the Laysan teal to survive when rabbits were introduced in 1903 and wreaked havoc on the vegetation for the next twenty years, before their eventual extermination (Caspers 1981). Nevertheless, the birds have a tenuous hold on existence, for the only other flock of Laysan teals disappeared in the late nineteenth century when their habitat, a lagoon on Lisiansky island, 224 km northwest of Laysan, dried up, destroying the food chain upon which the birds depended.

Sabkhas

The term "sabkha" is a transliteration of the Arabic word for "salt flat" and was recommended by Kinsman (1969) as the standard form in the English geological literature. As is so often the case, however, an original definition became corrupted by misunderstanding or misuse, and in this case the word has become virtually synonymous with "supratidal evaporitic flat," as Sonnenfeld (1984:158) would have it. There are, however, supratidal (coastal) sabkhas, and there are continental sabkhas associated with playa lakes in endorheic basins, and Purser (1985) would like to restrict the term to any environment where the substratum is affected by capillary evaporation that leads to the precipitation and preservation of sulfates or more soluble minerals of the evaporitic suite. He would convert the original descriptive term into one connoting process, a process that leads to the formation of a hypersaline brine within, and periodically on top of, the local sediments.

From the above considerations, we are aware that sabkhas, either continental or paralic, grade laterally into lagoons and lakes. The biological productivity of the exposed surface on the margin of the lagoon or lake can be thought of as one facet of the biological productivity of the water body as a whole; and as must be clear from all the examples cited above, hypersaline environments—*all* hypersaline environments—are very productive. Among lagoons, Shark Bay and Spencer Gulf have extensive microbial mats on their shores in intertidal and supratidal zones; Laguna Ojo de Liebre and its neighbor in Baja California, Laguna Guerrero Negro, have cyanobacterial mats in an extensive area of salt marshes and sabkha flats (Javor and Castenholz 1984). Several small anchialine pools within extensive sabkhas along the northern gulfs of the Red Sea have flourishing microbial mats within and around their margins (Friedman et al. 1985; Gerdes et al. 1985), and the sabkhas themselves have a large number of different types of mat (Gerdes et al. 1985). Around the Persian Gulf, there are other examples of flourishing algal and microbial communities in hypersaline lagoons and supratidal flats, including the now famous sabkhas of Abu Dhabi (Kendall and Skipwith 1969; Purser 1985).

This survey of hypersaline environments—lakes, salinas, lagoons, and sabkhas—leads us to two conclusions: first, that hypersalinity is no impediment to biological activity; second, that on the contrary, hypersalinity encourages exceptional biological productivity. The causes and the consequences of these conclusions now need to be examined.

ECOLOGY OF THE HYPERSALINE ENVIRONMENT

A well-understood concept in biology declares that the diversity of species endemic to a particular ecological niche diminishes as the harshness of the environment increases. Gerdes and Krumbein (1984) describe this as "Thienemann's rule." A simple illustration can be drawn from life in the polar regions of our planet: few metazoans can tolerate such extreme climatic conditions, and as a result, biological diversity decreases. The same effect can be seen in hypersaline environments, for few aquatic organisms have a mechanism for retaining their body fluids in the presence of brines more saline than seawater. At a concentration of 300‰, a sodium chloride brine exerts an osmotic pressure of more than 100 atmospheres (Baas-Becking 1928), and the effect of this upon species that inhabit the oceans is to cause them to desiccate rapidly and expire. Knowledge of the general inverse correlation between salinity and diversity was used by Williams (1981) to encourage biologists to study the whole ecosystem in salt lakes, for in these environments fish and macrophytes are absent, so that the work of the biologist attempting to untangle the complexity of trophic relationships is simplified.

This principle can be illustrated from a number of the hypersaline locales described in the preceding section. In Laguna Madre, as salinity increases from that of normal seawater to about 75‰, the number of species of fish in the water decreases from 67 to 10 (figure 5.3; Copeland 1967). In the oceanic zone near Shark Bay, there are 83 species of benthonic Foraminifera, but only 16 species survive in the bay, where salinities are between 56‰ and 70‰ (Davies 1970). In the Sea of Azov, from which the Sivash receives its inflow, there are 137 species of benthos, 108 species of zooplankton, and 188 species of phytoplankton, whereas in the southern, most saline end of the lagoon, there are 2 species of benthos, 9 species of zooplankton, and 16 species of phytoplankton (Zenkevitch 1963). It is precisely

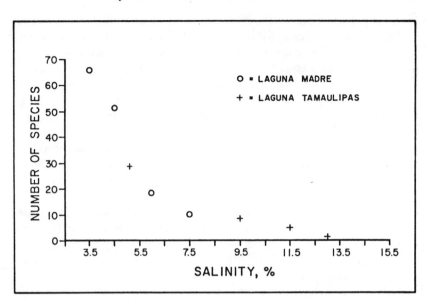

Figure 5.3
Abundance of species of fish in Laguna Madre, Texas, and Laguna
Tamaulipas, Mexico (data from Hildebrand 1958 and Copeland 1967).

because there is a relationship between salinity and survivability among
the phytoplankton that we need to understand the reasons for the
impoverishment.

As evaporation causes the salinity of seawater to increase, concurrent changes take place in both the chemistry and physics of the brine,
so that several factors may have a bearing upon the ability of organisms to survive:

1. Hypersalinity must be considered to be the most important influence. Marine invertebrates are in osmotic equilibrium with seawater, and if any organism is to survive in water more saline than
   seawater, its physiological functioning must include a mechanism
   for preventing desiccation by osmosis. Javor and Castenholz (1984)
   showed by experiment how drastically invertebrates without such
   a mechanism are affected by immersion in hypersaline water.
2. Accompanying the increase in salinity is an increase in alkalinity.
   The pH of normal seawater is 8.2, but the highest pH achieved
   is 9.0, at a salinity of twice the normal value (Copeland 1967).

3. At a pH of about 9, the concentration of carbon dioxide in the brine is nearly zero (Moberg et al. 1934), so that algae that are dependent upon carbon dioxide for photosynthesis must perish.
4. As salinity increases, the specific heat of the brine decreases (Kaufmann 1960), so that for a given amount of heat input, a unit volume of hypersaline water will show a greater increase in temperature than will the same volume of less saline water. The increase in temperature may be inimical to some species.
5. The solubility of oxygen in a brine decreases progressively as salinity increases and as the temperature increases (Copeland 1967; Hite 1970; Kinsman et al. 1973), and this affects the survival of some organisms.
6. Most hypersaline milieux develop in tropical climatic belts within 40° of the equator, and as a consequence, the surface waters are subjected to high light intensity, to which some organisms are sensitive (Larsen 1980; Williams 1981). The red coloration so frequently seen in strong brines is due to the pigment that protects halophilic bacterial cells from the effects of light. A corollary effect of that protection is an increase in the temperature of the brine, as the great quantity of pigment in the bacterial bloom absorbs light (Larsen 1980). This exacerbates the problem described in 4. above.
7. Most marine organisms are adjusted to the balance of anions and cations in seawater. Because precipitation of any salt changes the ratio of one ion to another, the biota can be affected (Larsen 1980).

Concurrent with the impoverishment in the number of species in the hypersaline environment is an enrichment in the number of individuals, a process that forms the second part of Thienemann's rule (Gerdes and Krumbein 1984). The above descriptions of life in evaporitic hypersaline waters are expressions of the abundant populations of the species that can survive the rigors of this environment. The maximum biomass appears to occur at about two or three times the concentration of seawater; at higher salinities, the number of organisms per unit volume decreases, reaching almost zero at concentrations at which the salts of potassium and magnesium precipitate.

More significant than biomass, however, is primary productivity, the rate at which energy is stored by photosynthesis in the form of organic substance. The "biomass" or "standing crop" forms only a small part of the annual production, especially for species that create a new generation every few days. In the Caspian Sea, the annual crop of phytoplankton is about 300 times greater than the standing crop (Bordovskiy 1965). We can reiterate that rates of primary productivity in hypersaline environments are greater than those for the most productive parts of the world's oceans.

The reasons for the burgeoning productivity of phytoplankton are related to the chemical and physical changes that take place in the brine as it becomes concentrated by evaporation:

1. Unless the body of water is completely sealed off from the ocean and is therefore desiccating, any water that is evaporated is replaced by further inflow, which brings more salts. Just as these salts are left behind in the brine, so too are the essential nutrients of biological activity, compounds of nitrogen and phosphorus, and they become concentrated in proportion to the salinity of the water. The actual concentration required for the support of life is very low—only a few milligrams of phosphorus (as phosphate) per cu m, and about eight times that amount by weight of nitrogen (as nitrate; Raymont 1963). Only slight differences in concentration separate the minimum requirements from the optimum conditions, and as a consequence, slight increases in concentrations of these nutrients can produce rapid and spectacular "blooms" of phytoplankton.

2. Because of the decreased diversity in the biota, surviving species have little or no competition from other species for nutrients, for space, or for any other requirements for existence.

3. The environmental changes described above affect the grazers and parasites that under other circumstances would inhibit the development of large populations of the species that they victimize. This can be seen very clearly on the Gavish Sabkha, on the southern part of the Sinai peninsula (Gerdes and Krumbein 1984). *Pirenella conica*, a marine gastropod, can tolerate the conditions of this extreme coastal environment, but only up to a salinity of 70‰. As a result, microbial mats, upon which the gastropod grazes, develop only in areas of higher salinity, and are inhibited where large populations of *Pirenella* are present. The same relationship has been determined by observation and experiment upon microbial mats in Laguna Guerrero Negro, Baja California, Mexico (Javor and Castenholz 1984). *Cerithidea californica*, a marine gastropod,

cannot tolerate a salinity greater than that of seawater, and so inhabits tidal channels and intertidal flats adjacent to the channels. Higher intertidal flats are drier and unsuitable for the cerithids, and it is on those surfaces that microbial mats persist. Within the channels and on the moist, lower intertidal areas, there are no mats.

The effect of these environmental stresses would appear to be the maintenance of a simple food chain, from algal and microbial primary producers to a few halotolerant grazers, and occasionally, as in the case of flamingoes and the Laysan duck, to vertebrates. And, in truth, such a chain can be seen in salt lakes, in some lagoons, and in solar ponds. But, just as all generalizations are highlighted by exceptions, or by more complicated examples of the simple truth, so biological activity in hypersaline environments maintains more than a simple food chain. This we are learning from detailed studies of the bacterial populations in evaporitic brines.

Estimates of the content of the biomass in the northern arm of the Great Salt Lake reveal that the contribution from bacteria is more than eleven times the contribution from algae (Post 1981). Despite that revelation, there have been relatively few detailed analyses of the entire bacterial population in hypersaline settings, and therefore we do not yet know the ecological relationships among the microscopic inhabitants of hypersaline water. From studies of Solar Lake, where 149 distinct microscopic "morphotypes" were isolated, many of them positively identified as known genera, it is apparent that microbial life in these settings is extremely complicated (Hirsch 1980). A similar conclusion can be derived from perusal of the work of Cornée (1984) in the salinas of Santa Pola, Spain and Salin-de-Giraud, France. If the bacterial contribution to the organic matter may be the most significant quantitatively, and if it can be preserved after sedimentation, then it should play a major role in determining the composition of the oil that is ultimately produced. The determination of the contribution, preservation, and effect of bacterial organic matter in the formation of oil from evaporitic source beds is a path of research upon which we are only now beginning to tread.

DESTRUCTION AND PRESERVATION OF
ORGANIC MATTER

Organic matter that is generated in the aquatic environment has several possible fates. It might encounter the unusual circumstances that will allow it to be preserved intact, but far greater is the probability that it will be partially or completely destroyed. The following description of the processes of destruction of organic matter is a necessary prelude to the understanding of the circumstances in which it could be preserved. If the evaporitic hypersaline environment did not have the capability of preserving at least a part of the organic matter generated by biological activity in the brines, then no further discussion of evaporitic source beds would be necessary. Because it is not possible to actually observe the destructive processes as they are taking place, the description that follows is a model, the necessary antithesis of our view of sedimentation and preservation of organic material in the deep hypersaline evaporitic milieu.

Destruction of Organic Matter Within the Water Column

In a deep-water evaporitic embayment, no less than in any other deep body of water, three zones would contribute to the partial or complete destruction of organic matter: the oxygenated water column from the surface downwards; the anoxic water column at greater depth; and the anoxic sediment column beneath the interface of sediment and water. This latter zone would have had two subzones: an upper zone of sulfate reduction, and an underlying zone of methanogenesis.

The Oxygenated Water Column. The oxygenated water column would probably have been the primary zone of destruction in the evaporitic embayment, as it is in the open ocean (Bordovskiy 1965), the Black Sea (Deuser 1971), and lakes (Wetzel 1975). This part of the water column in the evaporitic embayment would usually have been thin, probably no more than tens of meters, extending from the surface into the chemocline (halocline). The thinner the zone, the lesser would have been its capacity as a destroyer of organic matter. Furthermore, throughout much of the oxygenated water mass, because the concentration of oxygen in hypersaline brine would have been less than that in normal marine water, the destructive capacity of this zone would have been reduced even further.

The oxygen-carrying capacity of hypersaline brine is inversely proportional to its salinity (Kinsman et al. 1973). Hypersaline marine brines at the point of precipitation of gypsum, for example, can hold only about one-half of the approximately 8.5 ppm of oxygen that can be held by normal marine water. Because of this characteristic, the oxygen concentration of the oxygenated water mass would have diminished progressively away from the portal. Because the likelihood of destruction of organic matter would probably have been directly proportional to the oxygen concentration in the brine, the greatest potential for destruction would have been in the vicinity of the portal, and the least potential in the far reaches of the embayment.

Generally, as the food supply increases in a marine environment, the number of grazers (chiefly zooplankton) increases correspondingly. The same relationship probably holds for the marine mesosaline environment, although in that particular case, it would be the number of individuals, not the number of species, that would increase. We have described the abundance of the ubiquitous brine shrimp, *Artemia salina,* and also the abundance of cerithid gastropods associated with areas of prolific activity of mat-forming microorganisms. Zooplankton like *Artemia* would have flourished in the mesosaline oxic waters of the evaporitic embayment, and would have played an important role in preventing complete destruction of phytoplankton debris. In accordance with Stoke's law, the comparatively large fecal pellets of zooplankton would have sunk at a relatively rapid velocity through the oxygenated zone. The organic debris encapsulated in pellets would therefore be removed quickly from contact with aerobic saprophytic bacteria, the principal remineralizers, whereas the smaller individual organic fragments, sinking far more slowly, may not have survived the downward journey.

The Anoxic Water Mass. The transition to complete anoxia would have been completed within the halocline, so that at greater depth, within the highly saline bottom waters, an assemblage of saprophytic anaerobic microorganisms would have fed upon the sinking algal and bacterial debris. Simple organic compounds within this debris, as well as similar compounds formed by associated microorganisms, would have been metabolized by sulfate-reducing bacteria to form the by-products carbon dioxide and hydrogen sulfide, both of which would have dissolved in the brine.

The demarcation near the top of the halocline between reducing conditions below and oxidizing conditions above would have been the site for a reaction between hydrogen sulfide and oxygen. The reaction would be aided enzymatically, the agents being colorless chemosynthetic sulfur-oxidizing bacteria, or, if the halocline were in the photic zone, colored photosynthetic sulfur bacteria. Elemental sulfur would have accumulated within or outside of the bacterial cells, ultimately settling onto the floor of the embayment, where some of it might be attacked by sulfur-oxidizing bacteria that use nitrate as an oxidizing agent. These bacteria, along with unicellular algae and other bacteria, would be the principal source of organic matter that ultimately became entombed within the sediments.

Destruction of Organic Matter Within the Sediment Column

Zone of Sulfate Reduction. This zone at the bottom of the water column would have extended a few centimeters or tens of centimeters into the sediments, and any organic matter therein or on the floor of the embayment would have been subject to attack by sulfate-reducing bacteria, the organic matter being oxidized by sulfate anions. Because the depositional rate in such a setting as this would have been very low, it is possible that the organic matter might have remained in this zone for hundreds of years. Nevertheless, it would not necessarily have been subject to oxidation during all that time.

The reason for this is the nature of the sediments on the embayment floor, beneath mesosaline surface waters. Usually they would have been fine-grained carbonates (or marls) with restricted permeability that would have prevented free movement of brine into the sediments. In the absence of burrowing metazoans, or metazoans of any kind to stir up the sediment, sulfate could have moved downwards only by molecular diffusion. The replacement of any sulfate consumed would therefore have been a slow process.

Most of the easily oxidized organic material would probably have been consumed within the oxygenated water and in the anoxic water column, thereby depriving the sulfate-reducers of foodstuffs, further slowing down the destructive process of sulfate reduction. Under favorable ecological conditions, other foodstuffs—amino acids, fatty acids, and molecular hydrogen—would probably have been generated by an assemblage of associated bacteria by processes of hydrolysis and fermentation. These would have allowed sulfate reduction (and thereby oxidation of organic matter) to proceed unabated. However, the waters immediately above the floor of the embayment and within the sediments on the floor are the remnants of the highly saline brine that formed within the surface waters at the far reaches of the basin. The salinity of this brine might well have exceeded the ecological limits of many species of hydrogen-forming and acid-forming bacterial saprophytes. That being the case, no new foodstuffs would have been supplied to replace the easily oxidized organic matter already lost, and oxidation of the remaining organic matter by sulfate-reducing bacteria would have diminished greatly.

Zone of Methanogenesis. At a depth at which sulfate was no longer of sufficient concentration for sulfate-reducing bacteria to function, the methane-forming bacteria, the methanogens, would have operated. They do not function where sulfate-reducing bacteria are active (Abram and Nedwell 1978), but like the sulfate-reducers, they can utilize only hydrogen and simple organic compounds such as acetic acid. These compounds are formed by an associated assemblage of bacteria that hydrolyze and ferment organic compounds. The oxygen necessary for oxidation would have come from the organic matter itself and from carbonate species derived from solution of carbonate minerals. We are uncertain how these associated bacteria would have functioned under the conditions of hypersalinity that would have existed in the basin and environment we are describing. Certainly, degradation of organic matter in this zone would have proceeded slowly as compared to degradation of such matter in the overlying zones, if for no other reason than the fact that most of it that could have been consumed readily by the saprophytic microbes would already have been consumed in the overlying zones.

Methane that formed would probably have diffused into the overlying zones of sulfate reduction, where it would have been used as a foodstuff by the sulfate-reducing bacteria (Reeburg 1983). Probably only a small fraction (if any) would have reached the oxygenated water column, and there it would have been rapidly oxidized by bacterial agents.

Formation of Kerogen

Because primary productivity would have been great, because the processes of destruction in the water column and in the sediment column would have been generally inefficient, and because the rate of accumulation of mineral matter would have been slow, a sediment would have formed that was rich in organic matter. The beginning of this process can be seen today in sediments of the Dead Sea. Nissenbaum et al. (1972) compared organic constituents in sediment from beneath oxygenated waters and in sediment from beneath deeper hypersaline and anoxic water. Although the content of organic carbon in both cases was lower than that found in sediments from reducing environments in other lakes or from marine shelves, the sediment from beneath the anoxic water had a much higher concentration of hydrocarbons, fatty acids, amino acids, humic and fulvic acids, and chlorophyll derivatives than did the sediment from beneath the oxygenated water. Under oxidizing conditions, the organic matter in the sediment was rapidly converted into an insoluble nonextractable complex, whereas that in the anoxic environment was preserved. Most of this organic matter would ultimately have been incorporated into kerogen molecules, organic polymers of high molecular weight and complex structure. The kerogen would have formed from the degradation product of algal cells—humic acids—and a variety of microbial byproducts (Type II of Tissot and Welte 1984). Once chemically attached to kerogen, these byproducts, such as acetate anion, would no longer have been subject to microbial attack. In the deepwater evaporite environment, the kerogen that would have formed would have been rich in hydrogen. After sufficient burial and thermal exposure, it would have yielded oil.

PHYSICAL ENVIRONMENT OF THE DEEP-WATER
EVAPORITIC SETTING

The absence of a modern marine basin in which evaporites are precipitating from deep water allows us free intellectual rein to formulate a conceptual model of this particular environment. There is an obvious need to understand the sedimentological and chemical features of the setting in which, as we believe, many major source rocks originated.

We define "deep" as those depths at which wave action no longer stirs the sediments. On the open shelf this "depth to wave base" varies widely, but is usually in the range of 10 to 25 m, perhaps even deeper on shelves marked by conditions of high energy. For a number of reasons, however, wave action might not have penetrated so deeply in evaporitic basins. In the latitudinal range in which evaporation exceeds precipitation and thereby allows a suitable embayment to develop and contain a brine more saline than seawater, winds would be of lower frequency and intensity than those in more polar or more equatorial latitudinal belts. Also, within the basin itself, the concentrated brine would be more viscous than would seawater, and the presence of a marked halocline would be an impediment to free mixing throughout the water column. We can deduce, therefore, that wave action might not penetrate as far as it does on normal marine shelves, and we can state that in our terms, a deep evaporitic basin is one in which the brine has a depth of about 10 to 20 m or more.

Although several types of deep-water evaporitic environments may exist, one type may be connected to the ocean (or to another source of marine water) by a restricted opening—a portal. If the portal were to close, the environment would, at least initially, be that of a lacustrine deep-water evaporitic setting, similar to the Dead Sea. An essential feature of our conception of this basin is a sill at the portal separating the evaporite environment within from the normal seawater without (figure 5.4). The sill is a shallow submarine barrier, and implicit in this description is that water depth within the embayment is greater than water depth at the sill, and that water depth in the open-marine basin is the same as or deeper than that of the sill. The Red Sea is an example of such a basin, for at the strait of

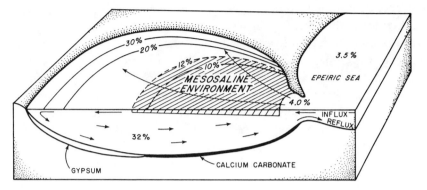

Figure 5.4
Simplified model of evaporitic basin with deep water (modified from Briggs and Pollack 1967). Contours are lines of isosalinity; arrows represent flow of water; cross-hatched zone is "mesosaline" area, which is the region of greatest biological productivity.

Bab-el-Mandeb, a sill separates the elongate embayment from the Indian Ocean.

As water in an evaporitic basin undergoes evaporation, salts would remain behind and the specific gravity of the near-surface water would increase. If the underlying water were less dense, the near-surface water would sink until it lost its integrity by mixing with water of lower specific gravity; or it would sink until at some depth it reached water of the same specific gravity; or it would sink until it reached the sea floor. Dense brine that reached the sea floor would flow down any incline until it reached water of the same specific gravity, or until it reached a low on the floor of the embayment, where it would accumulate as a brine pool. Here the pool would persist until displaced upward by introduction of an even denser brine, or until dissipated by molecular diffusion. In such a basin, the salinity of the near-surface waters would increase progressively away from the portal, and the most saline surface water would have been in the shallow shoals in the far reaches, whence it would drain into the depths of the basin down the sloping floor or in channels, to collect in discrete pools in depressions on the basin floor.

The drainage of hypersaline surface water into basinal deeps would have been intermittent. Brines formed during summer would sink to a level determined by their specific gravity, which itself would be a function of the intensity of evaporation during that season. Those brines immediately above the sediment-water interface in the deepest part of the embayment would have been emplaced during years with the most severe aridity, when the evaporation rate was at a maximum. They would be preserved because currents induced by winds would be restricted to the water mass above the halocline; only molecular diffusion, a slow process, could cause the deep brines to dissipate. The hypersaline mantle on the floor of the embayment would be important, if not essential, for the preservation of any organic matter deposited.

In a deep-water evaporite setting in which the depth of the sill was below wave base, the volume of dense brine in the basin deep would ultimately fill the basin to the "spill point." Then as the halocline rose above the depth of the sill, dense brine would flow out of the embayment as a refluxing current beneath the influxing current of normal marine water. This situation occurs in the Mediterranean Sea and the Red Sea today. In the Red Sea the most saline brine (40.6‰) has not yet filled the embayment to sill depth, and water of lesser salinity (39‰) flows as a reflux current over the sill and cascades into the depths of the Indian Ocean (figure 5.5; Dietrich et al. 1980). If a sill were absent, hypersaline water would flow as a reflux current out of the embayment, without forming a brine pool with its associated halocline. This is the situation in the Holocene Persian Gulf, where the more saline brine drains into the Gulf of Oman without first having resided in a brine pool.

Eventually, the halocline would stabilize at the depth of the "spill point," and would be the boundary between two distinct water bodies. The two would be related by the mineral and organic matter that originated in the upper layer and sank through the halocline into the denser waters below. The halocline would also provide an effective barrier to downward migration of oxygen from the near-surface waters, and to upward migration of hydrogen sulfide, if present, from deep waters below the halocline. Only in the shoals in the far reaches of the basin where the surface waters were most saline and where the halocline was absent could oxygen-bearing waters move into the depths. As we have discussed above, however, hypersaline waters are

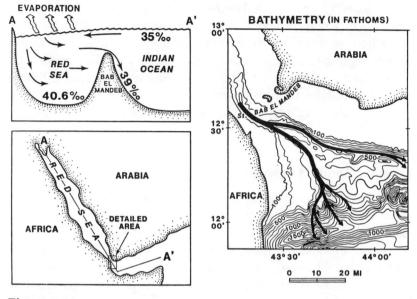

Figure 5.5
Refluxing current from the Red Sea. Maximum depth of the Red Sea is approximately 2,600 m, and depth of the sill is 175 m (after Dietrich et al. 1980 and Einsele and Werner 1972).

ineffectual in transporting large amounts of oxygen in solution.

Hypersaline waters may also have been derived from shallow lagoons surrounding the principal embayment. The Delaware Basin during Guadalupian times may have been an ancient example (Newell et al. 1953). In such a setting, a hypersaline brine pool might have formed at depth in the primary embayment as brines from fringing lagoons spilled over through gaps in, for example, a reef tract, while at the same time the salinity of the near-surface water might have been close enough to that of seawater to support flourishing assemblages of calcareous algae and invertebrates on the fringes of the primary embayment. In this model, the sill need not have been shallow nor the portal narrow, for the hypersalinity would not depend upon restriction of the principal embayment. The hypersaline waters would ultimately have filled the basin to the spill point, even if the sill were deep (or very deep), and even if the opening to the embayment were broad (or very broad). Under these conditions productivity in the embayment would probably have been lower than that in a hypersaline body of water, but the brine would have enhanced preservation of any organic matter that rained into deep water. The fringing lagoons would have trapped terrigenous sediments, thus contributing to a low sedimentation rate within the embayment and producing conditions that might have led to formation of good source rocks for oil.

Example: Todilto Limestone

The Middle Jurassic (Lucas et al. 1985) Todilto Formation occurs chiefly in southwestern Colorado and in northwestern New Mexico, and consists of a Limestone Member, which is about three m thick over much of its distribution, overlain by a Gypsum Member, which is of variable thickness, up to about 30 m. In an area of about 6,660 sq km in the southeastern part of the San Juan Basin, the Todilto Limestone Member is the apparent source rock for six known accumulations of oil within the immediately underlying Entrada Sandstone (Vincelette and Chittum 1981).

The reasoning by which the Todilto can be identified as the source depends upon the elimination of other possibilities. Extracts from Cretaceous shale units in the San Juan Basin have chemical characteristics that do not match those of oils from the Entrada accumulations. Furthermore, nothing within the Entrada itself, nor in the Triassic Chinle immediately below, would appear to have source potential; and of the units above the Entrada—the Todilto, the Jurassic Summerville Formation and the Jurassic Morrisson Formation—only the first-named could be a source (Vincelette and Chittum 1981). Furthermore, oil from the Media Field, the first Entrada discovery in the southeastern San Juan Basin, correlates with extracts from the Todilto Limestone.

Although we are dealing with one type of source rock, there may be considerable differences among members of the one type, both in the details of the original environment and in the chemistry of the product. In the case of the Todilto the quantity of organic matter in the limestone unit is lower than what we have come to expect of evaporitic sources ($\approx 1\%$), the depositional environment of the evaporitic limestone was unusual, and the type of oil has characteristics unlike those generated by other evaporitic carbonates.

The Limestone Member is distinctly laminated; couplets have a thickness of about 0.15 mm, and are composed of calcium carbonate and organic matter. Silt grains, probably wind-borne, were deposited within and on top of the thin (approximately 0.01 mm) organic laminae. Individual laminae can be correlated over distances of at least 3.7 km (Anderson and Kirkland 1960). In spite of their thinness, the organic laminae are extremely cohesive, and individual "sheets," which invariably contain abundant fragments of vascular plants, can be "lifted" from the dissolved remains of the limestone matrix.

Two species of fossil fish, *Hulettia americana* and *Todiltia schoewei*, and a species of fossil insect, an aquatic Hemipteran belonging to the family Naucoridae, have been found at many localities within the Limestone Member. Ecological deductions based on both the fish and the insects have been used to support a lacustrine environment for the Todilto, in marked contrast to deductions based upon the ratio of strontium-87 to strontium-86 in the carbonates. The isotopic data establish clearly that cations forming the Limestone Member were derived mainly from a marine source. Furthermore, our data on sulfur isotopes of the Gypsum Member (12 analyses from various localities) and the data of Ridgley and Goldhaber (1983) on other gypsum samples from the Gypsum Member indicate that the sulfate anions had a marine source; that is, they have a sulfur isotopic value expected for Middle Jurassic seawater.

Lucas et al. (1985) consider the Todilto to have been deposited in a salina formed within the erg that became the Entrada Sandstone along the western margin of the basin. The evaporites, however, may have been deposited in the far reaches of the Sundance Sea, on a broad shelf 30 to 60 m deep, where no physical barrier restricted circulation. We need describe neither a narrow strait (for example, see Harshbarger et al. 1957) nor a restriction caused by seepage of marine water through a sand barrier to explain evaporite deposition in the Todilto, for the movement of water over great distances from the open sea, as was the case here, might allow salinity to increase, while reflux and mixing would be inhibited by friction. This model of deposition of evaporites on an open shelf was originated by Arkhangel'skaya and Grigor'yev (1960) to explain formation of Cambrian evaporites on the Siberian Platform.

The ecological details outlined above for other evaporitic environments would have applied for the Todilto. Nutrients would have become concentrated, growth of algae would have been enhanced, and organic debris would have rained onto the anoxic seafloor. Oil in the Entrada reservoirs has a high pour point (50° to 90°F), and probably contains a substantial fraction of high-molecular-weight normal alkanes (waxes). These are not derived from marine algae, nor are they the degradation products of sulfate-reducing bacteria. They may have been bacterial in origin, or they may have been derived from waxes of vascular plants, probably conifers, that were borne as particles by wind and stream from highlands to the south of the Todilto sea. The plant waxes, long-chain linear compounds, have great resistance to degradation, and frequently contribute to organic matter accumulating in lakes. In a situation like this they could have made up a significant portion of the organic debris and therefore could have had a major influence on the nature of the hydrocarbon product that evolved from this particular evaporitic source rock. The source was a marine evaporite, but the oil appeared with a lacustrine cast.

Example: Paradox Basin, Utah and Colorado

During the Pennsylvanian, in late Atokan time and throughout the Desmoinesian, approximately 2,137 m (7,000 ft) of cyclical evaporites and shales were deposited in the Paradox Basin of southeastern Utah and southwestern Colorado. The basin developed on the southwestern side of the Uncompahgre Uplift, an elongate highland oriented northwest-southeast (figure 5.6). Rapid subsidence in the basin, concurrent uplift of the highlands on the northeast, and mild positive movements on the southwest and northwest margins of the trough allowed for the development of euxinic and evaporitic conditions that persisted until the basin was filled to sill depth by Late Pennsylvanian time. Thereafter, normal marine conditions prevailed. (This account of source beds in the Paradox Basin is based largely upon the work of Hite and Buckner 1981 and Hite et al. 1984).

The evaporites are restricted to the Hermosa Formation, which can be divided into three members: the Lower Member, 122 m (400 feet) of carbonates, shales and anhydrites in the southwest, but only 15.3 m

(50 ft) thick beneath halite in the center of the basin; the Paradox Member, defined by the first occurrence and last occurrence of halite in the sequence, and consisting of 29 evaporitic cycles with interbedded "shales"; and the Upper Member of shallow-water carbonates varying in thickness from 305 to 519 m (1,000 to 1,700 ft), and grading upwards into the Permian Cutler Formation (figure 5.7). The

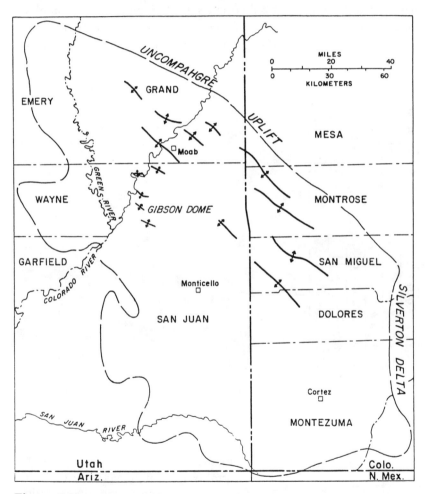

Figure 5.6
Boundaries of the Paradox Basin (redrawn from Hite et al. 1984). Location of salt anticlines shown by standard symbols.

stratigraphy is not as simple as the reader might be led to believe from the above description, for some carbonates of the Lower Member are the time equivalents of halite cycles in the Paradox Member. Furthermore, the evaporite cycles vary considerably in areal extent, the "zero halite" isopleth for cycles 6, 9, 13, 18, and 19 extending much further to the south and west than for the other cycles.

The evaporite cycles are bounded by disconformities and, from the base upwards, consist of anhydrite; silty dolomite; calcareous black "shale"; silty dolostone; anhydrite; and halite, with or without potash salts. Even within the halites, there are minor cycles of halite

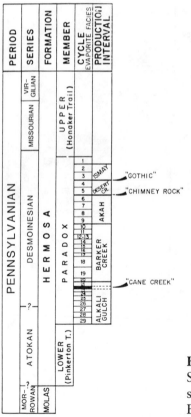

Figure 5.7
Stratigraphic nomenclature of Pennsylvanian sequence in the Paradox Basin (Hite et al. 1984).

with anhydrite, or of halite with anhydrite and potassium salts. The black shales are believed to be time-synchronous units of wide regional extent and therefore of great utility in basinal correlations. They can be traced from within the evaporitic sections into the carbonate facies in both the Lower Member and Upper Member of the Hermosa Formation, and they define intervals that are significant producers of hydrocarbons from within those carbonates. Three of the shales are important source rocks in cycles 2, 3, and 5. Members of the petroleum industry have given specific but informal names to the latter two, the "Gothic Shale" in cycle 3 and the "Chimney Rock Shale" in cycle 5, because they underlie major reservoirs elsewhere in the basin. In addition, the black shale of cycle 22 has been named the "Cane Creek Shale" from its exposure in the mine, now flooded, of the Texas Gulf Sulphur Company in the Cane Creek anticline near Moab, Utah.

The environment of deposition of the Hermosa Formation varied from a shallow shelf bathed in normal marine waters on the southwestern edge of the basin, to a large fan delta complex, the Silverton fan delta, in the southeastern area around Durango, Colorado, to deep hypersaline conditions within the basin itself. The cyclicity in the evaporites and in their coeval carbonate and clastic sediments was controlled by glacio-eustatic variations in sea level (Hite et al. 1984).

Mineralogical analyses of the black shales in the evaporite sequence show that the term "shale" is a misnomer, for the rocks contain at least 30% and sometimes more than 50% carbonate—calcite near the southern margins, but dolomite in the deeper, distal (and more saline) reaches. Clay minerals and clay-sized quartz make up the remainder of the shales. Consistently high values of organic carbon are found in the shales associated with the evaporites of the Paradox Member, in which values of 5% TOC are usual and values greater than 10% are known. The highest values of TOC were found in shales in the Upper Member of the Hermosa Formation, from which more than 20% TOC was recorded. The organic matter is a mixture of autochthonous material generated within the evaporitic brines, normal marine organic matter swept into the basin as part of the inflow that replaced evaporitic losses, and terrestrial organic matter from the Uncompahgre Uplift and other surrounding land.

Core from shales in the Hermosa Formation encountered in the Department of Energy Gibson Dome no. 1 (see figure 5.6) was subjected to geochemical examination in order to determine the quality and state of maturity of kerogen therein (Hite et al. 1984). The entire Lower Member, the entire Paradox Member, and 458 m (1,500 ft) of the Upper Member were mature enough to have generated oil, but the generated hydrocarbons had apparently been retained only in those shales that were completely interbedded with halite. Only where the shales were interbedded with porous carbonates had the hydrocarbons been expelled from the source. An estimate of potential of the Gothic Shale indicated that it had the capacity to generate almost 5,000 bbls per acre, and from this we can gauge the enormous source-rock potential of the multiple shales in the evaporitic sequence deposited in the hypersaline environment of the Paradox Basin.

That potential has been partly realized, for the "shales" have been responsible for a cumulative production of about 400 million bbls of oil and about 1 TCF of gas in fields such as Aneth and Ismay. It is the irony of this basin that the conditions that led to the formation of remarkable source beds also led in this case to the deposition of extensive, impermeable evaporites that would ultimately ensure that most of the products of those source beds would never be released.

Example: Saudi Arabia

From Late Carboniferous time until well into the Tertiary, a very shallow carbonate platform occupied large areas of the Middle East. The platform became differentiated into basins and shelves during the Middle Jurassic, with evaporites intermittently occupying the central depressions. It was in this setting that source rocks of the world's richest petroleum habitat were formed; it was here that perhaps half the world's oil was generated from evaporitic sources. The following account of the geology and geochemistry of one of those sources is based upon the published work of Murris (1980), Ayres et al. (1982), and Okla (1986).

The Arabian Basin was one of the bathymetric depressions that developed on the carbonate platform during Late Callovian and Oxfordian time (figure 5.8). Within the depression a sequence of carbonates and evaporites was deposited (figure 5.9), but where the se-

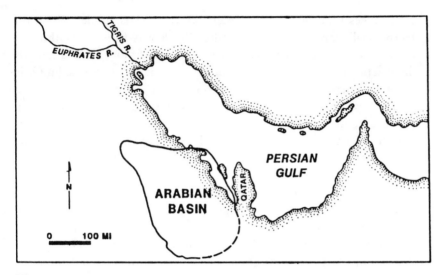

Figure 5.8
Location of the Arabian Basin.

quence is exposed at the surface, only limestones can be seen (Okla 1986). The oldest unit that shows signs of the differentiation of the carbonate platform (or "ramp") is the Tuwaiq Mountain Formation, which accumulated under euxinic conditions in the basin, but in aerated, high-energy environments on the adjoining shelves. This unit was succeeded by the Hanifa Formation, laminated lime muds and marls in the basin, grainstones on the surrounding shelves. Thereafter, from Late Oxfordian time onwards, carbonates and evaporites were deposited in shallow-marine to supratidal environments throughout the area, and Jurassic sedimentation concluded with four depositional cycles of shallowing-upward marine carbonates overlain by anhydrite. These cycles include the grainstones of the Arab Formation (A, B, C, and D), in which most of the oil in Jurassic reservoirs in Saudi Arabia is found. The final evaporite in those cycles is the Hith Anhydrite, 153 m (500 ft) thick and the massive seal that prevented vertical movement of oil generated from Jurassic source rocks.

The Hanifa Formation of the basinal depression is the principal source of the oil in the Arab reservoirs, and from several lines of evidence, it was deposited under hypersaline conditions, under circumstances that bear a close resemblance to those under which the Hermosa Formation of the Paradox Basin was formed. Within the Hanifa Formation, there are discrete laminae of anhydrite and early diagenetic nodules of anhydrite, direct statements of hypersalinity. In addition, there are laminae of fibrous calcite, whose morphology is also indicative of elevated salinity. This elevation of salinity was intermittent, for the environment was capable of supporting normal marine life (Ayres et al. 1982), and the sporadic presence of burrows and trails in the sediments indicates that anoxia, too, was not a permanent feature of the basin floor.

Our estimate of the nature of the sedimentological regime is similar to the one described above: marine water entered the embayment through a restricted portal, and evaporation produced elevated salinities. Periodically anoxia developed in the bottom waters, and organic matter raining down through the water column was preserved in fine carbonate muds. More than 100 m of source-rock facies, with total organic carbon exceeding 1%, accumulated in the Arabian Basin; in some sections the values exceeded 5%.

Kerogen in the source facies is predominantly hydrogen rich, well suited for generation of large amounts of petroleum. Calculations of maturation reveal that these source rocks started releasing their petroleum in the early Tertiary. The petroleum has characteristics of oils generated from carbonate sources: high sulfur content, low concentration of light naphthenic compounds, a pristane-to-phytane ratio of less than one, and a predominance of even-numbered normal alkanes (Palacas 1983). In this case, the carbonates were formed in an environment of elevated salinity induced by evaporitic conditions.

CONCLUDING REMARKS

We are satisfied that we have demonstrated the efficacy of the evaporitic hypersaline regime as a source of large volumes of petroleum. The environment is enormously productive biologically, and in the conditions most reasonable for the setting, a large amount of the or-

---

Figure 5.9
Stratigraphic column for the Arabian Basin (Ayres et al. 1982).

Time		Rock Unit		Lith.	Source Rock
Cretaceous	Maestrichtian	Aruma			
	—?— Campanian				
	—?— Santonian				
	—?— Coniacian				
	—?— Turonian	Wasia	Mishrif		
	—?— Cenomanian		Rumaila		
			Ahmadi		
	—?— Albian		Wara		
			Mauddud		
			Safaniya		
			Khafji		
	—?— Aptian	Thamama	Shu'aiba		
	—?— Barremian		Biyadh		
	—?— Hauterivian		Buwaib		
	—?— Valanginian		Yamama		
	—?— Berriasian		Sulaiy		//////
Jurassic	—?— Tithonian (Portlandian)	Hith			
		Arab	A		
			B		
	—?—		C		
			D		
	Kimmeridgian	Jubaila			
	—?—	Hanifa			//////
	Oxfordian				
	—?—	Tuwaiq Mountain			//////
	Callovian	Dhruma			
	—?—				//////
	Bathonian				
	—?—				//////
	Bajocian				
	—?—				
	Toarcian	Marrat			//////
	—?—				

ganic matter can be preserved, entombed in fine-grained inorgani-
cally precipitated carbonates. Thereafter, the inexorable process of
maturation guaranteed generation of oil of high quality. We have
described three basins where such a source, and such events, have
taken place, but by no means are those the only three. Others come
to mind after but a few moments recollection: in reefs of the Michigan Basin there are oils generated from the Silurian evaporitic carbonates; the Devonian Basin of western Canada includes evaporitic
sources; the Jurassic Smackover Formation contributed oil to traps
in interior basins of the Gulf Coast region, and that oil may well have
come from hypersaline facies therein; the Permian Zechstein of Europe is not to be excepted, nor the multiple evaporites in the Williston Basin.

The next step toward complete comprehension of evaporitic source
rocks must be the definition of the chemical characteristics of their
product, and a detailed evaluation of all the biological contributors
to those chemical features. The role of algal organic matter may be
understood, but the role of bacteria is only now being charted. After
the definition of the chemical characteristics, it will be possible to
identify the product before the source, and thereby evaluate the
stratigraphic sequence in any basin for the point of origin of the oil.
After that, an analysis of potential pathways of migration could reveal other prospects that by a different approach might have remained inviolable. Research in the petroleum industry, whether it be
into subjects as different as seismic stratigraphy or source beds, is
always aimed at the prospect, the ultimate target. This essay on evaporitic source beds was written with that same thought in mind.

REFERENCES

Arkhangel'skaya, N. A. and V. N. Grigor'yev. 1960. Formation of halogenic
zones in marine basins illustrated by the example of the Lower Cambrian
evaporite basin of the Siberian Platform. Trans. from *Izvestiya Acad. Sci.
USSR, Geol. Series* 4:41–54.
Abram, J. W. and D. B. Nedwell, 1978. Inhibition of methanogenesis by
sulfate-reducing bacteria competing for transfer hydrogen. *Arch. Microbiol.* 117:88–92.

Allen, R. P. 1956. *The Flamingoes: Their Life History and Survival.* Nat. Audubon Soc. Research Report no. 5. New York.

Anderson, R. Y. and D. W. Kirkland. 1960. Origin, varves, and cycles of the Jurassic Todilto Formation, New Mexico. *Am. Assoc. Petrol. Geol. Bull.* 44:37–52.

Ayres, M. G., M. Bilal, R. W. Jones, L. W. Slentz, M. Tartir, and A. O. Wilson. 1982. Hydrocarbon habitat in main producing areas, Saudi Arabia. *Am. Assoc. Petrol. Geol. Bull.* 66:1–9.

Baas-Becking, L. G. M. 1928. On organisms living in concentrated brine. *Tijdsch. Ned. Dierk. Ver.*, series 3, pp. 6–9.

Baas-Becking, L. G. M. 1931. Historic notes on salt and salt-manufacture. *Scientific Monthly* 32:434–446.

Bauld, John. 1984. Microbial mats in marginal marine environments: Shark Bay, Western Australia, and Spencer Gulf, South Australia. In Y. Cohen, R. W. Castenholz, and H. O. Halvorson, eds., *Microbial Mats: Stromatolites*, pp. 39–58. New York: Liss.

Beyth, M. 1980. Recent evolution and present stage of Dead Sea brines. In *Hypersaline Brines and Evaporitic Environments*, pp. 155–166. See Nissenbaum 1980.

Bordovskiy, O. K. 1965. Accumulation of organic matter in bottom sediments. *Marine Geology* 3:33–82.

Briggs, L. I., and H. N. Pollack. 1967. Digital model of evaporite sedimentation. *Science* 155:453–456.

Brock, T. D. 1979. Ecology of saline lakes. In M. Shilo, ed., *Strategies of Microbial Life in Extreme Environments*, pp. 29–47. Berlin: Dahlem Konferenzen.

Carpelan, L. H. 1957. Hydrobiology of the Alviso salt ponds. *Ecology* 38:375–390.

Caspers, H. 1957. Black Sea and Sea of Azov. In J. W. Hedgpeth, ed., *Ecology*, pp. 803–890. Vol. 1 of *Treatise on Marine Ecology and Paleoecology*. Geol. Soc. Amer. Mem. no. 67. New York.

Caspers, H. 1981. On the ecology of hypersaline lagoons on Laysan Atoll and Kauai Island, Hawaii, with special reference to the Laysan duck, *Anas laysanensis Rothschild. Hydrobiologia* 82:261–270.

Copeland, B. J. 1967. Environmental characteristics of hypersaline lagoons. *Contr. Marine Sci.* 12:207–218.

Cornée, A. 1983. Sur les bactéries des saumures des sédiments de marais salants méditerranéens: Importance et role sédimentologique. Documents du GRECO vol. 52, no. 3. Paris: Laboratoire de géologie du Museum.

Cornée, A. 1984. Etude préliminaire des bactéries des saumures et des sédiments des salins de Santa Pola (Espagne). Comparaison avec les marais salants de Salin-de-Giraud (Sud de la France). *Revista D'Investigacions Geologiques* 38/39:109–122.

Davies, G. R. 1970. Algal-laminated sediments, Gladstone Embayment, Shark

Bay, Western Australia. In *Carbonate Sedimentation and Environments, Shark Bay, Western Australia,* pp. 165–205. Am. Assoc. Petrol. Geol. Mem. no. 13. Tulsa, Okla.: AAPG.

Davis, J. S. 1973. The importance of microorganisms in solar salt production. In A. L. Coogan, ed., *Fourth Symposium on Salt,* 2:369–372. Cleveland: Northern Ohio Geol. Soc.

Dembicki, H., Jr., W. G. Meinschein, and D. E. Hattin. 1976. Possible ecological and environmental significance of the predominance of even-carbon number C20-C30 n-alkanes. *Geochim. et Cosmochim. Acta* 40:203–208.

Deuser, W. G. 1971. Organic-carbon budget of the Black Sea. *Deep-Sea Research* 18:995–1004.

Dietrich, G., K. Kalle, W. Krauss, and G. Siedler. 1980. *General Oceanography: An Introduction.* 2d ed. New York: Wiley.

Eardley, A. J. 1938. Sediments of Great Salt Lake, Utah. *Am. Assoc. Petrol. Geol. Bull.* 22:1305–1411.

Einsele, G. and F. Werner. 1972. Sedimentary processes at the entrance of the Gulf of Aden/Red Sea. *Geol. Jahrbuch* 10:39–62.

Eugster, H. P. 1985. Oil shales, evaporites, and ore deposits. *Geochim. et Cosmochim. Acta* 49:619–635.

Foree, E. G. and P. L. McCarty. 1970. Anaerobic decomposition of algae. *Environ. Sci. Technology* 4:842–849.

Friedman, G. M. 1985. Gulf of Elat (Aqaba): Geological and sedimentological framework. In *Hypersaline Ecosystems,* pp. 39–71. See Friedman and Krumbein 1985.

Friedman, G. M. and W. E. Krumbein, eds. 1985. *Hypersaline Ecosystems: The Gavish Sabkha.* New York: Springer-Verlag.

Gerdes, G. and W. E. Krumbein. 1984. Animal communities in Recent potential stromatolites of hypersaline origin. In Y. Cohen, R. W. Castenholz, and H. O. Halvorson, eds., *Microbial mats: Stromatolites,* pp. 59–82. New York: Liss.

Gerdes, G., Y. Spira, and C. Dimenta. 1985. The fauna of the Gavish Sabkha and the Solar Lake—A comparative study. In *Hypersaline Ecosystems,* pp. 322–345. See Friedman and Krumbein 1985.

Gibor, A. 1956. The culture of brine algae. *Biol. Bull.* 111:223–229.

Grabau, A. W. 1920. *Principles of Salt Deposition.* Vol. 1 of *Geology of the Non-metallic Mineral Deposits Other than Silicates.* New York: McGraw-Hill.

Gross, M. G. 1972. *Oceanography.* Englewood Cliffs, N.J.: Prentice-Hall.

Halbouty, M. T., A. A. Meyerhoff, R. E. King, R. H. Dott, Sr., D. H. Klemme, and T. Shabad. 1970. World's giant oil and gas fields, geologic factors affecting their formation, and basin classification. In M. T. Halbouty, ed., *Geology of Giant Petroleum Fields,* pp. 502–555. Am. Assoc. Petrol. Geol. Mem. no. 14, Tulsa, Okla.: AAPG.

Hammer, U. T. 1981. Primary production in saline lakes. *Hydrobiologia* 81:47–57.

Harshbarger, J. W., C. A. Repenning, and J. H. Irwin. 1957. *Stratigraphy of the Uppermost Triassic and Jurassic Rocks of the Navajo Country.* U.S. Geological Survey Prof. Paper no. 291. Washington, D.C.

Hedberg, H. D. 1967. Geologic controls of petroleum genesis. *Proceedings 7th World Petrol. Cong.* 2:3–11.

Hedgpeth, J. W. 1947. The Laguna Madre of Texas. 12th North Amer. Wildlife Cong. Trans., pp. 364–380.

Hedgpeth, J. W. 1957. Estuaries and lagoons, part 2: Biological aspects. In J. W. Hedgpeth, ed., *Ecology,* pp. 693–749. Vol. 1 of *Treatise on Marine Ecology and Paleoecology.* Geol. Soc. Amer. Mem. no. 67. New York.

Hildebrand, H. H. 1958. Estudios biologicos preliminares sobre la Laguna Madre de Tamaulipas. *Ciencia* 17:7–9.

Hirsch, P. 1980. Distribution and pure culture studies of morphologically distinct Solar Lake microorganisms. In *Hypersaline Brines and Evaporitic Environments,* pp. 41–60. See Nissenbaum 1980.

Hite, R. J. 1970. Shelf carbonate sedimentation controlled by salinity in the Paradox Basin, southeast Utah. In J. L. Rau and L. F. Dellwig, eds., *Third Symposium on Salt,* 1:48–66. Cleveland: Northern Ohio Geol. Soc.

Hite, R. J. and D. H. Buckner. 1981. Stratigraphic correlations, facies concepts, and cyclicity in Pennsylvanian rocks of the Paradox Basin. In D. L. Wiegand, ed., *Geology of the Paradox Basin,* pp. 147–159. Rocky Mount. Assoc. Geologists 1981 Field Conference. Denver.

Hite, R. J., D. E. Anders, and T. G. Ging. 1984. Organic-rich source rocks of Pennsylvanian age in the Paradox Basin of Utah and Colorado. In J. Woodward, F. F. Meissner, and J. L. Clayton, eds., *Hydrocarbon Source Rocks of the Greater Rocky Mountain Region,* pp. 255–274. Denver: Rocky Mount. Assoc. Geologists.

Hutchinson, G. E. 1967. *Introduction to Lake Biology and the Limnoplankton.* Vol. 2 of *A Treatise on Limnology.* New York: Wiley.

Javor, B. J. 1983. Nutrients and ecology of the Western Salt and Exportadora de Sal saltern brines. In B. C. Schreiber, and H. L. Harner, eds., *Sixth Symposium on Salt,* 1:195–206. Alexandria, Va.: Salt Institute.

Javor, B. J. and R. W. Castenholz. 1984. Productivity studies of microbial mats, Laguna Guerrero Negro, Mexico. In Y. Cohen, R. W. Castenholz, and H. O. Halvorson, eds., *Microbial Mats: Stromatolites,* pp. 149–170. New York: Liss.

Jones, R. W. 1984. Comparison of carbonate and shale source rocks. In J. G. Palacas, ed., *Petroleum Geochemistry and Source Rock Potential of Carbonate Rocks,* pp. 163–180. Am. Assoc. Petrol. Geol. Studies in Geol. no. 18. Tulsa, Okla.: AAPG.

Kaufmann, D. W., ed. 1960. *Sodium Chloride.* New York: Reinhold.

Kendall, C. G. St. C. and P. A. d'E. Skipwith. 1969. Recent algal mats of a Persian Gulf lagoon. *Jour. Sed. Pet.* 38:1040–1058.

Kinsman, D. J. J. 1969. Modes of formation, sedimentary associations, and diagnostic features of shallow-water and supratidal evaporites. *Am. Assoc. Petrol. Geol. Bull.* 53:830–840.

Kinsman, D. J. J., M. Boardman, and M. Borcsik. 1973. An experimental determination of the solubility of oxygen in marine brines. In A. Coogan, ed., *Fourth Symposium on Salt,* 1:325–327. Cleveland: Northern Ohio Geol. Soc.

Kirkland, D. W. and R. Evans. 1981. Source-rock potential of evaporitic environment. *Am. Assoc. Petrol. Geol. Bull.* 65:181–190.

Kirkland, D. W., J. P. Bradbury, and W. E. Dean. 1983. The heliothermic lake—A direct method of collecting and storing solar energy. *Arch. Hydrobiol./Suppl.* 65:1–60.

Larsen, H. 1980. Ecology of hypersaline environments. In *Hypersaline Brines and Evaporitic Environments,* pp. 23–39. See Nissenbaum 1980.

Logan, B. W. and D. E. Cebulski. 1970. Sedimentary environments of Shark Bay, Western Australia. In B. W. Logan, G. R. Davies, J. F. Read, and D. E. Cebulski, *Carbonate Sedimentation and Environments, Shark Bay, Western Australia,* pp. 1–37. Am. Assoc. Petrol. Geol. Mem. no. 13. Tulsa, Okla.: AAPG.

Lotze, F. 1957. *Steinsalz und Kalisalze.* Berlin: Gebruder Borntraeger.

Lucas, S. G., K. K. Kietzke, and A. P. Hunt. 1985. The Jurassic System in east-central New Mexico. In *New Mexico Geological Society Guidebook to the Santa Rosa–Tucumcari Region,* pp. 213–242. New Mexico Geological Society.

Malek-Aslani, M. 1980. Environmental and diagenetic controls of carbonate and evaporite source rocks. *Gulf Coast Assoc. Geol. Societies Trans.* 30:445–458.

Mason, D. T. 1968. Aspects of the limnology of Mono Lake, California (abs.). In *Symposium: Endorheic Lakes and Isolated Ocean Basins.* Am. Soc. Limnology and Oceanography, June 24–29. Logan, Utah.

Melack, J. M. 1981. Photosynthetic activity of phytoplankton in tropical African soda lakes. *Hydrobiologia* 81:71–85.

Moberg, E. G., D. M. Greenberg, R. Revelle, and E. C. Allen. 1934. The buffer mechanism of sea water. *Scripps Inst. Oceanography Tech. Ser. Bull.* 3:231–278.

Moody, J. D. 1959. Discussion. In Relationship of primary evaporites to oil accumulation. *5th World Petrol. Cong. Proc.,* section 1, pp. 134–138.

Murris, R. J. 1980. Middle East: Stratigraphic evolution and oil habitat. *Am. Assoc. Petrol. Geol. Bull.* 64:597–618.

Newell, N. D., J. K. Rigby, A. G. Fisher, A. J. Whiteman, J. E. Hickox, and J. S. Bradley. 1953. *The Permian Reef Complex of the Guadalupe Mountains Region, Texas and New Mexico.* San Francisco: Freeman.

Nissenbaum, A., ed. 1980. *Hypersaline Brines and Evaporitic Environments.* Vol. 28 of *Developments in Sedimentology.* New York: Elsevier.

Nissenbaum, A., M. J. Baedecker, and I. R. Kaplan. 1972. Organic geochemistry of Dead Sea sediments. *Geochim. et Cosmochim. Acta* 36:709–727.

Oehler, D. Z., J. H. Oehler, and A. J. Stewart. 1979. Algal fossils from a Late Precambrian hypersaline lagoon. *Science* 205:388–390.

Okla, S. M. 1986. Litho- and microfacies of Upper Jurassic carbonate rocks outcropping in central Saudi Arabia. *Jour. Petrol. Tech.* 9:195–206.

Orr, W. L. and A. G. Gaines. 1974. Observations on rate of sulfate reduction and organic matter oxidation in the bottom waters of an estuarine basin: The upper basin of the Pettaquamscutt River (Rhode Island). In B. Tissot and F. Bienner, eds., *Advances in Organic Geochemistry,* pp. 791–812. Paris: Tecnip.

Palacas, J. G. 1983. Carbonate rocks as sources of petroleum: Geological and chemical characteristics and oil-source correlations. *11th World Petrol. Cong. Proc.,* 2:31–43.

Parker, R. H. 1959. Macro-invertebrate assemblages of central Texas coastal bays and Laguna Madre. *Am. Assoc. Petrol. Geol. Bull.* 43:2100–2166.

Perrodon, M. A. 1968. Introduction a la géologie des évaporites. *Rev. Assoc. Franc. Tech. Pétrol.,* no. 190, pp. 21–33.

Peterson, J. A. and R. J. Hite. 1969. Pennsylvanian carbonate-evaporite cycles and their relation to petroleum occurrence, southern Rocky Mountains. *Am. Assoc. Petrol. Geol. Bull.* 53:884–908.

Phleger, F. B. and G. C. Ewing. 1962. Sedimentology and oceanography of coastal lagoons in Baja California, Mexico. *Geol. Soc. Amer. Bull.* 73:145–182.

Por, F. D. 1985. Anchialine pools—Comparative hydrobiology. In *Hypersaline Ecosystems.* pp. 136–144. See Friedman and Krumbein 1985.

Post, F. J. 1981. Microbiology of the Great Salt Lake north arm. *Hydrobiologia* 81:59–69.

Purser, B. H. 1985. Coastal evaporite systems. In *Hypersaline Ecosystems,* pp. 72–102. See Friedman and Krumbein 1985.

Quinn, B. G. 1966. Biology of the Great Salt Lake. In *The Great Salt Lake,* pp. 25–34. Utah Geol. Soc. Guidebook—Geology of Utah, no. 20. Salt Lake City: Utah Geol. Soc.

Raymont, J. E. G. 1963. *Plankton and Productivity in the Oceans.* New York: Macmillan.

Reeburg, W. S. 1983. Rates of biogeochemical processes in anoxic sediments. *Ann. Rev. Earth Planet. Sci. 1983* 11:269–298.

Richardson, M., M. A. Arthur, and B. J. Katz. 1986. Miocene synrift evaporites of the Red Sea: Their deposition and hydrocarbon source potential (abs.). *Am. Assoc. Petrol. Geol. Bull.* 70:638–639.

Ridgley, J. L. and M. Goldhaber. 1983. Isotopic evidence for a marine origin of the Todilto Limestone (abs.). *Abstracts with Program, 36th Ann. Geol. Soc. Amer. Rocky Mtn. Section and 79th Ann. Cordilleran Section Meeting* 15(5):414.

Sassen, R., J. A. Nunn, C. H. Moore, F. C. Meendsen, and V. K. Iliff. 1986. Models for hydrocarbon generation and destruction, eastern Jurassic Smackover Formation. *Am. Assoc. Petrol. Geol. Bull.* 70:644.

Simmons, E. G. 1957. An ecological survey of the upper Laguna Madre. *Inst. Marine Sci. Pub.* 4:156–200.

Sonnenfeld, P. 1984. *Brines and Evaporites.* New York: Academic Press.

Sonnenfeld, P. 1985. Evaporites as oil and gas source rocks. *Jour. Petrol. Geol.* 8:253–271.

Strakhov, N. M. 1970. *Principles of Lithogenesis.* New York: Plenum Press.

Ten Haven, H. L., J. W. de Leeuw, and P. A. Schenk. 1985. Organic geochemical studies of a Messinian evaporite basin, northern Apennines (Italy), part 1: Hydrocarbon biological markers for a hypersaline environment. *Geochim. et Cosmochim. Acta* 49:2181–2191.

Thomas, M. 1982. Approche géochimique du systéme sédimentaire des marias salants de Salin-de-Giraud (Sud de la France). *Géol. Médit.* 9 (4):487–500.

Thomas, J. C. and D. Geisler. 1982. Peuplements benthiques a Cyanophycées des marais salants de Salin-de-Giraud (Sud de la France). *Géol. Médit.* 9 (4):391–412.

Tissot, B. P. 1981. Generation of petroleum in carbonate rocks and shales of marine or lacustrine facies and its geochemical characteristics. In J. F. Mason, ed., *Petroleum Geology in China,* pp. 71–82. Tulsa, Okla.: Pennwell Books.

Tissot, B. P. and D. H. Welte. 1978. *Petroleum Formation and Occurrence.* New York: Springer-Verlag.

Tissot, B. P. and D. H. Welte. 1984. *Petroleum Formation and Occurrence.* 2nd ed. New York: Springer-Verlag.

Vincelette, R. R. and W. E. Chittum. 1981. Exploration for oil accumulations in Entrada Sandstone, San Juan Basin, New Mexico. *Am. Assoc. Petrol. Geol. Bull.* 65:2546–2570.

Warren, J. K. 1986. Shallow-water evaporitic environments and their source rock potential. *Jour. Sed. Pet.* 56:442–454.

Weeks, L. G. 1961. Origin, migration, and occurrence of petroleum. In G. B. Moody, ed., *Petroleum Exploration Handbook,* pp. 5-1 to 5-50. New York: McGraw-Hill.

Wetzel, R. G. 1975. *Limnology.* Philadelphia: Saunders.

Williams, W. D. 1981. Inland salt lakes: An introduction. *Hydrobiologia* 81:1–14.

Woolnough, W. G. 1937. Sedimentation in barred basins, and source rocks of oil. *Am. Assoc. Petrol. Geol. Bull.* 21:1101–1157.

Zahl, P. A. 1967. Life in a "dead" sea—Great Salt Lake. *Nat. Geog. Mag.* 132:252–263.
Zenkevitch, L. 1963. *Biology of the Seas of the U.S.S.R.* New York: Interscience.

6

Applications of Stable Isotope Geochemistry to the Study of Evaporites

CATHERINE PIERRE

Stable isotopes are very subtle intrinsic markers of the external parameters of a system; application of stable isotope geochemistry to sedimentological studies of evaporites provides specific information on the ambient conditions of deposition and diagenesis, as illustrated by the literature on this subject. As a result, the study of stable isotopes is undoubtedly essential to the solution of many sedimentary problems in evaporite systems. It remains fundamental, however, that prior to any geochemical investigation, the material must be carefully studied along many lines of investigation, for petrography, mineralogy, biological content, and so forth, to precisely define and know its sedimentological framework.

Stable isotope geochemistry can be useful in the recognition of the ambient parameters of naturally occurring mineral-water sedimen-

Catherine Pierre is a Researcher in the Department of Dynamic Geology, Université Pierre et Marie Curie, Paris.

tary systems. One of the characteristics of evaporite environments is the large variability in the free solutions of ambient parameters— salinity, temperature, pH, Eh, dissolved oxygen content—which are controlled primarily by the mass-balance between the solution input and the water loss by evaporation, and by the biological activity (mainly algae and bacteria). In response to these external modifications, chemical and isotopic compositions of the solutions also change, and reflect the geochemical picture of the evaporite system. Another peculiar characteristic of evaporite settings is that the interstitial environment of sedimentary deposits often becomes the place of early diagenetic reactions, where vertical or lateral chemical gradients are existent in the pore fluids. Such situations are created either when allochthonous solutions percolate through the sediments or when *in situ* bacterial activity occurs. Similarly, rock-solution interactions during late diagenesis of evaporites are induced by migration of fluids throughout the sediments. These may occur over a range of low to high temperature-pressure conditions, depending on whether the sediments are uplifted or buried after deposition.

The following article presents general considerations concerning the isotope geochemistry of water-carbonate-sulfate that apply in both terrestrial and marine evaporitic systems. This presentation will show why stable isotope measurements are often the necessary complement to classical sedimentological studies in present-day and fossil evaporites. The article briefly reviews key reactions and factors critical to the use of stable isotope geochemistry in the evaluation of evaporites and discusses the interactive systems related to evaporite deposition and diagenesis; in its final section, specific examples of selected studies are given to illustrate the utility of isotope geochemistry in evaporative sedimentary problems.

STABLE ISOTOPES IN THE WATER-CARBONATE-SULFATE SYSTEMS: NATURAL VARIATIONS AT THE EARTH'S SURFACE

The Isotopic Reactions

Two types of reactions are responsible for the variations in isotopic compositions in natural compounds: chemical reactions, where the starting and the end products have different chemistry and between which there is an exchange of isotopes; and pure isotopic exchange reactions, where the chemical system remains unchanged.

Equilibrium and Kinetic Isotopic Reactions. Each isotopic reaction must satisfy the Law of Mass Action: at equilibrium (chemical and isotopic), a constant α, called the *isotopic fractionation factor,* relates the isotopic concentrations of the products to those of the reactants for a given temperature.

However, several isotopic reactions within the earth's surface environment may occur out of equilibrium; thus isotopic transfers between the components of the reaction are mainly controlled by kinetic effects. For such reactions, the α value at any temperature is never constant and depends on constraints external to the system. For instance, in the biogeochemical reactions which are catalyzed and rate-controlled by the enzymatic activity of microorganisms, the isotope fractionations between initial and end products vary as a function of environmental conditions, such as quantity and quality of nutritive substrate and competition with other microorganisms (which may produce poisonous compounds).

The Isotopic Fractionation Factor; The Isotopic Enrichment Factor. The general equation between α and temperature is given as the form: $1{,}000 \ln \alpha = A\, 10^6\, T^{-2} + B$, where A and B are constants, and temperature T is given in degrees Kelvin. This shows that α is a reverse function of temperature, i.e., that the isotopic fractionation between the product and the reactant of a reaction increases when the temperature decreases.

Using the δ notation for the isotopic compositions of the reactant A and product B of the reaction $A \rightleftharpoons B$ gives

$$\alpha_{A-B} = \frac{1{,}000 + \delta_A}{1{,}000 + \delta_B}$$

where

$$\delta = \left(\frac{R\ \text{sample}}{R\ \text{reference}} - 1\right)\cdot 10^3$$

and where

$$R = \frac{\text{heavy isotopes}}{\text{light isotopes}}$$

SMOW is used for oxygen and hydrogen compounds (Craig 1961a); PDB for oxygen and carbon compounds (Craig 1957); and C.D.T. for sulfur compounds (Jensen and Nakai 1962).

The isotopic enrichment factor $\epsilon = (\alpha - 1) \cdot 10^3$ is also commonly used in its simplified algebraic form.

It is easy to demonstrate that $1{,}000 \ln \alpha_{A-B} \cong \delta_A - \delta_B \cong \epsilon_{A-B}$

The Reservoir Effect: Isotopic Behavior in Open and Closed Systems. The reservoir effect describes the special case of isotopic evolution in small-sized reactant reservoirs which diminish continuously as the reaction proceeds. The Rayleigh's distillation equation gives, at any time, the isotopic composition of the reactant as a function of the residual fraction of the reactant reservoir:

$$\delta - \delta_0 = -\epsilon \ln f$$

In this equation, δ and δ_0 are respectively the isotopic compositions of the reactant at t and $t = 0$ (starting reactant); ϵ is the isotopic enrichment factor of the reaction; f is the residual fraction of reactant at time t.

The isotopic behavior of the reactant and product reservoirs is illustrated in figure 6.1. *Open system conditions* mean that the instantaneous product of the reaction escapes from the system: there, at any time, the isotopic difference between the reactant and the product is constant and equal to ϵ. *Closed system conditions* mean that the product accumulates until all the reactant has reacted: there, at any time, the isotopic difference between the reactant and the product changes, and the final isotopic composition of the accumulated product is equal to the initial isotopic composition of the reactant.

Numerous geochemical and biogeochemical reactions are subjected to reservoir effects: the partial condensation of clouds to form rain; the partial bacterial reduction of sulfate from pore waters; and fractionated crystallization of minerals, to name a few.

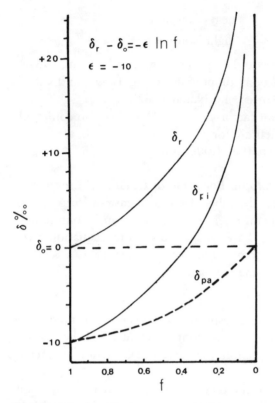

Figure 6.1
Rayleigh's distillation in open and closed systems. In both the cases, at each time, the isotope composition of the reactant reservoir is δ_r. In open system conditions, the isotope composition of the instantaneous product is $\delta_{p,i}$. In closed system conditions, the isotope composition of the accumulated product is $\delta_{p,a}$. When the remaining fraction f of reactant is zero, the isotope composition of the accumulated product is equal to the isotope composition of the starting reactant δ_0.

The Water System

In any geochemical or biogeochemical reaction that takes place at earth surface temperatures, water is always present. Water is fundamental because, even when not directly involved in the reaction, it acts as a link between the reactants and the products. Water serves to transport solid, gaseous, and ionic species, and commonly reacts with these species; water is also necessary to the life functions of organisms; finally, liquid water is the medium in which most exchanges of energy during chemical and biochemical reactions occur.

The isotopic composition of surface waters depends mainly on the evaporation-condensation processes during which the changes of physical state of water occur. Other isotope variations in water may arise when a small water reservoir exchanges isotopes with a larger reservoir of gas or rock.

The Evaporation-Condensation Cycle of Water. Ocean water represents the major constituent ($\approx 98\%$) of the hydrosphere, and was chosen as the reference SMOW (Standard Mean Ocean Water; Craig 1961a) for δ measurements of waters and other oxygen and hydrogen compounds.

The δ value of mean seawater is therefore equal to zero on the SMOW scale. Because the water vapor pressure is larger for $H_2^{16}O$ molecules than for $HD^{16}O$ and $H_2^{18}O$ molecules, the atmospheric water vapor, which is generated above the oceans, is depleted in heavy isotopes (i.e., δ < 0) relative to the seawater reservoir. Consequently, precipitation, which originates from the condensation of atmospheric water vapor, generally has negative δ values.

The most important δ variations in atmospheric precipitation are related to temperature effects (i.e., altitude and latitude effects) because the isotopic fractionation factor between the condensed phases (liquid and solid water) and the vapor increases as temperature drops (Merlivat 1978); as a result, the more ^{18}O-^2H–depleted precipitation in the world is centered in the high-latitude regions (Yurtsever 1975; Craig and Gordon 1965).

Other δ variations in precipitation are attributed to the continental effect, which corresponds to the progressive inland decrease of δ values in meteoric waters. A typical case of reservoir effect is illustrated where the fractionated condensation of clouds that advance above continents yields precipitation more and more depleted in heavy isotopes relative to the initial reservoir of atmospheric water vapor.

On a worldwide scale, the oxygen and hydrogen isotopic composition of meteoric waters obeys a general law of the form: $\delta^2H = s\,\delta^{18}O + d$ (Craig 1961b; Dansgaard 1964; Fontes 1976; Craig and

Gordon 1965). This results from the fact that condensation occurs at equilibrium when water vapor is saturated. In most oceanic areas, the equation expressing this condition is: $\delta^2H = 8\ \delta^{18}O + 10$.

However, the parameters s and d may vary locally due to kinetic effects. The value of s increases when the relative humidity and/or the temperature of ocean's surface increases. The parameter d, called *deuterium excess,* depends on the relative air humidity, which is close to zero for 100% humidity and up to 30 for 50% humidity (Merlivat and Jouzel 1979). In the water cycle, evaporation is the counterpart of condensation. As mentioned previously, during water evaporation, light water molecules preferentially leave the liquid to form vapor; by mass-balance effect, the remaining fraction of water becomes enriched in heavy isotopes relative to the initial water body. Evaporated marine waters thus have positive δ values, while evaporated continental waters may exhibit negative to positive δ values depending on their initial δ values and on their degree of evaporation.

Experimental and natural data show how complex the isotopic behavior of water is during the evaporation process. Because water evaporation occurs when the relative air humidity is lower than 100%, it is basically a nonequilibrium process. This implies large kinetic effects that are due to the molecular diffusion of the water's isotopic species from the water surface towards the air (Merlivat 1978). The ratio of ^2H and ^{18}O enrichment factors in evaporated waters becomes lower than the equilibrium value of 8; it commonly ranges between 6 and 3 and depends on the relative air humidity (Craig et al. 1963).

Two other factors act during evaporation to limit the isotopic enrichment in water. The isotopic exchange between water and vapor occurs at any time at the liquid-air boundary until the system reaches steady state (Craig et al. 1963; Craig and Gordon 1965).

The presence of dissolved salts in water lowers the water activity of the solution. That is, the tendency of a chemical species to react is limited, due to the presence of an elevated concentration of various ions. The coupling of these factors explains the differences in the isotopic evolution pattern of evaporated salt solutions that have various chemistries and are affected by different ambient conditions during evaporation (figure 6.2). This has two consequences in the isotopic evolution of saline solutions subjected to evaporation. Decreasing the thermodynamic activity of water also means that the activity of

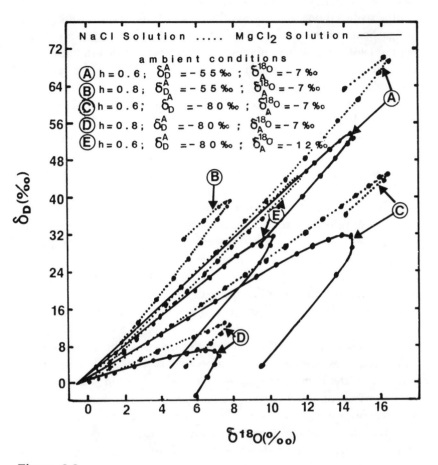

Figure 6.2
Evaporation curves for NaCl and MgCl$_2$ brines on δD^c vs $\delta^{18}O^a$ plots
(D is deuterium, c is concentration, a is activity). Initial conditions
and parameters for these calculations: salt concentration = 0.5 mo-
lal; $\delta_{L,0}{}^c$ = 0.0‰ for both D and ^{18}O (L = bulk liquid); ρ/ρ_{18} =
ρ/ρ_D = 0.982; $\rho_{i,L}/\rho_D$ = $\rho_{i,L}/\rho_{18}$ = 0.19; $\alpha^{18}O$ = 0.9908; αD = 0.9269
(ρ = viscosity, i.e., salinity; $\alpha^{18}O$ and αD = oxygen and hydrogen
isotope fractionation factors between vapor and pure water). Dashes
on curves signify a decrement of 0.05 in the fraction of residual water
(Sofer and Gat 1975).

heavy and light water molecules decreases. Maximum isotopic enrichments in saline solutions are thus lower than in pure water, other things being equal (Sofer and Gat 1972, 1975; Gonfiantini 1966; Fontes 1966, 1976). The highest $\delta^{18}O$ values recorded in natural environments have been measured in extreme arid regions, in a marine pond along the Gulf of Elat ($\delta^{18}O = 11‰$ in the Solar Lake; Aharon et al. 1977), and in an endoreic lake of the Sahara ($\delta^{18}O = 30‰$; Fontes and Gonfiantini 1967a).

Furthermore, hydration of ions, by substracting water molecules from the solution, introduces additional isotopic fractionation effects in the water that are specific to each ionic species (Götz and Heinzinger 1973; Sofer and Gat 1972, 1975).

Generally, the isotopic composition of water in saline solution subjected to evaporation shows a typical two-step evolution. During the first step, the solution is subjected only to the isotope exchange with the vapor: δ values in water increase progressively, up to the stage where salts begin to precipitate, so that a pseudo–steady state is reached. During the second step, both the decrease of water activity and the hydration of ions act to progressively lower the δ values in concentrating brines (figure 6.2). Reported on a δ^2H-$\delta^{18}O$ diagram, the δ values of evaporating saline solutions describe two curves of different slopes, connected by a loop that is more or less open, depending on the nature of ions present in the solution (figure 6.3).

The shapes of these curves are slightly modified, however, when the δ values are expressed either on an activity scale or on a concentration scale (Sofer and Gat 1975). The δ values reported on an activity scale represent the isotope compositions of the free water of the solutions, while the δ values reported on a concentration scale represent the isotope compositions of the entire water (i.e., free water plus the hydration water of ions) of the solutions. Therefore, the expression of water δ values on an activity scale is more convenient for studies on the evaporation process, while the expression of water δ values on a concentration scale is better for evaluating eventual mixing of solutions.

Simple equations for converting δ concentration to δ activity scales have been obtained experimentally by Sofer and Gat (1972, 1975) using solutions of known molar concentrations.

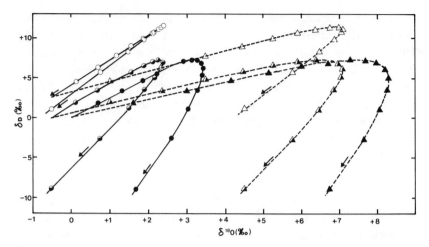

Figure 6.3
The isotopic content of an evaporating $MgCl_2$ solution, plotted in different δ scales (given in SMOW): open symbols = δD activity vs $\delta^{18}O$ activity; half-shaded symbols = δD concentration vs $\delta^{18}O$ activity; filled symbols = δD concentration vs $\delta^{18}O$ concentration. Initial conditions and parameters as in figure 6.2. Dashed line curves: isotope composition of vapor, $\delta D_A = -80.0‰$; $\delta^{18}O_A = -7‰$; relative air humidity, $h = 0.8$. Solid lines: isotope composition of vapor, $\delta D_A = -80.0‰$; $\delta^{18}O_A = -12.0‰$; relative air humidity, $h = 0.8$. Spacing of symbols on the curves corresponds to a decrement of the residual water fraction by 0.05 (Sofer and Gat 1975).

The term $\qquad \Delta\delta = \dfrac{\delta_c - \delta_a}{\delta_a + 10^3} \cdot 10^3$

Concentrations are given in molalities (i.e., $m.kg^{-1}$ water):

$\Delta\delta^{18}O = 1.11\, m_{Mg^{2+}} + 0.47\, m_{Ca^{2+}} - 0.16\, m_{K^+}$

$\Delta\delta\,^2H = -6.1\, m_{Ca^{2+}} - 5.1\, m_{Mg^{2+}} - 2.4\, m_{K^+} - 0.4\, m_{Na^+}$

Special mention has to be given to the isotope behavior of water that is incorporated in the crystal lattice of hydrated minerals. For gypsum, crystallized in the temperature range between 17°C and 57°C, experimental and natural data have shown that the crystal water is enriched in ^{18}O by 4‰ and depleted in ^2H by 15‰ to 20‰, relative to the mother water where crystallization has occurred (Fontes 1965; Fontes and Gonfiantini 1967b; Gonfiantini and Fontes 1963; Sofer 1978). It is therefore possible to reach the isotope compositions of parent solutions of primary gypsum deposits that remain protected from diagenetic alteration. However, gypsum is often subject to diagenetic recrystallization, as when the sediment is exposed to solutions alternatively diluted and concentrated, or when the passage of gypsum to an anhydrite phase (due to high temperature or high salinity conditions) is followed by rehydration with surrounding solutions.

Gas-Water and Mineral-Water Isotope Exchange Reactions. Isotope exchange reactions, between water and gas or minerals of the rock matrix, occur in closed systems where the water/gas or water/rock mass-ratio becomes low. These reactions are greatly enhanced when the thermodynamic equilibrium conditions of the gas-mineral-water system change so that minerals may dissolve or crystallize and exchange isotopes with the surrounding fluids. The pressure-temperature increase during burial is commonly evoked to account for the increasing water $\delta^{18}O$ values—decreasing silicate and carbonate $\delta^{18}O$ values with increasing depth (figure 6.4). Similar $\delta^{18}O$ increases in groundwaters are measured in geothermal areas and are due to high-temperature water-rock exchanges (Gonfiantini et al. 1973).

Oxygen isotopic depletion and hydrogen isotopic enrichment in groundwaters are attributed respectively to exchanges of waters with large amounts of CO_2 and H_2S which are products either of volcanic activity (CO_2 and H_2S) or of bacterial sulfate reduction (H_2S).

Knowledge of the isotopic characteristics of these modified waters is fundamental in the reconstruction of the diagenetic history of sedimentary deposits that have been deeply buried or have encountered hydrothermal conditions.

The Carbonate System

The chemical reactions describing the equilibrium of the carbonate, and for which the equilibrium constants are temperature and pressure-dependent, are the following:

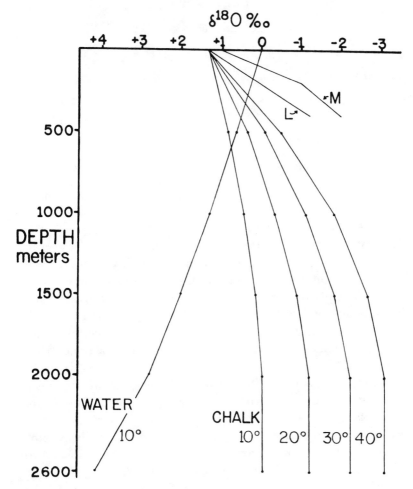

Figure 6.4
Variation in the $\delta^{18}O$ of chalk (on the PDB scale) with depth for four geothermal gradients (in °C/Km, as calculated to result from compaction-induced burial diagenesis in seawater). Variation in the $\delta^{18}O$ of water which remains after compaction (on the SMOW scale) for a geothermal gradient of 10°C/Km is also shown. Natural examples obtained by JOIDES (line L from Lawrence 1973; line M from Matter et al. 1975) undergo more rapid chemical change, which must be caused by fluids other than modified interstitial (connate) water (Land 1980).

$$CO_2 + H_2O \rightleftharpoons H_2CO_3$$

$$H_2CO_3 \rightleftharpoons HCO_3^- + H^+$$

$$HCO_3^- \rightleftharpoons CO_3^{2-} + H^+$$

For pH lower than 9.0, alkalinity is mainly due to the presence of HCO_3^- ions.

When pH becomes higher than 9.0, the ions CO_3^{2-} are dominant:

$$CO_2 + OH^- \rightleftharpoons HCO_3^-$$

$$HCO_3^- + OH^- \rightleftharpoons CO_3^{2-} + H_2O$$

The dissolution and precipitation of carbonates from aqueous solutions at earth surface temperatures are thus basically dependent on pH and CO_2 partial pressure.

Normal seawater is a buffered solution, characterized by a rather constant pH value, close to 8.1. In contrast, fresh waters are rarely buffered and exhibit a wide range of pH values. Furthermore, in saline solutions, pH decreases in response to the strong hydration of ions (especially of the divalent ions such as CO_3^{2-}) and to the decrease of water activity (Krumgalz 1980). For instance, pH values close to 7.0 are often measured in halite-saturated marine brines (Pierre 1982).

In natural environments, the activity of living aquatic organisms and microorganisms either yields CO_2 by respiration and by organic matter decay, or uses CO_2 for photosynthesis. The resulting variations in partial pressure of CO_2 are responsible respectively either for lowering the pH or for increasing the pH of the solutions. Variations of pH and CO_2 partial pressure are readily detected in the pore solutions of organic-rich sediments where CO_2 is either released during bacterial sulfate reduction or consumed during methanogenesis (Nissenbaum et al. 1972). Other significant changes in the CO_2 partial pressure of the solutions are due to evaporation that takes place either in the surficial waters of a basin or in pore waters of the sediments above the water table (i.e., in the vadose zone).

It is during the dissolution-precipitation of carbonates that oxygen and carbon isotopes are transferred between the different reservoirs, water, $\Sigma\ CO_2$ (total dissolved inorganic carbon), and solid carbonate. The kinetics of these isotopic reactions are often sufficiently rapid, relative to the crystallization rate, to ensure isotopic equilibrium between the aqueous and solid components. Nevertheless, in saline solutions, the rates of isotopic exchange reactions are considerably reduced (Sofer and Gat, 1972); kinetic isotope effects may thus arise in brines, especially when carbonate crystallization results from a rapid supersaturation of the solution. Other kinetic isotope fractionation effects take place during the production of biogenic carbonates, and are mainly due to the metabolism of organisms.

The Oxygen Isotope Behavior of Carbonates. The oxygen isotope fractionation factor α between a solid carbonate and water depends on the nature of the cations which enter the crystal lattice: at the same temperature, the larger the cation, the lower the α value (table 6.1). Because for each carbonate mineral, α depends also on temperature, it is therefore possible to relate the $\delta^{18}O$ values of carbonate and water with the temperature of crystallization, if the oxygen isotope equilibrium is realized.

The carbonate-water, oxygen isotope temperature scale established by Epstein et al. (1953) and revised by Craig (1965) for earth surface temperatures is a simplified equation that relates the three variables $\delta^{18}O_{water}$ – temperature – $\delta^{18}O_{CaCO_3}$:

Table 6.1
Oxygen Isotope Fractionation Factors of Some Carbonate-Water Equilibrium Reactions

Mineral	*1,000 ln* $\alpha_{carb\text{-}water}$	$\alpha^{18}O_{25°C}$
Witherite		
$BaCO_3$	$2.57 \cdot 10^6 T^{-2} - 4.23$	1.0250^a
Strontianite		
$SrCO_3$	$2.69 \cdot 10^6 T^{-2} - 3.24$	1.0274^a
Calcite	$2.78 \cdot 10^6 T^{-2} - 2.89$	
$CaCO_3$		1.02882^a
Dolomite	$2.62 \cdot 10^6 T^{-2} - 2.17$	1.03218^b
$Ca,Mg(CO_3)_2$	$3.06 \cdot 10^6 T^{-2} - 3.24$	1.03171^c
	$3.20 \cdot 10^6 T^{-2} - 1.5$	1.03510^d

[a]O'Neil et al. 1969   [b]Fritz and Smith 1970   [c]Matthews and Katz 1977   [d]Northrop and Clayton 1966.

$$t = 16.9 - 4.2 \ (\delta^{18}O_{CaCO_3} - \delta^{18}O_{water}) + 0.13 \ (\delta^{18}O_{CaCO_3} - \delta^{18}O_{water})^2$$

In this equation, temperature is given in degrees Celsius, and the $\delta^{18}O$ values of water and carbonate are referred to the same reference (PDB). Therefore, the present-day marine $CaCO_3$ minerals have $\delta^{18}O$ values close to zero (PDB) when crystallization occurs near 17°C. Each increase of 1 $\delta^{18}O$ unit for $CaCO_3$ corresponds either to the same 1 $\delta^{18}O$ increase for water, or to a decrease in temperature of about 4.2°C. The reverse applies to decreasing $\delta^{18}O$ values of $CaCO_3$.

It is thus expected that $CaCO_3$ minerals precipitated in progressively evaporated waters will exhibit increasing $\delta^{18}O$ values. Conversely, crystallization or diagenesis in meteoric waters will produce carbonates depleted in ^{18}O relative to the marine reference.

The discrepancy that appears between different authors for the $\alpha \ ^{18}O_{dolomite-water}$ experimental values is primarily due to the impossibility of crystallizing true dolomite in laboratory conditions at low temperatures (except poorly ordered assemblages, such as the "protodolomites" synthesized by Fritz and Smith 1970 in the 25°–79°C range). However, the experimental data of Fritz and Smith (1970), as well as the extrapolations at low temperatures of the high-temperature data of Northrop and Clayton (1966) and Matthews and Katz (1977) show that dolomite might be enriched in ^{18}O by a few parts per thousand (3‰ to 6‰ at sedimentary temperatures) relative to a cogenetic calcite. Numerous sedimentological and isotopic studies have dealt with the problem of sedimentary dolomites, and many of them in evaporitic environments (Persian Gulf sabkhas, Coorong lagoons of Australia, Deep Springs Lake of California, lagoons of Baja California).

A review of the dolomite question was given recently by Land (1980). Following him, two major points, at least, have to be emphasized concerning the isotope geochemistry of dolomite. First, dolomitization is a dissolution-precipitation process during which the carbonate ions (originating eventually from a dissolved $CaCO_3$ precursor) equilibrate at the ambient temperature with water for ^{18}O and $\Sigma \ CO_2$ for ^{13}C; this means that if coexisting calcium carbonate and dolomite are not strictly cogenetic, the differences between their $\delta^{18}O$ and $\delta^{13}C$ values cannot be identified with the $\epsilon^{18}O$ and $\epsilon^{13}C$ values between dolomite and calcite. This especially has to be the case for carbonate crystallization within evaporitic settings that are affected by large variations of ambient parameters even during early stages of diagenesis.

Second, it is possible that the $\alpha^{18}O$ dolomite-water value is not constant at a given low temperature, due to an additional and variable kinetic isotope fractionation effect during crystallization. Studies of the crystallochemistry of dolomite have shown that kinetic barriers control the rate of nucleation of cation-ordered dolomite. The effect of the barrier of hydration of Mg^{2+} ions may be diminished when strongly solvated ions (such as Li^+) are also present in the solutions (Gaines 1974, 1980). Although dolomite formation requires solutions with a molar ratio $Mg^{2+}/Ca^{2+} > 1$, the Ca^{2+} concentration needs to be low because this ion preferentially occupies the Mg^{2+} sites in the dolomite crystal lattice (Lippmann 1973). The competition of crystal growth between calcite and dolomite may be suppressed when either high alkalinities or high SO_4^{2-} concentrations in the initial solutions lower the Ca^{2+} concentrations in the resulting dolomitizing solution, by precipitation of $CaCO_3$ or $CaSO_4$ minerals (Lippmann 1973). This explains why in many evaporitic successions, dolomite is commonly associated with gypsum and anhydrite.

The Carbon Isotope Behavior of Carbonates. The carbon isotope composition of carbonates is directly related to the $\delta^{13}C$ value of the total dissolved inorganic carbon reservoir ($\Sigma\, CO_2 = CO_2 + HCO_3^-$ $+ CO_3^{2-} + H_2CO_3$) of the solution where crystallization takes place. The temperature changes modify only slightly the $\delta^{13}C$ values of carbonate minerals (0.04‰ by °C). However, different ^{13}C fractionations occur during crystallization; for instance, at 25°C, aragonite is enriched by 1.8‰ relative to calcite (Rubinson and Clayton 1969), and dolomite is heavier by about 2‰ relative to cogenetic calcite, as predicted from high-temperature extrapolations (Sheppard and Schwarcz 1970).

The reservoir of dissolved inorganic carbon in any solution may originate from several sources: dissolution of atmospheric CO_2, dissolution of preexisting carbonate rocks, or as a product of the weathering (by atmospheric CO_2) of silicate rocks, oxidation of organic compounds, and dissolution of the CO_2 produced during respiration of living organisms. Differences in the mass-balance between these components and the eventual output of carbon from the aqueous system (degassing of CO_2 by evaporation, photosynthesis and production of organic matter, carbonate precipitation) are responsible for the large variations measured in the carbon isotope composition of ΣCO_2 in natural waters.

Owing to the different carbon isotope fractionations that occur between the components of the carbonate system (table 6.2), the $\delta^{13}C$ value of the reservoir of total dissolved inorganic carbon equilibrated with recent atmospheric CO_2 ($\delta^{13}C = -6.4‰$ vs PDB for the preindustrial period; Craig and Keeling 1963) is close to $+2‰$ ($\pm 1‰$ due to local temperature variations) at earth surface temperatures. This δ value corresponds to the average measured in surficial ocean waters.

The input and output to the ΣCO_2 reservoir of the products of organic activity that are highly ^{13}C-depleted (Craig 1953) account for the most changes of the $\delta^{13}C$ values of the aqueous inorganic carbon reservoir. Large supplies of organically derived CO_2 lower the $\delta^{13}C$ value of the ΣCO_2 reservoir. Such situations are common in ocean-bottom waters due to oxidation of organic matter at depth (Kroopnick 1974; Kroopnick et al. 1970), and in many rivers, lakes, and groundwaters, where CO_2 originates mainly from degradation products of terrestrial soils (Deines 1980; Buchardt and Fritz 1980). The contribution of CO_2 of organic origin becomes significant in the pore

Table 6.2
Carbon Isotope Enrichment Factors of the Individual Equilibrium Reaction in the Carbonate System

Reaction A − B	ϵ_{A-B}	$\epsilon 25°C$
CO_2 gas − CO_2 aqueous	$373\ T^{-1} - 0.19$	1.06^a
CO_2 gas − HCO_3^-	$-9,483\ T^{-1} + 23.89$	-7.93^b
CO_2 aqueous − HCO_3^-	$-9,866\ T^{-1} + 24.12$	-8.99^b
CO_2 gas − CO_3^{2-}	$-9,037\ T^{-1} + 22.73$	-7.59^c
$CaCO_3$ − HCO_3^-	$-4,232\ T^{-1} + 15.10$	0.90^d
CO_2 gas − calcite	$-2.988 \cdot 10^6\ T^{-2}$ $+7.6663 \cdot 10^3\ T^{-1}$ -2.4612	-10.38^e

[a]Vogel et al. 1970 [b]Mook et al. 1974 [c]Thode et al. 1965 [d]Rubinson and Clayton 1969 [e]Bottinga 1969.

waters of organic-rich sediments where anaerobic conditions favor the activity of sulfate-reducing bacteria and methanogenic bacteria. The complete set of chemical reactions that are involved during organic matter diagenesis has been synthesized by Irwin et al. (1977). The general expression of bacterial sulfate reduction,

$$2 \, CH_2O + CaSO_4 \dashrightarrow CaCO_3 + CO_2 + H_2S + H_2O$$

shows that carbonate and sulfide are the by-products of the reaction. The carbonate formed during this process is characterized by low $\delta^{13}C$ values which reflect the organic origin of the carbon. A few examples of these diagenetic carbonates, which exhibit $\delta^{13}C$ values down to $-25‰$, have been described in evaporitic formations at peculiar levels interbedded with bituminous layers or in the vicinity of oil fields (Dessau et al., 1959, 1962; Feely and Kulp 1957; Kirkland and Evans 1976; Pierre 1974; Pierre and Rouchy 1985, 1986a, 1988; McKenzie 1985). Light hydrocarbons, such as methane, may also serve as substrate to sulfate-reducing bacteria:

$$CH_4 + CaSO_4 \dashrightarrow CaCO_3 + H_2S + H_2O$$

In these cases, authigenic carbonates present very low $\delta^{13}C$ values (down to $-60‰$) because the CH_4 itself is highly ^{13}C-depleted (Cheney and Jensen 1965; Hathaway and Degens 1969).

As soon as bacterial sulfate reduction ends (i.e., when the SO_4^{2-} level is zero in pore solutions), bacterial methanogenesis arises by CO_2 reduction and fermentation of organic matter:

$$CO_2 + 4 \, H_2 \dashrightarrow CH_4 + 2 \, H_2O$$

$$2 \, CH_2O \dashrightarrow CO_2 + CH_4$$

The partition of carbon isotopes between CO_2 and CH_4 that occurs during partial CO_2 reduction ($\epsilon \, ^{13}C_{CO_2-CH_4} = 50‰$; Nissenbaum et al. 1972) or during organic matter fermentation ($\epsilon \, ^{13}C_{CO_2-CH_4} = 80‰$; Nakai 1960; Rosenfeld and Silverman 1959) leads to large increases in $\delta^{13}C$ values of the produced CO_2, while methane becomes highly depleted in ^{13}C ($-90‰$ to $-45‰$) relative to the original source of carbon. Very high $\delta^{13}C$ values (up to $30‰$) are typical of these authigenic carbonates crystallized in methane-generating sediments. Numerous examples of such carbonates (mainly dolomite) are documented in the diatomite–organic-rich formations such as the Miocene Monterey Formation of California and younger deposits off southern California and the western and eastern coasts of Baja California (Kelts and McKenzie 1982; Pisciotto and Mahoney 1981; Friedman and Murata 1979; Murata et al. 1967, 1969). It is noteworthy to underline that such ^{13}C-rich authigenic carbonates are characteristics of anoxic organic sediments, strictly devoid of sulfate, because methane-forming bacteria cannot coexist with sulfate-reducing bacteria.

At greater burial depth, abiotic reactions of thermal decomposition of organic matter take place: the CO_2 that is produced exhibits low $\delta^{13}C$ values ($\geqq -25$‰), as does the associated CH_4. Furthermore, since the α $^{13}C_{CO_2-CH_4}$ decreases as temperature increases (Bottinga 1969), the $\delta^{13}C$ values of thermogenic CH_4 increase with downhole temperatures (Deines 1980).

Some moderate $\delta^{13}C$ changes in carbonates may be due to a reservoir effect if the system becomes closed or semiclosed relative to CO_2. The progressive crystallization of carbonates in a system strictly closed for CO_2 leads to the progressive decrease of $\delta^{13}C$ values in both the remaining $\Sigma\,CO_2$ reservoir and in the instantaneous carbonate product (Deuser and Degens 1967). When the system is semiclosed for CO_2 (no input, but an output of CO_2), the $\delta^{13}C$ values of the $\Sigma\,CO_2$ as well as of the precipitated carbonate increase due to the withdrawal of gaseous CO_2 (Deines et al. 1974). These two processes may explain a few local $\delta^{13}C$ variations in carbonates found in restricted environments, such as in pore fluids or in stratified water masses.

The Sulfate System

The oxygen and sulfur isotope fractionation which occurs during crystallization between aqueous and solid sulfate is constant, at least at earth surface temperatures. For gypsum–aqueous sulfate equilibrium, the mean values of the isotope enrichment factors are respectively:

$$\epsilon^{18}O_{gypsum-SO_4^{2-}} = 3.5‰ \text{ (Lloyd 1968)}$$

and

$$\epsilon^{34}S_{gypsum-SO_4^{2-}} = 1.65‰ \text{ (Thode and Monster 1965).}$$

The $\delta^{18}O$ and $\delta^{34}S$ values of sulfate evaporites are thus directly related to the $\delta^{18}O$ and $\delta^{34}S$ values of the aqueous sulfate reservoir where crystallization has occurred.

The different origins and mass-balance between inputs and outputs of sulfur compounds that contribute to the formation of any aqueous sulfate reservoir justify the large ^{18}O and ^{34}S variations observed in evaporitic sulfates.

Present and Past Isotope Composition of the Mean Oceanic Sulfate. At the present time, the oxygen and sulfur isotope compositions of the ocean pool of aqueous sulfate are rather constant: $\delta^{18}O = 9.5$ ± 0.5‰ vs SMOW (Longinelli and Craig 1967; Lloyd 1967, 1968); $\delta^{34}S = 20 \pm 0.5$‰ vs C.D.T. (Thode et al. 1961; Thode 1970; Zak et al. 1980). The $\delta^{18}O$ value of oceanic sulfate is lower by about 20‰ than one expects for the SO_4^{2-}-water oxygen isotope equilibrium (Lloyd 1968; Mizutani and Rafter 1969). This is explained by the very low rate of isotope exchange between water and sulfate (Lloyd 1967; Teis 1956). However, the increasing kinetics of this reaction can be driven in acid solutions even at low temperatures (Hoering and Kennedy 1957).

The composition of the oceans was apparently different than today during some periods in the past, when extensive deposition or weathering of sulfate or sulfide minerals had modified the global sulfur cycle, and had thus changed the isotope composition of oceanic sulfate (Thode et al. 1961; Thode and Monster 1965; Holser and Kaplan 1966; Holser 1977; Claypool et al. 1980). The ^{18}O and ^{34}S age curves of marine sulfate have been established by a compilation of isotope data on gypsum and anhydrite from marine evaporitic successions of various localities in the world. The largest variations are for ^{34}S ($10 < \delta^{34}S$ vs C.D.T. < 35), while the range of $\delta^{18}O$ does not exceed 10‰ (figure 6.5).

Except for the Precambrian period, for which measurements are as yet too scarce, the evolution of $\delta^{34}S$ values through time of marine sulfate records a few major geochemical steps. From the Lower Cambrian to the end of Silurian, the $\delta^{34}S$ values decrease quite gradually from 32‰ to 17‰. The Devonian period is characterized by an abrupt shift towards higher $\delta^{34}S$ values, rising to 35‰ during the Frasnian time. Then, the $\delta^{34}S$ values decrease again, to $\cong 10$‰ in the Upper

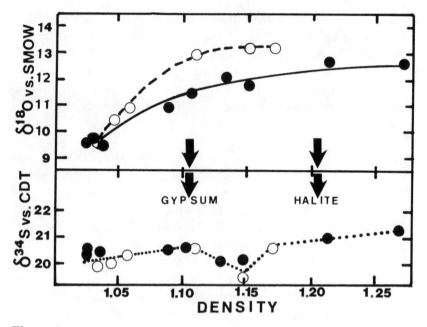

Figure 6.5
Oxygen and sulfur isotopic behavior of aqueous sulfate during evaporation of seawater in the salt pans at Salin-de-Giraud, western Mediterranean (open circles = solutions collected in July 1976; filled circles = solutions collected in June 1979; Pierre 1985).

Permian–Lower Triassic. A very sharp $\delta^{34}S$ increase occurs at the end
of the Lower Triassic (i.e., the Rotliegendes event; Holser 1977). Finally, the present-day $\delta^{34}S$ value of 20‰ is reached progressively
starting at the beginning of Tertiary, after some local pulses.

The Causes of Isotope Changes in Sedimentary Sulfates. When
studied in more detail and on a local scale and within each marine
evaporitic basin, during each period the δ values of sulfates fluctuate
more or less around mean values. This fluctuation depends upon inorganic and organic processes that interact to affect the isotope mass-
balance of sulfur compounds. These reactions include, on the one
hand, the dissolution-precipitation of sulfate minerals, and on the
other hand, the oxidation-reduction of sulfur compounds, monitored
by bacterial activity.

1. Dissolution-precipitation of sulfate minerals: This process is very sensitive in closed or semiclosed basins, because solutions are not continuously supplied. Solid sulfate may thus be dissolved intermittently by diluted waters, and may recrystallize again if the solutions are not drained out of the basin. In such systems, owing to the small and finite size of the aqueous SO_4^{2-} reservoir, and to the ^{18}O and ^{34}S enrichments in favor of the solid sulfate relative to the aqueous sulfate, the $\delta^{18}O$ and $\delta^{34}S$ values of the precipitated sulfate and of the remaining aqueous sulfate decrease progressively, following the typical pattern of reservoir effect (Holser and Kaplan 1966; Pierre 1982).

2. Bacterial oxidation-reduction of sulfur compounds: These processes have the largest impact on the variations of $\delta^{18}O$ and $\delta^{34}S$ values of sulfate, because microorganisms select preferentially light isotope species during metabolic activity.

   a. *The bacterial reduction of sulfate to sulfide* is a multistep reaction; each step is affected by an individual isotope fractionation, the amplitude of which is controlled by the kinetics of the reaction. A slow rate of reaction takes place in direct and opposite ways, while a rapid reaction is necessarily one way.

   The sulfur isotope behavior of the sulfate-to-sulfide pathway has been described in detail by Rees (1973). The first step of SO_4^{2-} incorporation into the bacterial cell leads to a 3‰ enrichment in ^{34}S of the internal SO_4^{2-}. The second step of formation of an enzyme-SO_4^{2-} complex has a zero sulfur isotope fractionation. The third and fourth steps, which correspond to the breakdown of S-O bonds, are responsible for the largest isotope fractionation effects: sulfite and sulfide are each depleted in ^{34}S by 25‰ (at maximum) relative to their parent product. Therefore, a maximum sulfur isotope enrichment of 47‰ has to be expected between sulfate and sulfide formed during bacterial reduction of sulfate under open system conditions for both SO_4^{2-} and S^{2-} compounds (Jorgensen 1979). However, many natural and experimental values of $\epsilon^{34}S_{sulfate-sulfide} \cong 25$‰ indicate that the last step from sulfite to sulfide is very rapid and is thus accompanied by near-zero sulfur isotope fractionation (Harrison and Thode 1958; Kemp and Thode 1968; Kaplan and Rittenberg 1964; Zak et al. 1980). During this multistep reaction, sulfate-sulfite and water may equilibrate with respect to oxygen if the kinetics of the second and third reactions are faster than the following reaction (Mizutani and Rafter 1973).

Most experimental and natural data predict that the remaining aqueous sulfate, from a partial bacterial-sulfate reduction, is enriched in both ^{34}S and ^{18}O relative to the original source of aqueous sulfate. During this reaction the ratio $\epsilon^{34}S/\epsilon^{18}O$ is often close to 4 (Rafter and Mizutani 1967; Mizutani and Rafter 1969; Zak et al. 1980); however, the wide range of this ratio ($-7.8 < \epsilon^{34}S/\epsilon^{18}O < 23.5$) emphasizes the effect of the kinetics of reaction on the extent of the isotope fractionation effect between sulfur compounds and water (Mizutani and Rafter 1973).

b. *The inorganic or organic oxidation of sulfide to sulfate* produces a near-zero sulfur isotope fractionation between the initial reduced and the final oxidized sulfur compounds (Nakai and Jensen 1964; Mizutani and Rafter 1969). The first experiments of Lloyd (1967 1968) and Mizutani and Rafter (1969) had shown that the sulfide-to-sulfate oxidation process is a multistep reaction where oxygen incorporated in the sulfate is derived from water and dissolved molecular oxygen in the ratio 2:1. Assuming that the kinetic isotope effects during the incorporation of water oxygen and molecular oxygen into sulfate are respectively 0‰ and -8.7‰ and are constant (Lloyd 1967, 1968), one could propose a simplified relationship between the oxygen isotope compositions of water-dissolved oxygen and sulfate:

$$\delta^{18}O_{SO_4^{2-}} = 0.32\ (\delta^{18}O_{dissolv.O_2} - 8.7) + 0.68 \cdot \delta^{18}O_{water}$$

Transposing this equation to present-day normal seawater conditions—$\delta^{18}O_{water} = 0$‰ vs SMOW (Craig 1961a); $\delta^{18}O_{atmosph.O_2} = 23.5$‰ vs SMOW (Kroopnick and Craig 1972)—would give $\delta^{18}O = 4.7$‰ vs SMOW for the sulfate formed by sulfide oxidation in seawater.

Recent experimental and natural isotope data on sulfates from sulfide oxidation have reviewed the problem of kinetics of oxygen incorporation in the sulfate from the water and molecular oxygen (Taylor et al. 1984; Van Everdingen and Krouse 1985). Using the example of pyrite oxidation to sulfate, Taylor et al. (1984) proposed two types of reaction.

The first reaction uses oxygen only from water:

$$FeS_2 + 14\,Fe^{3+} + 8\,H_2O \longrightarrow 15\,Fe^{2+} + 2\,SO_4^{2-} + 16\,H^+ \quad (1)$$

The second reaction uses oxygen from both water and molecular oxygen in the ratio 1:7:

$$FeS_2 + \frac{7}{2}\,O_2 + H_2O \longrightarrow Fe^{2+} + 2\,SO_4^{2-} + 2\,H^+ \quad (2)$$

The rates at which reaction (1) and reaction (2) contribute respectively to sulfate production depend on their relative kinetics and on the availability of molecular oxygen in the aqueous environment. The reevaluation of the kinetic isotope effects due to incorporation in the sulfate of water oxygen and molecular oxygen gives $\epsilon^{18}O$ values of 4.1 and -11.2 respectively (Taylor et al. 1984). The $\delta^{18}O$ values of sulfate from sulfide oxidation can thus be evaluated using the following equation (Van Everdingen and Krouse 1985):

$$\delta^{18}O_{SO_4^{2-}} = f\,(\delta^{18}O_{water} + 4.1)$$

$$+ (1 - f)[0.875\,(\delta^{18}O_{dissolv.O_2} - 11.2) + 0.125\,(\delta^{18}O_{water} + 4.1)]$$

in which f is the fraction of SO_4^{2-} produced by the first reaction (i.e., in which oxygen comes solely from water).

Using the same proportion for oxygen atoms from water (0.68) and from molecular oxygen (0.32) as Lloyd, the calculation from the above equation gives $\delta^{18}O = 7.4\permil$ for the sulfate formed by sulfide oxidation in normal seawater (i.e., $2.7\permil$ difference from Lloyd's equation). Further isotope measurements will check and probably refine this new equation, especially for evidence of eventual variations of the kinetic isotope fractionations during oxygen incorporation from water and dissolved oxygen.

The role of sulfur redox reactions in the oxygen and sulfur isotope transfers between sulfur compounds and water has been documented in marine salt pans from the southern Mediterranean coast of France (Fontes and Pierre 1978; Pierre and Fontes 1982; Pierre 1982, 1985). There, the free solutions of these shallow basins are oxygenated, while the interstitial solutions of the organic-rich bottom sediments are re-duced and subjected to bacterial-sulfate reduction. Concentration gradients of sulfate and sulfide are thus extant between the free and the interstitial solutions, and cause the diffusive transfer of sulfur components through the sediment-water interface. The $\delta^{18}O$ and $\delta^{34}S$ values of the aqueous SO_4^{2-} of free solutions are thus controlled by the direction and the intensity of the diffusive fluxes of oxidized and reduced sulfur compounds between the free solutions and the inter-stitial solutions.

During summer months, when marine-derived waters undergo pro-gressive evaporation in the basins, the $\delta^{18}O$ values of water and aqueous SO_4^{2-} both increase, while $\delta^{34}S$ values of the aqueous SO_4^{2-} remain quite stable, up to the stage of gypsum saturation (figure 6.6). This points to oxygen isotope exchange reaction between water and aqueous SO_4^{2-} which occurs during the oxidation in free solutions of the up-ward diffusive sulfide from the bottom sediment. Steady-state con-ditions for sulfur isotopes mean that the downward diffusive flux of sulfate is balanced by the upward diffusive flux of sulfide. At higher salinities, the δ values of aqueous SO_4^{2-} are stabilized by the removal of sulfate ions from the solutions by gypsum precipitation. Further-more, the formation on the basin floor of gypsum and halite crusts lowers the rates of the upward and downward diffusive fluxes of sulfur compounds.

During winter months, the massive dilution by rainwaters of the solutions in the basin causes the reversal of the sulfate diffusive flux: the upward diffusion of SO_4^{2-} enriched in ^{18}O and ^{34}S by partial bac-terial sulfate reduction in the pore solutions is responsible for the simultaneous increases of $\delta^{18}O$ and $\delta^{34}S$ values of the aqueous SO_4^{2-} of the free solutions (figure 6.6).

EXAMPLES OF THE UTILITY OF STABLE ISOTOPE GEOCHEMISTRY IN EVAPORITE STUDIES

The preceding section has demonstrated the special application of stable isotopes in the recognition of ambient parameters of natural mineral-water sedimentary systems. This second section presents a few selected examples from the literature where isotope geochemistry applied to past or present evaporites has provided the essential an-

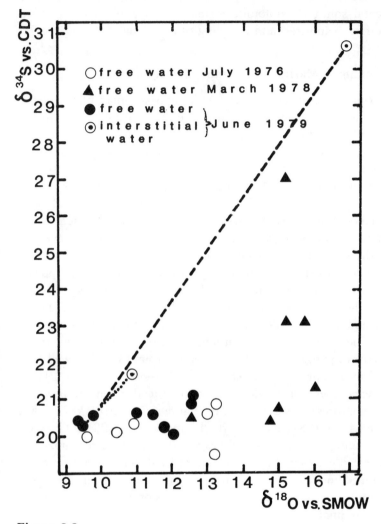

Figure 6.6
^{18}O and ^{34}S variations in aqueous sulfate of free and interstitial solutions collected in the salt pans at Salin-de-Giraud, western Mediterranean, during July 1976, March 1978, and June 1979. The dashed lines join the δ values measured in the interstitial solutions and in the overlying free solutions (Pierre 1985).

swer to sedimentary problems, such as early mixing of marine and continental solutions, dolomite formation, carbonate replacements after sulfate evaporites, and recycling of evaporites.

Early Mixing of Marine and Continental Solutions

The mixing of solutions of different chemistries may play a determinant role during sedimentation and early diagenesis in the crystallization/dissolution of minerals, especially because this process commonly provides an easy way to pass through the chemical barrier of mineral saturation. In other respects, the mineral paragenesis and the ratio of abundance of minerals in evaporitic sequences had to differ from one another owing to differences in the chemical composition between the components of the mixtures. Numerous examples of isotope studies in recent and fossil evaporites have evidenced the importance of such mixing of solutions.

Highly demonstrative examples are provided by modern evaporitic settings where solutions can be easily collected. Such instances are found in the evaporite flats that border the Mormona and Ojo de Liebre lagoons on the western coast of Baja California, northwestern Mexico. There, the mixing of marine-derived brines at different stages of concentration together with continental groundwaters is well documented by the progressive decreasing $\delta^{18}O-\delta^2H$ values in a landward direction, whereas capillary evaporation tends to enrich the water of these mixtures in heavy isotopes (figure 6.7; Pierre 1982, 1983; Pierre and Fritz 1984; Pierre, Ortlieb, and Person 1984). These effects of mixing of solutions and evaporation are also recognized in the water that enters within the crystal lattice during the precipitation of gypsum. Similar conclusions have been deduced from stable isotope compositions of brines from the sabkhas of the Persian Gulf and of northern Sinai (McKenzie et al. 1980; Gat and Levy 1978).

In fossil evaporites, only the minerals themselves can serve to evaluate the isotope composition of water where crystallization occurred, because parent solutions are no longer available (figure 6.8). Carbonates are often used to ascertain the oxygen isotope composition of water, but it is sometimes difficult to judge the true value for $\delta^{18}O$ of water due to uncertainties in temperature estimations and the diagenetic masking of the isotopic record. The only way to directly ob-

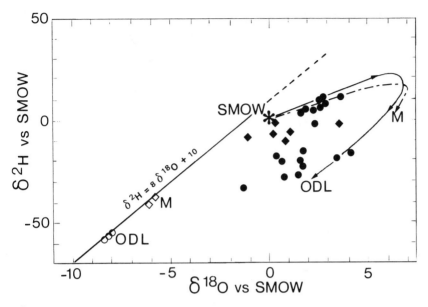

Figure 6.7
^{18}O-^2H relationship in brines from the supratidal evaporite flats of Mormona Lagoon (filled rhombs) and of the Ojo de Liebre Lagoon (filled circles). The δ values of local groundwaters of the Mormona area (open rhombs) and of the Ojo de Liebre area (open circles) are well disposed on the meteoric water line ($\delta^2H = 8\delta^{18}O + 10$) defined by Craig (1961). The two curves (M and ODL) represent the seawater evaporation path in the Mormona and Ojo de Liebre lagoons. Both $\delta^{18}O$ and δ^2H values are expressed on a concentration scale. (Data from Pierre 1982, 1983; Pierre, Ortlieb, and Person 1984; Pierre and Fritz 1984.)

tain the parent solutions is to extract water that is trapped either in the crystal lattice of hydrated minerals or in primary fluid inclusions of minerals. In gypsum facies that have been protected from late diagenetic alteration, it is possible to ascertain the water present at the time of gypsum crystallization. For instance, stable isotope analyses of the water of hydration from Messinian gypsum from Mediterranean land outcrops and deep-sea cores reveal undoubted mixing of marine and continental solutions at the time of gypsum deposition (Pierre 1974, 1982; Pierre and Fontes 1978, 1979; Bellanca and Neri

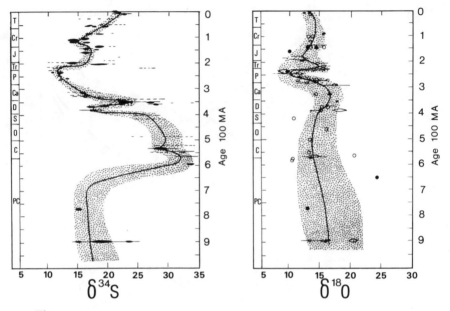

Figure 6.8
Sulfur and oxygen isotope age curves for marine sulfate. The heavy lines are the best estimates for $\delta^{34}S$ and $\delta^{18}O$ values of sulfate minerals in equilibrium with the world ocean-surface sulfate at that date. The shaded areas are the estimate of the uncertainty of these curves (from Claypool et al. 1980).

1986). Similarly, the oxygen and hydrogen isotope compositions of fluid inclusions extracted from Permian halite of the Palo Duro Basin in Texas indicate the probable mixing of evaporite brines with meteoric water in at least a few levels, either at the stage of primary halite crystallization or during later recrystallization events (Knauth and Beeunas 1986). Mixing of marine and continental solutions has thus not only taken place in present day coastal evaporitic settings, but it also occurred in giant evaporitic basins of the past during stages of low stand levels of water.

The Formation of Supratidal Dolomite. Evaporitic supratidal flats represent one of the preferential sites where physico-chemical conditions are highly propitious for dolomite formation, mainly because the coprecipitation of calcium sulfate minerals (gypsum or anhydrite) from the concentrated solutions significantly depresses the Ca^{2+} and SO_4^{2-} concentrations, which are thought to inhibit nucleation of dolomite crystals when their levels are too high.

Marine sabkhas along the Persian Gulf coast are well documented by numerous sedimentary studies; they give the best example of massive dolomitization of calcium carbonate sediments by brines that are largely marine-derived. The progressive effects of diagenesis during sabkha progradation have been well described by the isotopic survey of dolomitized carbonate sediments sampled along a 7 km transect across the sabkhas of Abu Dhabi (McKenzie 1981). From a general point of view, the $\delta^{18}O$ values (1.35 to 3.90‰ vs PDB) of dolomites are compatible with crystallization in isotopic equilibrium for oxygen with ambient conditions (i.e., summer temperatures and $\delta^{18}O_{water}$). The rather constant and high $\delta^{13}C$ values of dolomites (2.8 ± 0.6‰ vs PDB) indicate that the total dissolved inorganic carbon was mainly equilibrated with atmospheric CO_2. Increasing diagenesis toward the inner edge of the sabkha is evidenced in the dolomitized sediments by the larger size of dolomite rhombs and by the higher degree of order in the dolomite crystal lattice. Furthermore, in the oldest sediments, the 3.2‰ difference of $\delta^{18}O$ values between coexisting dolomite and calcium carbonate falls within the range of experimental equilibrium dolomite/calcite ^{18}O fractionation which has been evaluated for similar temperatures. This means that ^{18}O reequilibration of calcium carbonate with pore solutions probably occurred during progressive diagenesis, so that calcium carbonate and dolomite may be considered as a mineralogical pair (i.e., the minerals are cogenetic).

The evaporitic supratidal flats along the Ojo de Liebre lagoon are established in a siliciclastic context. The site of dolomite formation is the southern landward margin of the evaporitic complex (Pierre 1982; Pierre, Ortlieb, and Person 1984). The dolomite sediments, where calcium carbonate is totally lacking, are only present below the water table; these observations suggest that dolomite crystallizes directly from the pore solutions within the sandy matrix. Chemical and isotope data clearly demonstrate that interstitial solutions consist of a mixture of Mg^{2+}-rich, marine-derived brines with calcium bicarbonate–bearing continental waters, that is subjected to concen-

tration by evaporation through the capillary fringe of sediments. Dolomites cover a relatively narrow range of high $\delta^{18}O$ values (2.3 to
5.1‰ vs PDB) and of low $\delta^{13}C$ values (-5.3 to -2.4‰ vs PDB; figure
6.9). The range of measured values of $\alpha\ ^{18}O_{dol\text{-}water}$ (1.0326 to 1.0354)
is very close to the range of theoretical values of $\alpha\ ^{18}O_{dol\text{-}water}$ (1.0321
to 1.0369) calculated from the equations of Northrop and Clayton
(1966), Fritz and Smith (1970), and Matthews and Katz (1977) for
the same range of *in situ* temperatures. This evidence of dolomite-
water equilibrium conditions for ^{18}O gives support to the hypothesis
of primary precipitation of dolomite at this site. It is also assumed
that the rather low $\delta^{13}C$ values of dolomites point to the contribution
to the T.D.I.C. reservoir of biogenic CO_2 of terrestrial origin that is
provided by continental groundwaters. In this case, the siliceous matrix is passive relative to the pore solutions that are the unique vector
for ion transport and chemical interaction. The mixing of solutions

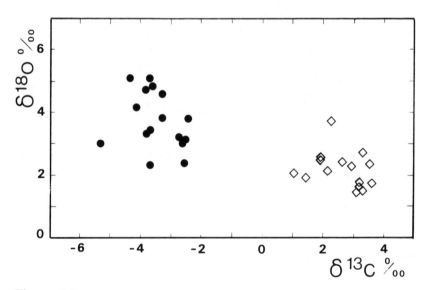

Figure 6.9
$\delta^{18}O$ and $\delta^{13}C$ values of dolomites from the Persian Gulf sabkhas
(open rhombs) and from the supratidal evaporite flats of the Ojo de
Liebre lagoon (filled circles). (Data from McKenzie 1981; Pierre, 1982;
Pierre, Ortlieb, and Person 1981, 1984.)

of different chemistries also seems essential to reach the ideal thermodynamic conditions to precipitate dolomite.

In spite of their similarities of morphology, the evaporitic supratidal flats from the Persian Gulf coast and from the Ojo de Liebre lagoon are the sites of two different mechanisms of dolomite authigenesis, largely evidenced by stable isotope investigations. The Persian Gulf dolomites develop by replacive crystal growth after *in situ* dissolution by Mg^{2+}-rich and CO_2-bearing marine-derived brines of the calcium carbonate sedimentary matrix; the Ojo de Liebre dolomites are emplaced displacively within the siliceous sedimentary matrix where the inflowing Mg^{2+}-rich marine-derived brines are mixed with Ca^{2+}-HCO_3^-–bearing continental groundwaters.

The Origin of Massive Carbonate Replacement
After Sulfate Evaporites

The origin of the large carbonate bodies replacing gypsum/anhydrite deposits in the Permian Castile Formation of Texas and New Mexico, as well as in the Messinian "Gessoso Solfifera" Formation of Central Sicily, was mainly illuminated by stable isotope studies (figure 6.10; Kirkland and Evans 1976; Dessau et al. 1959, 1962; Pierre 1974, 1982; McKenzie 1985). The very low $\delta^{13}C$ values of these carbonates expressed the organic origin of the carbon, and made obvious that these carbonates were formed after the reduction of sulfate by bacteria using organic compounds or hydrocarbons as an energy source. Furthermore, the oxygen isotope composition of these carbonates shows that diagenesis may have occurred either early or late. In the "Gessoso Solfifera" Formation, the mechanism developed quite contemporaneously with sedimentation, so that carbonates crystallized in evaporated waters, explaining the very high $\delta^{18}O$ values (figure 6.10). In the Castile Formation, diagenetic carbonates that exhibit mostly very low $\delta^{18}O$ values (figure 6.10) belong to a late diagenetic event, when continental groundwaters and migrating hydrocarbons have invaded and passed through the sedimentary deposits of the Delaware Basin.

Other examples of diagenetic carbonates after sulfate evaporites, observed in the cap rocks of diapirs from the Gulf Coast (Feely and Kulp 1957; Cheney and Jensen 1965) and in the Middle Miocene

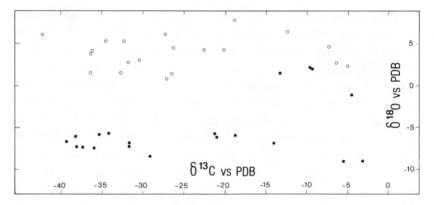

Figure 6.10
$\delta^{18}O$ and $\delta^{13}C$ values of carbonates replacing sulfate evaporites in the Permian Castile Formation of the Delaware Basin (filled circles), and in the Upper Miocene "Gessoso Solfifera" Formation of the central Sicilian Basin (open circles). (Data from Kirkland and Evans 1976; Dessau et al. 1959, 1962; Pierre 1974, 1982.)

evaporites of Egypt (Pierre and Rouchy 1986a, 1988), have given very similar isotope data, and are also interpreted as the result of bacterial sulfate reduction.

This mechanism of carbonate diagenesis emphasizes the essential link between evaporites and hydrocarbons or organic compounds. It is also more than likely that further isotope studies on carbonates in many evaporitic series will reveal similar diagenetic origin.

The Recycling of Evaporites

A major characteristic of evaporites is their high solubility, so that exposure of a sedimentary series would produce significant dissolution features by fresh-water weathering of the more soluble salts (mainly halite and gypsum/anhydrite deposits). Such effects may thus largely affect the overlying sedimentary layers, which commonly undergo collapse, fracture, and brecciation phenomena. Furthermore, if the solutions bearing the products of dissolution then collect in a restricted basin (either marine or continental), where conditions are favorable to evaporite formation, the dissolved salts are recycled in the new evaporitic deposits. The detection of such recycling processes is not always easy by means of petrography, because sedimentary facies as well as mineral paragenesis in the evaporite series will describe the evaporation pattern rather than the origin of the ions of the solutions. In that respect, oxygen and sulfur isotope compositions of the sulfate minerals provide good markers of the origin of the SO_4^{2-} ion, which enters into the constitution of many evaporitic minerals (gypsum, anhydrite, glauberite, polyhalite, bloedite, epsomite, kainite, kieserite). Good examples of recycling of Triassic sulfate during the Tertiary have been evidenced in the Granada Basin, the Ebro Basin, and in the Paris Basin.

The Granada Basin, in central Spain, is an intramontane trough isolated within the Betic Cordillera. Massive gypsum deposits with unfossiliferous marly and detrital intercalations overlie marine marls and carbonates of Tortonian age. They are covered by lacustrine sediments intercalated with a few lignite levels. Oxygen and sulfur isotope compositions of the various sedimentary facies of gypsum occur in a very narrow range ($16.3 < \delta^{18}O$ vs SMOW < 17.7; $15.8 < \delta^{34}S$ vs CDT < 17.8), remarkably out of the range of values characteristic for marine gypsum of late Tertiary age (figure 6.11; Rouchy and Pierre 1979; Pierre 1982). The mixing of marine Tertiary sulfate with reoxidized sulfide or native sulfur seems rather improbable as the cause of these unusual isotopic values, because reduced sulfur compounds are completely absent from these sediments and from their surroundings. On the other hand, gypsiferous deposits of Keuper age outcrop at the northern and eastern margins of this basin, as well as within the basin itself, where Triassic salt diapirs have pierced the overlying beds. The values measured in the Tertiary gypsum of the Granada Basin do, however, fall within the range of the values characteristic of the Keuper marine gypsum (figure 6.11). It is therefore assumed that during late Miocene time, the Granada Basin became isolated from the ocean and was an endoreic trough. It was fed by continental waters that dissolved the outcropping Triassic gypsum and were ultimately concentrated by evaporation to precipitate gypsum having Triassic rather than Tertiary isotopic characteristics.

Another example of chemically recycled gypsum is found in the Upper Bartonian gypsum of the Paris Basin (Middle Eocene), the origin of which was a much-debated question. Recently, oxygen and sulfur isotope studies have added strong arguments in support of the

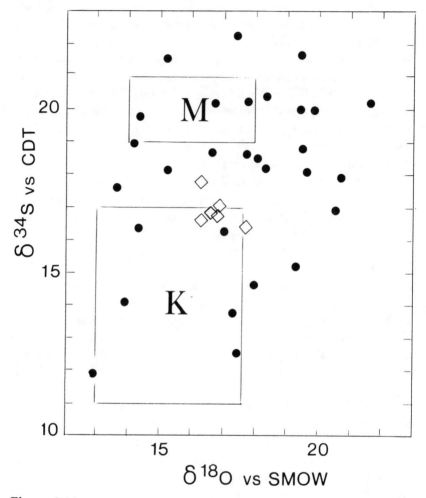

Figure 6.11
$\delta^{18}O$ and $\delta^{34}S$ values of the Upper Eocene gypsum of the Paris Basin (filled circles) and of the Upper Miocene gypsum of the Granada Basin (open rhombs). The areas K and M correspond to the range of δ values measured in marine sulfate evaporites of Keuper and Muschelkalk ages. (Data from Fontes and Letolle 1976; Rouchy and Pierre 1979; Pierre 1982.)

probable continental origin of the solutions that were supplied into the Paris Basin at the end of Eocene. The δ values of the upper Bartonian gypsum generally fall outside of the domain defined by the δ values of marine Tertiary gypsum, whereas they fit very well within the range of values known for Permo-Triassic marine gypsum (figure 6.11; Fontes and Nielsen 1966; Fontes and Letolle 1976). Although a minor contribution of Tertiary marine sulfate cannot be excluded, especially at the beginning of the evaporitic sequence, the main source of sulfate ions was apparently from the Permo-Triassic gypsum layers leached by continental waters at the eastern margins of the Paris Basin.

CONCLUSIONS

Stable isotopes are good indicators of the origin of water and aqueous ionic species, and consequently of sedimentary and authigenic minerals. In this way isotopic investigations can provide the key that enables us to unlock the geochemical evolution pattern of an evaporitic basin through space and time. Thus isotopes constitute a powerful tool in the recognition of the depositional and diagenetic histories of any evaporite, either marine or continental. In spite of these attractive prospects for environmental reconstructions, isotope studies in evaporites remain too scarce, and a fuller and more complete analysis is required to obtain a better geochemical panorama of evaporitic systems.

It was not the aim of this paper to provide a universal guide for the use of stable isotopes in evaporite studies, because the rule in evaporitic settings is variability. A few mechanisms that are frequently inherent in such depositional systems are the corollaries of the restriction of evaporitic environments. For instance, the effects of the diagenesis of organic matter during bacterial sulfate reduction and methanogenesis are relevant to the striking relationship between evaporites and hydrocarbons. Furthermore, the large variations in the water budget within all evaporitic basins are responsible for important changes of the water level; during periods of low water level, marginal areas are exposed to subaerial diagenesis with continental waters, while during high stands the entire basin may be flooded by marine waters and undergo another type of diagenesis.

Finally, apart from these environmental considerations, the application of stable isotopes to the stratigraphy of marine sulfate evaporites is of major interest either when the series are devoid of other stratigraphic markers, or when sulfates have been recycled as an aqueous phase by weathering or as a solid phase by tectonics. However, due caution must be exercised to avoid being misled by diagenetically altered isotope values.

REFERENCES

Aharon, P., Y. Kolodny, and E. Sass. 1977. Recent hot brine dolomitization in the "Solar Lake," Gulf of Elat: Isotopic, chemical, and mineralogical study. *J. Geology* 85:27–48.

Ault, W. V. and J. L. Kulp. 1959. Isotope geochemistry of sulfur. *Geochim. Cosmochim. Acta* 16:201–235.

Bellanca, A. and R. Neri. 1986. Evaporite carbonate cycles of the Messinian, Sicily: Stable isotopes, mineralogy, textural features, and environmental implications. *Jour. Sed. Petrol.* 56:614–621.

Berner, R. A. 1971. *Principles of Chemical Sedimentology*. New York: McGraw-Hill.

Bottinga, Y. 1969. Calculated fractionation factors for carbon and hydrogen isotopic exchange in the system calcite-carbon dioxide-graphite-methane-hydrogen-water vapor. *Geochim. Cosmochim. Acta* 33:49–64.

Buchardt, B. and P. Fritz. 1980. Environmental isotopes as environmental and climatological indicators. In P. Fritz and J. C. Fontes, eds., *Handbook of Environmental Isotope Geochemistry* 1:473–504. New York: Elsevier.

Cheney, E. and M. L. Jensen. 1965. Stable carbon isotopic composition of biogenic carbonates. *Geochim. Cosmochim. Acta* 29:1331–1346.

Claypool, G. E., W. T. Holser, I. R. Kaplan, H. Sakai, and I. Zak. 1980. The age curves of sulfur and oxygen isotopes in marine sulfate and their mutual interpretation. *Chem. Geology* 28:199–260.

Clayton, R. N., B. F. Jones, and R. A. Berner. 1968. Isotope studies of dolomite formation under sedimentary conditions. *Geochim. Cosmochim. Acta* 32:415–432.

Clayton, R. N., H. C. W. Skinner, R. A. Berner, and M. Rubinson. 1968. Isotopic compositions of recent South Australian lagoonal carbonates. *Geochim. Cosmochim. Acta* 32:983–988.

Craig, H. 1953. The geochemistry of the stable carbon isotopes. *Geochim. Cosmochim. Acta* 3:53–93.

Craig, H. 1957. Isotopic standards for carbon and oxygen and correction factors for mass spectrometric analyses of carbon dioxide. *Geochim. Cosmochim. Acta* 12:133–149.

Craig, H. 1961a. Standard for reporting concentrations of deuterium and oxygen-18 in natural waters. *Science* 133:1833–1834.

Craig, H. 1961b. Isotopic variations in meteoric waters. *Science* 133:1702–1703.

Craig, H. 1965. The measurement of oxygen isotope paleotemperatures. In E. Tongiorgi, ed., *Stable Isotopes in Oceanographic Studies and Paleotemperatures,* pp. 3–24. Pisa: C.N.R., Laboratorio di Geologia Nucleare.

Craig, H. and L. I. Gordon. 1965. Deuterium and oxygen-18 variations in the ocean and marine atmosphere. In E. Tongiorgi, ed., *Stable Isotopes in Oceanographic Studies and Paleotemperatures,* pp. 161–182. Pisa: C.N.R., Laboratorio di Geologia Nucleare.

Craig, H., L. I. Gordon, and Y. Horibe. 1963. Isotopic exchange effects in the evaporation of water. *J. Geophys. Res.* 68:5079–5087.

Craig, H. and C. D. Keeling. 1963. The effects of atmospheric N_2O on the measured isotopic composition of atmospheric CO_2. *Geochim. Cosmochim. Acta* 27:549–551.

Dansgaard, W. 1964. Stable isotopes in precipitation. *Tellus* 16:436–468.

Deines, P. 1980. The isotopic composition of reduced organic carbon. In P. Fritz and J. C. Fontes, eds. *Handbook of Environmental Isotope Geochemistry,* 1:329–406. New York: Elsevier.

Deines, P., D. Langmuir, and S. H. Russell. 1974. Stable carbon isotope ratios and the existence of a gas phase in the evolution of carbonate groundwaters. *Geochim. Cosmochim. Acta* 38:1147–1164.

Dessau, G., R. Gonfiantini, and E. Tongiorgi. 1959. L'origine dei giacimenti solfiferi italiani alla luce delle indagini isotopiche sui carbonati della serie gessoso-solfifera della Sicilia. *Boll. Servizio Geol. It.* 81:313–348.

Dessau, G., M. L. Jensen, and N. Nakai. 1962. Geology and isotopic studies of Sicilian sulfur deposits. *Econ. Geology* 57:410–438.

Deuser, W. G. and E. T. Degens. 1967. Carbon isotope fractionation in the system CO_2 (gas)–CO_2 (aqueous)–HCO_3 (aqueous). *Nature* 215:1033–1035.

Epstein, S., R. Buchsbaum, H. A. Lowenstam, and H. C. Urey. 1953. Revised carbonate water isotopic temperature scale. *Bull. Geol. Soc. Am.* 62:417–425.

Feely, H. W. and J. L. Kulp. 1957. Origin of Gulf Coast salt-dome sulfur deposits. *Bull. Amer. Assoc. Petrol. Geol.* 41:1802–1853.

Fontes, J. C. 1965. Fractionnement isotopique dans l'eau de cristallisation du sulfate de calcium. *Geol. Rundschau* 55:172–178.

Fontes, J. C. 1966. Intérêt en géologie d'une étude isotopique de l'évaporation. Cas de l'eau de mer. *C. R. Acad. Sci.* (Paris) 263:1950–1953.

Fontes, J. C. 1976. *Isotopes du milieu et cycles des eaux naturelles: Quelques aspects.* Paris: Thèse Sciences.

Fontes, J. C., P. Fritz, J. Gauthier, and G. Kulbicki. 1967. Minéraux argileux, éléments traces et compositions isotopiques ($^{18}O/^{16}O$ et $^{13}C/^{12}C$) dans les formations gypsifères de l'Eocène supérieur et de l'Oligocène de Cormeilles-en-Parisis. *Bull. Centre Rech. Pau–SNPA* (1967): 315–366.

Fontes, J. C., P. Fritz, and R. Letolle. 1970. Composition isotopique, minéralogigue, et genèse des dolomies du Bassin de Paris. *Geochim. Cosmochim. Acta* 34:279–294.

Fontes, J. C. and R. Gonfiantini. 1967a. Comportement isotopique au cours de l'évaporation de deux bassins sahariens. *Earth Planet. Sci. Letters* 3:258–266.

Fontes, J. C. and R. Gonfiantini. 1967b. Fractionnement isotopique de l'hydrogène dans l'eau de cristallisation du gypse. *C. R. Acad. Sci.* (Paris) 265:4–6.

Fontes, J. C. and R. Letolle. 1976. ^{18}O and ^{34}S in the Upper Bartonian gypsum deposits of the Paris Basin. *Chem. Geology* 18:285–295.

Fontes, J. C. and H. Nielsen. 1966. Isotopes de l'oxygène et du soufre dans le gypse parisien. *C. R. Acad. Sci.* (Paris) 262:2685–2687.

Fontes, J. C. and J. P. Perthuisot. 1971. Faciès minéralogiques et isotopiques des carbonates de la Sebkah el Melah (Zarzis, Tunisie): Les variations du niveau de la Méditerranée orientale depuis 40,000 ans. *Rev. Geogr. Phys. Geol. Dynam.* 13:299–314.

Fontes, J. C. and C. Pierre. 1978. Oxygen 18 changes in dissolved sulphate during sea water evaporation in saline ponds. In G. M. Friedman, ed., *10th International Congress on Sedimentology*, pp. 215–216. Jerusalem: I.A.S.

Friedman, I. and W. E. Hall. 1963. Fractionation of $^{18}O/^{16}O$ between coexisting calcite and dolomite. *J. Geology* 71:238–243.

Friedman, I. and K. J. Murata. 1979. Origin of dolomite in Miocene Monterey Shale and related formations in the Temblor Range, California. *Geochim. Cosmochim. Acta* 43:1357–1365.

Fritz, P. and D. G. W. Smith. 1970. The isotopic composition of secondary dolomites. *Geochim. Cosmochim. Acta* 34:1161–1173.

Gaines, A. M. 1974. Protodolomite synthesis at 100°C and atmospheric pressure. *Science* 183:518–520.

Gaines, A. M. 1980. Dolomitization kinetics: Recent experimental studies. In *Concepts and Models of Dolomitization*, pp. 81–86. Soc. Econ. Pal. Min. Special Publication no. 28. Tulsa, Okla.: SEPM.

Gat, J. R. and Y. Levy. 1978. Isotope hydrology of inland sabkhas in the Bardawil area, Sinai. *Limnol. Oceanogr.* 23:841–850.

Gonfiantini, R. 1966. Effetti isotopici nell'evaporazione di acque salate. *Atti Soc. Tosc. Sci. Nat.* (ser. A), 72:550–569.

Gonfiantini, R., S. Borsi, G. Ferrara, and C. Panichi. 1973. Isotopic composition of waters from the Danakil depression (Ethiopia). *Earth Planet. Sci. Letters* 18:13–21.

Gonfiantini, R. and J. C. Fontes. 1963. Oxygen isotopic fractionation of the water of crystallization of gypsum. *Nature* 200:644–646.

Götz, D. and K. Heinzinger. 1973. Sauerstoffisotopieeffekte und hydratstruktur von alkalihalogenid. Lösungen in H₂O und D₂O. *Z. Naturforsch* 28a:137–141.

Harrison, A. G. and H. G. Thode. 1958. Sulphur isotope abundance in hydrocarbon and source rocks in Uinta Basin, Utah. *Bull. Amer. Assoc. Petrol. Geol.* 42:2642–2649.

Hathaway, J. C. and E. T. Degens. 1969. Methane-derived marine carbonates of Pleistocene age. *Science* 165:690–692.

Hoering, T. C. and J. W. Kennedy. 1957. The exchange of oxygen between sulfuric acid and water. *J. Amer. Chem. Soc.* 79:56–60.

Holser, W. T. 1977. Catastrophic chemical events in the history of the ocean. *Nature* 267:403–408.

Holser, W. T. and I. R. Kaplan. 1966. Isotope geochemistry of sedimentary sulfates. *Chem. Geology* 1:93–135.

Irwin, H., C. Curtis, and M. Coleman. 1977. Isotopic evidence for source of diagenetic carbonates formed during burial of organic-rich sediments. *Nature* 269:209–213.

Jensen, M. L. and N. Nakai. 1962. Sulfur isotope meteorite standards, results, and recommendations. In M. L. Jensen, ed., *Biogeochemistry of Sulfur Isotopes*, pp. 30–35. Nat. Sci. Found. Symposium. New Haven: Yale University Press.

Jorgensen, B. B. 1979. A theoretical model of the stable sulfur isotope distribution in marine sediments. *Geochim. Cosmochim. Acta* 43:363–374.

Kaplan, I. R. and S. C. Rittenberg. 1964. Microbiological fractionation of sulfur isotopes. *J. Gen. Microbiol.* 34:195–212.

Kelts, K. and J. A. McKenzie. 1982. Diagenetic dolomite formation in Quaternary anoxic diatomaceous muds of D.S.D.P. Leg 64, Gulf of California. In I. Curray, D. Moore, et al., *Initial Reports of the Deep Sea Drilling Project, Leg 64*, pp. 553–569. Washington, D.C.: U.S. Government Printing Office.

Kemp, A. L. W. and H. G. Thode. 1968. The mechanism of the bacterial reduction of sulfate and of sulfide from isotope fractionation studies. *Geochim. Cosmochim. Acta* 32:71–91.

Kirkland, D. W. and R. Evans. 1976. Origin of limestone buttes, gypsum plain, Culberson County, Texas. *Bull. Amer. Assoc. Petrol. Geol.* 60:2005–2018.

Knauth, L. P. and M. A. Beeunas. 1986. Isotope geochemistry of fluid inclusions in Permian halite with implications for the isotopic history of ocean water and the origin of saline formation waters. *Geochim. Cosmochim. Acta* 50:419–433.

Kroopnick, P. 1974. The dissolved O₂-CO₂-¹³C system in the eastern equatorial Pacific. *Deep Sea Research* 21:211–227.

Kroopnick, P. and H. Craig. 1972. Atmospheric oxygen: Isotopic fractionation and solubility fractionation. *Science* 175:54–56.

Kroopnick, P., W. G. Deuser, and H. Craig. 1970. Carbon 13 measurements
on dissolved inorganic carbon at the North Pacific (1969) GEOSECS sta-
tion. *J. Geophys. Res.* 75:7668–7671.

Krumgalz, B. 1980. Salt effect on the pH of hypersaline solutions. In
A. Nissenbaum, ed., *Hypersaline Brines and Evaporitic Environments.*
Vol. 28 of *Developments in Sedimentology,* pp. 73–83. New York:
Elsevier.

Land, L. S. 1980. The isotopic and trace element geochemistry of dolomite: The state of the art. In *Concepts and Models of Dolomitization,*
pp. 87–110. Soc. Econ. Pal. Min. Special Publication no. 28. Tulsa, Okla.:
SEPM.

Lawrence, J. R. 1973. Interstitial water studies, Leg 15—oxygen and carbon
isotope variations in water, carbonates, and silicates from the Venezuela
Basin (Site 149) and the Aves Rise (Site 148). In I. MacGregor, B. C.
Heezen et al., *Initial Reports of the Deep Sea Drilling Project, Leg 20,*
pp. 891–899. Washington, D.C.: U.S. Government Printing Office.

Lippmann, F. 1973. *Sedimentary Carbonate Minerals.* Berlin, Heidelberg,
and New York: Springer-Verlag.

Lloyd, R. M. 1967. Oxygen 18 composition of oceanic sulfate. *Science*
156:1228–1231.

Lloyd, R. M. 1968. Oxygen isotope behavior in the sulfate-water system. *J.
Geophys. Res.* 73:6099–6110.

Longinelli, A. and H. Craig. 1967. Oxygen 18 variations in sulfate ions in
sea-water and saline lakes. *Science* 156:56–59.

Matter, A., R. G. Douglas, and K. Perch-Nielsen. 1975. Fossil preservation,
geochemistry, and diagenesis of pelagic carbonates from Shatsky Rise,
Northwest Pacific. In R. L. Larsen, R. Moberly et al., *Initial Reports of
the Deep Sea Drilling Project, Leg 32,* pp. 891–892. Washington, D.C.:
U.S. Government Printing Office.

McKenzie, J. A. 1981. Holocene dolomitization of calcium carbonate sediments from the coastal sabkhas of Abu Dhabi, U.A.E.: A stable isotope
study. *J. Geology* 89:185–189.

McKenzie, J. A. 1985. Stable isotope mapping in Messinian evaporitic carbonates of central Sicily. *Geology* 13:851–854.

McKenzie, J. A., K. J. Hsü, and J. F. Schneider. 1980. Movement of subsurface waters under the sabkha, Abu Dhabi, U.A.E., and its relation to
evaporative dolomite genesis. In *Concepts and Models of Dolomitization,*
pp. 11–30. Soc. Econ. Pal. Min. Special Publication no. 28. Tulsa, Okla.:
SEPM.

Matthews, A. and A. Katz. 1977. Oxygen isotope fractionation during the
dolomitization of calcium carbonate. *Geochim. Cosmochim. Acta* 41:1431–
1438.

Merlivat, L. 1978. *Le deuterium et l'oxygène 18 dans l'eau naturelle: Un moyen d'étude de processus physiques.* Paris-Orsay: Thèse Sciences.

Merlivat, L. and J. Jouzel. 1979. Global climatic interpretation of the deuterium–oxygen 18 relationship for precipitation. *J. Geophys. Res.* 84:5029–5033.

Mizutani, Y. and T. A. Rafter. 1969. Oxygen isotopic composition of sulphates. Part 4: Bacterial fractionation of oxygen isotopes in the reduction of sulphate and in the oxidation of sulphur. *New Zealand J. Sci.* 12:60–68.

Mizutani, Y. and T. A. Rafter. 1973. Isotopic behaviour of sulphate oxygen in the bacterial reduction of sulphate. *J. Geochem.* 6:183–191.

Mook, W. G., J. C. Bommerson, and W. H. Staverman. 1974. Carbon isotope fractionation between dissolved bicarbonate and gaseous carbon dioxide. *Earth Planet. Sci. Letters* 22:169–176.

Murata, K. J., I. Friedman, and B. M. Madsen. 1967. Carbon-13-rich diagenetic carbonates in Miocene formations of California and Oregon. *Science* 156:1484–1486.

Murata, K. J., I. Friedman, and B. M. Madsen. 1969. *Isotopic Composition of Diagenetic Carbonates in Marine Miocene Formations of California and Oregon.* U.S. Geol. Survey Prof. Paper no. 614B. Washington, D.C.

Nakai, N. 1960. Carbon isotope fractionation in natural gas in Japan. *J. Earth Sci.* 8:176–180.

Nakai, N. and M. L. Jensen. 1964. The kinetic isotope effect in the bacterial reduction and oxidation of sulfur. *Geochim. Cosmochim. Acta* 28:1893–1912.

Nissenbaum, A., B. J. Presley, and I. R. Kaplan. 1972. Early diagenesis in a reducing fjord, Saanich inlet, British Columbia. Part 1: Chemical and isotopic changes in major components of interstitial water. *Geochim. Cosmochim. Acta* 36:1007–1027.

Northrop, D. A. and R. N. Clayton. 1966. Oxygen isotope fractionations in systems containing dolomite. *J. Geology* 74:174–196.

O'Neil, J. R., R. N. Clayton, and T. K. Mayeda. 1969. Oxygen isotope exchange between divalent metal carbonates. *J. Chem. Phys.* 51:5547–5558.

Pierre, C. 1974. *Contribution à l'étude sédimentologique et isotopique des évaporites Messiniennes de la Méditerranée: Implications géodynamiques.* Paris: Thèse 3ème cycle.

Pierre, C. 1982. *Teneurs en isotopes stables (^{18}O, ^{2}H, ^{13}C, ^{34}S) et conditions de genèse des évaporites marines: Applications à quelques milieux actuels et au Messinien de la Méditerranée.* Orsay, Paris-Sud: Thèse.

Pierre, C. 1983. Polyhalite replacement after gypsum at Ojo de Liebre Lagoon (Baja California, Mexico): An early diagenesis by mixing of marine brines and continental waters. In B. C. Schreiber and H. L. Harner, eds., *Sixth Symposium on Salt,* 1:257–265. Alexandria, Va.: Salt Institute.

Pierre, C. 1985. Isotopic evidence for the dynamic redox cycle of dissolved sulphur compounds between free and interstitial solutions in marine salt pans. *Chem. Geology* 53:191–196.

Pierre, C. and J. C. Fontes. 1978. Isotope composition of Messinian sediments from the Mediterranean Sea as indicators of paleoenvironments and diagenesis. In K. J. Hsü, L. Montadert et al., *Initial Reports of the Deep Sea Drilling Project, Leg 42A*, pp. 635–650. Washington, D.C.: U.S. Government Printing Office.

Pierre, C. and J. C. Fontes. 1979. Oxygène 18, carbone 13, deutérium et soufre 34: Marqueurs géochimiques de la diagenèse et du paléomilieu évaporitique du Messinien de la Méditerranée. *Bull. Mus. Natn. Hist. Nat.* (Paris), series 4, no. 1, section 3, paper no. 1, pp. 3–18.

Pierre, C. and J. C. Fontes. 1982. Etude isotopique des saumures et des gypses des marais salants de Salin-de-Giraud (Sud de la France). *Géol. Méditerr.* 9:479–486.

Pierre, C. and B. Fritz. 1984. Remplacement précoce de gypse par la polyhalite: L'exemple de la bordure sud-orientale de la lagune d'Ojo de Liebre (Basse California, Mexique). *Rev. Géol. Dyn. Géogr. Phys.* 25:157–166.

Pierre, C., L. Ortlieb, and A. Person. 1981. Formation actuelle de dolomite supralittorale dans les sables quartzo-feldspathique: Un exemple au sud de la lagune Ojo de Liebre (Basse California, Mexique). *C. R. Acad. Sci.* (Paris), vol. 293, sér. 2, pp. 73–79.

Pierre, C., L. Ortlieb, and A. Person. 1984. Supratidal evaporitic dolomite at Ojo de Liebre lagoon: Mineralogical and isotopic arguments for primary crystallization. *J. Sed. Petrol.* 54:1049–1061.

Pierre, C. and J. M. Rouchy. 1985. Carbonate pseudomorphs after gypsum/anhydrite deposits: Petrography and stable isotopes investigations in search of the mechanism of diagenesis (abs.). In: *6th European Regional Meeting I.A.S.*, pp. 361–362. April. Lerida, Spain.

Pierre, C. and J. M. Rouchy. 1986a. Sedimentological and geochemical diagnostic features of carbonate replacements after sulfate evaporites: Application to the example of the Middle Miocene of Egypt (abs.). In *Geochemistry of the Earth Surface and Processes of Mineral Formation*. pp. 28–29. March. Granada, Spain.

Pierre, C. and J. M. Rouchy. 1986b. Oxygen and sulfur isotopes in anhydrites from Givetian and Visean evaporites of northern France and Belgium. *Isotope Geoscience* 58:245–252.

Pierre, C. and J. M. Rouchy. 1988. The carbonate replacements after sulfate evaporites in the Middle Miocene of Egypt. *Jour. Sed. Petrol.* 58:446–456.

Pierre, C., J. M. Rouchy, A. Laumondais, and E. Groessens. 1984b. Sédimentologie et géochimie isotopique (^{18}O, ^{34}S) des sulfates évaporitiques givétiens et dinantiens du Nord de la France et de la Belgique: Importance pour la stratigraphie et la reconstitution des paléomilieux de dépôt. *C. R. Acad. Sci.* (Paris) 299:21–26.

Pisciotto, K. A. and J. J. Mahoney. 1981. Isotopic survey of diagenetic carbonates, Deep Sea Drilling Project, Leg 63. In R. S. Yeats, B. U. Haq et al., *Inital Reports of the Deep Sea Drilling Project, Leg 63*, pp. 595–609. Washington, D.C.: U.S. Government Printing Office.

Rafter, T. A. and Y. Mizutani. 1967. Preliminary study of variations of oxygen and sulphur isotope in natural sulphates. *Nature* 216:1000–1002.

Rees, C. E. 1973. A steady-state model for sulphur isotope fractionation in bacterial reduction processes. *Geochim. Cosmochim. Acta* 37:1141–1162.

Rosenfeld, W. D. and S. R. Silverman. 1959. Carbon isotope fractionation in bacterial production of methane. *Science* 130:1658–1659.

Rouchy, J.-M. and C. Pierre. 1979. Données sédimentologiques et isotopiques sur les gypses des séries évaporitiques messiniennes, d'Espagne méridionale et de Chypre. *Rev. Géol. Dyn. Géogr. Phys.* 21:267–280.

Rubinson, M. and R. N. Clayton. 1969. Carbon 13 fractionation between aragonite and calcite. *Geochim. Cosmochim. Acta* 33:997–1022.

Sheppard, S. M. F. and H. P. Schwarcz. 1970. Fractionation of carbon and oxygen isotopes and magnesium between coexisting metamorphic calcite and dolomite. *Contrib. Mineral. Petrol.* 26:161–198.

Sofer, A. 1978. Isotopic composition of hydration water in gypsum. *Geochim. Cosmochim. Acta* 42:1141–1150.

Sofer, Z. and J. R. Gat. 1972. Activities and concentrations of oxygen 18 in concentrated aqueous salt solutions: Analytical and geophysical implications. *Earth Planet. Sci. Letters* 15:232–238.

Sofer, Z. and J. R. Gat. 1975. The isotope composition of evaporating brines: Effect of the isotopic activity ratio in saline solutions. *Earth Planet. Sci. Letters* 26:179–186.

Taylor, B. E., M. C. Wheeler, and D. K. Nordstrom. 1984. Isotope composition of sulfate in acid mine drainage as measures of bacterial oxidation. *Nature* 308:530–541.

Teis, R. V. 1956. Isotopic composition of oxygen in natural sulfate (in English). *Geochemistry* (USSR), pp. 257–263.

Thode, H. G. 1970. Geochemistry of sulphur isotopes. In *McGraw-Hill Yearbook of Science and Technology*, pp. 351–353. New York: McGraw-Hill.

Thode, H. G. and J. Monster. 1965. Sulfur isotope geochemistry of petroleum, evaporites, and ancient seas. In *Fluids in Subsurface Environments*, pp. 367–377. Amer. Assoc. Petrol. Geol. Memoir no. 4. Tulsa, Okla.: AAPG.

Thode, H. G., J. Monster, and H. B. Dunford. 1961. Sulphur isotope geochemistry. *Geochim. Cosmochim. Acta* 26:159–174.

Thode, H. G., M. Shima, C. E. Rees, and K. V. Krishnamurty. 1965. Carbon

13 isotope effects in systems containing carbon dioxide, bicarbonate, carbonate, and metal ions. *Can. J. Chem.* 43:582–595.

Van Everdingen, R. O. and H. R. Krouse. 1985. Isotope composition of sulphates generated by bacterial and abiological oxidation. *Nature* 315:395–396.

Van Everdingen, R. O., M. A. Shakur, and H. R. Krouse. 1982. ^{34}S and ^{18}O abundances differentiate Upper Cambrian and Lower Devonian gypsum-bearing units, District of MacKenzie, N.W.T.: An update. *Can. J. Earth Sci.* 19:1246–1254.

Vogel, J. C., P. M. Grootes, and W. G. Mook. 1970. Isotope fractionation between gaseous and dissolved carbon dioxide. *Z. Phys.* 230:225–238.

Yurtsever, Y. 1975. Worldwide survey of stable isotopes in precipitation. *Rept. Sect. Isotope Hydrology—Vienna I.A.E.A.*, November. Vienna: International Atomic Energy Agency.

Zak, I., H. Sakai, and I. R. Kaplan. 1980. Factors controlling the $^{18}O/^{16}O$ and $^{34}S/^{32}S$ isotope ratios of ocean sulfates, evaporites, and interstitial sulfates from modern deep-sea sediments. In *Isotope Marine Chemistry*, pp. 339–373. Tokyo: Uchida Rokakuho.

7

The Origin of
Salt Structures

RALPH O. KEHLE

Both tectonic and nontectonic origins have been proposed for salt domes and other salt structures. Surveys of these theories are given by Lachmann (1911), Rogers (1918–19), Wolf (1923), DeGoyler (1925), and Nettleton (1955). Most workers presently believe that salt domes result from buoyancy—a type of gravity instability that arises where lower-density salt is buried beneath higher-density strata. According to this theory, when a minimum depth (p. 2–3) of burial is achieved, the salt flows upward through the overlying sediment. The analysis presented here shows that buoyancy may not be the most important factor in salt-dome growth. Rather it appears that the creation of salt sinks below thick overburden anomalies is more important. A similar conclusion is reached by Jackson and Talbot

Ralph O. Kehle is Chairman of the Board and Chief Executive Officer of Hershey Oil Company, Pasadena, California.

(1986). This mechanism functions regardless of sediment density and accounts for the initiation of structural growth at shallow burial depths.

The buoyancy theory, first proposed by Arrhenius (1912), is supported by the model work of Nettleton (1934, 1955), Dobrin (1941), and Parker and McDowell (1955). More recent work by Biot (1963a, 1963b, 1966), Biot and Ode (1965), and Danes (1964) demonstrated that buoyancy can cause salt domes. These analyses predict the influence of the mechanical properties and thickness of the salt and overlying sediments on the initial gross structure of a forming dome. With the aid of a computer, Fletcher (1967) traced the growth of such a dome from inception to maturity. His method predicts the origin, location, and type of both primary and secondary structures arising during salt-dome formation. The present analysis confirms this result, but finds buoyancy to be a subordinate mechanism.

In most previous experimental and theoretical work, the salt and sediment sequence is modeled as a horizontal sequence of uniform layers of fluid. It is commonly assumed that the fluids exhibit linear (Newtonian) mechanical properties and that the layer representing salt is less dense than overlying layers. The assumption of fluidlike behavior is supported by Chenevert (1968, personal communication), who was unable to detect a minimum differential stress for salt flow, and by Heard (1976). However, it is well known that salt creep is strongly nonlinear (that is, non-Newtonian; see for example Ode 1968, Heard 1976, and various articles in Hardy and Langer 1984). Moreover, uniformly dense, uniformly thick overburden is uncommon in nature, and this assumption in the model is potentially misleading.

In addition to a consensus that salt buoyancy is the principal mechanism in forming salt structures outside of orogenic regions, most investigators also believe that at least 900 to 1,200 m or 3,000 to 4,000 ft of overburden are required before dome growth begins (e.g., Balk 1947; Trusheim 1960; Gussow 1968). This may be construed as support for the buoyancy hypothesis, because the density of Gulf Coast Tertiary sediments increases with increasing thickness until it exceeds that of salt when depth of burial equals 762 to 914 m or 2,500–3,000 ft (Dickinson 1953).

The premise that salt movement does not begin until salt is buried to a depth exceeding several thousand feet is contradicted by structures in the interior salt basins of East Texas, northern Louisiana and central Mississippi. There, nondiastrophic, salt-supported structures began to grow beneath as little as 300 ft of overlying sediment (the Smackover Formation; see the seismic sections in Rosenkrans and Marr 1967). If these features resulted from buoyancy, then clearly much less than 1,220 m or 4,000 ft of overburden is required to initiate growth. Conversely, if large thicknesses of overburden are required for buoyant salt motion, many structures in the interior salt basins clearly are not caused by salt buoyancy. In this paper we attempt to evaluate the assumption of uniformly thick and dense overburden common to previous analyses; to resolve the contradiction regarding structures in the interior salt basins of the Gulf Coast; and to clarify the mechanics of salt-dome growth. A simple theory of salt flow is developed to accomplish these aims. In this study, we employ a few simple techniques of elementary hydraulics to establish the direction of salt flow, but not the details of the flow regime. This avoids the complicated mathematics used in previous analyses (Biot 1963a, 1963b, 1966; Biot and Ode 1965; Fletcher 1967). The presentation is made as follows:

In section 1, underlying principles are reviewed and a single principle is developed from which all remaining results are drawn: salt flows downhill if the gravity head exceeds an opposing pressure head. Uphill flow occurs only if the opposing pressure head exceeds the gravity head.

In section 2, two "end members" of possible salt-sediment configurations are investigated. In a model with a dipping salt layer and parallel overburden, salt always flows downhill, no matter how great the density and thickness of the overburden. In a model with a dipping salt layer and onlapping horizontal strata, the direction of salt flow between two elevations on the salt surface is determined by the average density of sediment in that elevation interval. Uphill flow occurs only if this average density exceeds that of salt.

In section 3, the results of section 2 are applied to geologic problems. Investigation of a horizontal salt and sediment sequence shows that dome-constructing uphill flow occurs only if there are irregularities of large dimension in the original salt surface and sediments onlapping the irregularity are more dense than salt. However, salt flow is calculated to be much greater from beneath positive anomalies on the sediment surface than into positive irregularities of the same size on the salt surface. This suggests that buoyant elevation of original highs on the salt surface is subordinate to movement away from local sediment loads. Examples of such loads are deltas, reefs, and turbidite fans.

In section 4, a discussion is given of the growth history of a salt structure initiated by delta progradation into a shallow salt basin (East Texas). Many young salt-dome terrains are characterized by a karst-like topography of closed depressions rather than by a collection of low-relief salt pillows. The lower continental slope, offshore Texas and Louisiana, exhibits this topography (Lehner 1969).

In section 5, the evolution of a group of Gulf Coast salt domes is discussed. The analysis presented in section 4 also shows that salt tectonics on the dipping flanks of basins is dominated by downhill salt flow, which masks potential buoyant salt tectonics. The evolution of this basinward flow of salt, encouraging sliding of the overlying sediments, is developed in section 5. In areas where this occurs, the structures that develop, though salt supported, are not salt motivated. Rather, they are the result of deformation of the upper plate of a gravity slide and are common to all gravity slides, whether underlain by salt or not. Structures of this type are abundant along the flanks of the interior salt basins in the northern Gulf of Mexico Province.

In section 6, the final section, case histories of several salt structures are presented.

1. SALT FLOW

Analysis of salt mechanics is typically predicated on the assumption that salt may be treated as a Newtonian fluid (Biot 1966; Biot and Ode 1965, Fletcher 1967), even though salt creep follows a highly nonlinear viscoelastic flow equation (Ode 1968; Heard 1975; Carter and Hansen 1983). If the goal of an analysis is a prediction of both growth rate and details of structural form, this discrepancy between actual and theoretical salt behavior may invalidate some previous conclusions.

This complication is avoided in the following analysis because the goal is to determine the direction of salt motion, if any, rather than the details of the flow regime. Elementary hydraulic theory is sufficient to determine the direction of flow no matter what the properties of the fluid. Even if salt possesses a yield point, the results are valid with the provision that no flow takes place in any direction unless the state of differential stress exceeds the limiting yield point.

Case: Motion in a Tilted Fluid Layer. Consider a motionless fluid confined between tilted parallel plates (figure 7.1). The fluid in the manometers at p_1 and p_2 stands at the same height above an arbitrary datum. No flow occurs because the hydraulic head within the fluid is everywhere the same, a fact illustrated by the common elevation of the fluid in the manometers show in figure 7.1. The manometer reading is comprised of two parts, the elevation of the tap above an arbitrary datum, called the gravity potential, and the height of the fluid column in the manometer, called the pressure head.

The total hydraulic head is the sum of the gravity potential and the pressure head. Fluid flow occurs when there is a difference in hydraulic head between points in a fluid, with flow occurring between places of high hydraulic head and places of low hydraulic head.

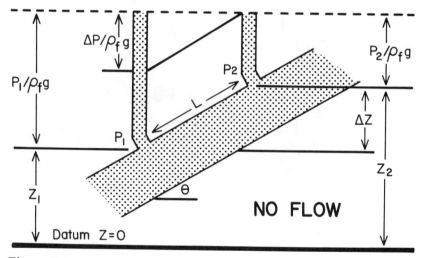

Figure 7.1
Tilted fluid layer, no flow condition. Fluid pressure is proportional to height of fluid column in manometers. Total hydraulic head, which equals pressure plus gravitational potential, is everywhere constant.

If the total fluid head is the same at the two points, then no flow occurs between the two points. Mathematically this is written:

$$Z_1 + P_1/\rho_f g = Z_2 + P_2/\rho_f g \tag{1a}$$

or equivalently:

$$\rho_f g Z_2 + P_2 = \rho_f g Z_1 + P_1 \tag{1b}$$

In these equations ρ_f is the density of the fluid, g the acceleration of gravity, Z the height above an arbitrary datum, and P the fluid pressure.

If a valve is opened at the lower end of the parallel plate system, the pressure at point p_1 will drop to say P_1^* and at p_2 to say P_2^*. The resulting manometer levels are shown in figure 7.2. Flow occurs downhill as indicated in figure 7.2, even though $P_1^* > P_2^*$; that is, flow is from a region of low pressure to a region of high pressure. This occurs because the decrease in gravity head between point p_1

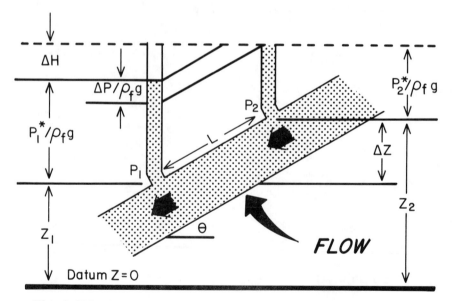

Figure 7.2
Tilted fluid layer, restricted downdip flow. Flow is from point of high hydraulic head (p_2) to lower hydraulic head (p_1) even though pressure at p_1 is greater than at p_2.

and p_2 exceeds the increase in fluid pressure. In this example, the net gravitational force which points downhill is larger than the differential fluid pressure which points uphill. Therefore, flow is downhill in the direction of increasing fluid pressure. What we have discovered is that fluids flow from regions of high total hydraulic head to regions of low total hydraulic head. If pressure is applied (say with a pump) to the lower end of the parallel plates of figure 7.1, and the pressure at p_1 is increased, the pressure at p_2 increases but by a lesser amount if flow occurs uphill (figure 7.3). As can be determined by inspection, the total head at p_1, ($\rho_f g Z_1 + P_1$), is greater than the total head at p_2, ($\rho g Z_2 + P_2$), by an amount $\Delta \hat{H}$. This illustrates an important point regarding salt-dome formation: to obtain uphill flow, the pressure drop in the uphill direction has to be greater than the gravity or hydrostatic head increase in that direction.

Another point that needs to be made is that the rate of fluid flow increases as an increase in the gradient fluid head regardless of the type of fluid. Coupled with the knowledge that fluids flow in the direction of decreasing fluid head (not pressure), this simple fact pro-

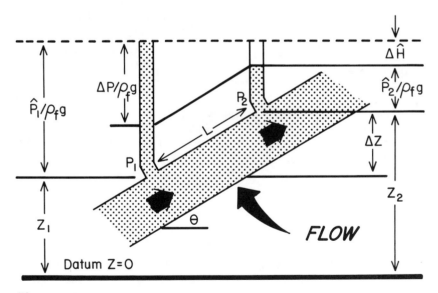

Figure 7.3
Tilted fluid layer, uphill flow. Pressure drop from p_1 to p_2 (ΔP) is greater than increase in gravitational potential from p_1 to p_2 ($\Delta Z \rho_f g$).

vides us with insight into several important questions on the initiation and growth of salt structures.

2. SALT-SEDIMENT CONFIGURATION: FLOW IN A SEDIMENTARY SALT LAYER

A salt layer located within a rock sequence behaves mechanically in the same manner as a fluid layer confined between parallel or subparallel plates, at least until deformation alters the original geometry of the strata. The gravity head within a salt layer is easily determined by choosing an arbitrary datum and measuring elevations within the salt mass relative to the datum. The pressure distribution (lithostatic pressure) at the salt-sediment interface, which depends on the thickness and density distribution of the overburden, is more difficult to determine. Actual geologic settings are varied; each yields a different pressure distribution. However, considerable insight into most real settings can be gained by examining two ideal "end members": tabular overburden layers of uniform thickness lying parallel to the salt surface, each layer with an arbitrarily assigned uniform density (figure 7.4); and horizontal formations of constant thickness, overlying an inclined layer of salt, each with a given depth of burial (figure 7.6). Although the salt layer is pictured as tabular, the following conclusions are independent of the nature of the lower salt surface.

Sediment Layers Parallel to Salt Surface

A salt layer lying on "basement" and overlain by a rock sequence whose beds parallel the salt surface is illustrated in figure 7.4, and the equivalent manometer readings are depicted in figure 7.5. This configuration models an undeformed salt layer in most geologic settings and is commonly applicable to concordant salt structures, e.g., salt pillows. Notice that the pressure head is the same at p_1 and p_2. This is because the overburden weight (lithostatic pressure) is the same at all localities; hence the pressure head is everywhere constant. This overburden pressure is given by:

$$P = g \left(\rho_1 h_1 + \rho_2 h_2 + \rho_3 h_3\right)/\cos\theta \tag{2}$$

where h_i and ρ_i, $i = 1 \ldots 4$, are the layer thickness and densities

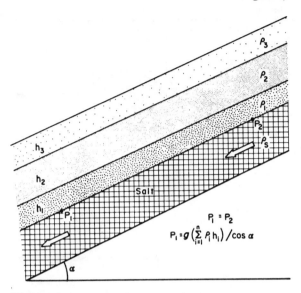

Figure 7.4
Schematic of dipping sedimentary rock sequence overlying salt layer.
Salt flow is downdip (Jackson and Galloway 1984).

respectively, g the acceleration of gravity, and θ the dip of the beds.
This result can be generalized to n layers, where the pressure equals:

$$P = g\left(\sum_{i=1}^{n}\rho_i h_i\right)/\cos\theta \tag{3}$$

Because the pressure head is everywhere the same, the difference
in total head between p_1 and p_2 is the difference in the gravity head
between these two points, $\rho_s\, g(Z_1 - Z_2)$, where ρ_s is the density of
salt. Because p_1 is lower than p_2, the hydraulic head diminishes downhill. The salt will flow downhill no matter what the thickness or density of the overburden.

This result does not imply that gravitational instabilities that result
in the formation of salt structures are eliminated by a slight incli-
nation of the rock sequence. If the above sequence had been horizontal, the analysis yields the conclusion that no flow would take
place in the salt, because both the pressure head and the gravity head
would be constant everywhere within a horizontal layer with uniform

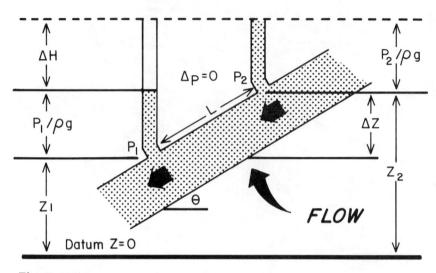

Figure 7.5
Tilted fluid layer model representing hydraulic condition existing in salt layer of figure 7.4. Fluid pressure (overburden weight) is constant, but total hydraulic head decreases downdip as gravitational potential decreases.

overburden load. The manner in which gravity instabilities arise in such sequences is discussed later (section 3).

Horizontal Sediment Layers Onlapping an Inclined Salt Bed

A second idealized configuration is an inclined salt layer onlapped by a horizontal sequence of rock layers, illustrated in figure 7.6. This configuration models a typical diapir and in some circumstances a salt pillow exhibiting large topographic relief. In the following analysis an assumption can be made that the deeper layers are progressively denser. A possible set of manometer readings corresponding to figure 7.6 is shown in figure 7.7, where it can be seen that downdip, the fluid head decreases to a minimum and then begins to increase again. The explanation for this is that down to a certain depth, the average density of the sedimentary rock is less than that of salt; below that depth the average density of the sedimentary column exceeds that of salt.

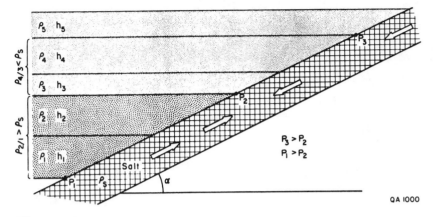

Figure 7.6
Schematic of dipping salt-sediment layer with onlapping horizontal sedimentary strata. Direction of salt flow depends on average density of sediments (Jackson and Galloway 1984).

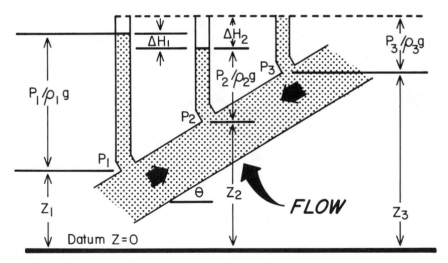

Figure 7.7
Tilted fluid layer model representing hydraulic conditions in salt layer of figure 7.6. Flow in fluid converges toward the center manometer tap.

This can be demonstrated by calculating the fluid head at each manometer tap, as follows:

$$\text{At } p_3, \ H_3 = P_3 + g\rho_s Z_3 = g(\rho_5 h_5 + \rho_s Z_3) \tag{4}$$

$$\text{At } p_2, \ H_2 = P_2 + g\rho_s Z_2 = g(\rho_5 h_5 + \rho_4 h_4 + \rho_3 h_3 + \rho_s Z_2) \tag{5a}$$

$$= g(\hat{\rho}_{35} T_{35} + \rho_s Z_2) \tag{5b}$$

$$\text{At } p_1, \ H_1 = P_1 + g\rho_s Z_1 = g(\rho_1 h_1 \tag{6a}$$

$$+ \rho_2 h_2 + \rho_3 h_3 + \rho_4 h_4 + \rho_5 h_5 + \rho_s Z_1)$$

$$= g(\hat{\rho}_{15} T_{15} + \rho_s Z_1) \tag{6b}$$

where $\hat{\rho}_{ij}$ are the average densities of a group of layers defined by

$$\hat{\rho}_{ij} = \left(\sum_{k=i}^{j} \rho_k h_k \right) / T_{ij} \tag{7a}$$

and T_{ij} is the total interval thickness defined by

$$T_{ij} = \sum_{k=i}^{j} h_k \tag{7b}$$

The two head drops ΔH_1 and ΔH_2 are obtained by subtracting H_1 from H_2 and H_2 from H_3,

$$\Delta H_1 = g\rho_s(Z_2 - Z_1) + (P_2 - P_1)$$

$$= g\rho_s(Z_2 - Z_1) - g(\rho_2 h_2 + \rho_1 h_1)$$

$$= g\rho_s(Z_2 - Z_1) - g\hat{\rho}_{12} T_{12} \tag{8a}$$

$$\Delta H_2 = g\rho_s(Z_3 - Z_2) + (P_3 - P_2)$$

$$= g\rho_s(Z_3 - Z_2) - g(\rho_4 h_4 + \rho_3 h_3)$$

$$= g\rho_s(Z_3 - Z_2) - g\hat{\rho}_{34} T_{34} \tag{8b}$$

where ρ_{12} and ρ_{34} are the average densities for layers 1–2 and 3–4 respectively. Comparison of figures 7.6 and 7.7 shows that:

$$Z_2 - Z_1 = h_2 + h_1 = T_{12} \tag{9a}$$

$$Z_3 - Z_2 = h_3 + h_4 = T_{34} \tag{9b}$$

so that equations 8a and 8b can be written:

$$\Delta H_1 = gT_{12}(\rho_s - \hat{\rho}_{12}) \tag{10a}$$

$$\Delta H_2 = gT_{34}(\rho_s - \hat{\rho}_{34}) \tag{10b}$$

This analysis demonstrates that the hydraulic head at the top of the salt decreases downdip so long as the average density of the sediments adjacent to the salt is less than that of salt. Where the sediment density exceeds the density of salt, the total hydraulic head increases downdip. Thus, salt flow is downhill between two elevations between which the average density of the sediments is less than that of salt; salt flow is uphill between two points if the average density of the sediments in the interval exceeds that of salt.

In figures 7.6 and 7.7 the average density of layers 3–5 is less than that of salt, and thus flow is downhill from p_3 to p_2. However, the average density of layers 1 and 2 is greater than that of salt, causing the salt to flow uphill from p_1 to p_2. Clearly this would result in a thickening of salt in the vicinity of p_2, forming a structure such as a salt pillow or an overhang in a growing salt diapir. This is a possible although unlikely mechanism for the formation of salt pillows. An additional conclusion derived from this analysis (equations 10a and 10b) is that total overburden load plays no role in determining either the direction or rate of salt motion. Only the distribution of density is important.

3. SALT FLOW IN NATURAL SETTINGS

The conclusions drawn from the previous section may be applied to problems of salt motion in natural settings. Although details of the flow field cannot be determined, by using this simple analysis considerable insight into the factors that control salt flow can be gained. Two types of settings are analyzed below: horizontal salt and sediment layers, with and without thickness variation; and an inclined salt layer with an overlying wedge of sediment that increases in thickness downdip. The former corresponds to the central "flat" portions of a salt basin, whereas the latter corresponds to the basin flanks.

Horizontal Salt-Sediment Sequences

Most analyses of salt-dome formation discuss the initiation of gravitational instability in a horizontal salt and sediment sequence. The salt is assumed to be less dense than the overlying strata, resulting in a gravity instability. Model experiments illustrate the growth of such instability, but they cast little light on how and why they form, although it is acknowledged that model salt domes rarely form unless a "seed" or nucleus is introduced into the model.

The following analysis offers an explanation of how a gravity instability initiates and why a seed is required. In particular, it demonstrates that salt-dome growth does not occur if a horizontal salt-sediment sequence possesses no initial irregularities; that if the salt surface is irregular, the direction of flow is away from highs and into lows until the density of the sediment onlapping the irregularities is greater than that of salt; that the hydraulic gradient induced by an uneven sediment load is much larger than that induced by buoyant forces in an equivalent setting; and that in a dipping salt-sediment sequence, regional downhill salt flow masks buoyant salt motion.

Case 1: No Flow with Perfect Horizontal Strata. In figure 7.4, the section above the salt layer is laterally uniform. Consequently the hydraulic gradient within the salt equals the gravity potential gradient because the overburden weight is everywhere the same. This is also true when all of the layers are horizontal, but additionally the gradient of the gravitational potential is zero because the strata are horizontal. Thus in uniform horizontal strata the hydraulic gradient is zero, no flow occurs, and no gravity instabilities (salt domes) form.

This result does not contradict previous theoretical analyses, all of which assume horizontal strata and all of which predict dome growth if a density inversion exists. This is because these earlier analyses have a built-in assumption that the original salt-sediment interface possesses irregularities of random size and shape, distributed randomly over the surface. These irregularities are the mathematical equivalents of the "seeds" used by experimenters to initiate dome growth in models. What the theories predict and the experiments demonstrate (see Biot 1963a, 1963b; Biot and Ode 1965; Nettleton 1934) is that a density inversion causes stresses across irregularities that in turn cause salt flow into, instead of away from, those irregularities. Unfortunately, these simple observations are commonly hidden in the mathematical complexity of the analysis.

Case 2: Flow Occurs when the Salt-Sediment Interface is Irregular. A schematic representation of a horizontal salt-sediment interface with an initial irregularity is given in figure 7.8. The vertical elevation of the irregularity may be considered as grossly exaggerated in the figure; in a later example the irregularity is assumed to be 30.5 m (100 ft) high and 8 km (5 mi) across. This implies a vertical exaggeration of about 50:1 in figure 7.8. For example, the bump in this salt surface could be the result of uneven deposition or postdepositional deflation or dissolution.

The elevation Z_1 of the top of the irregularity is of special impor-

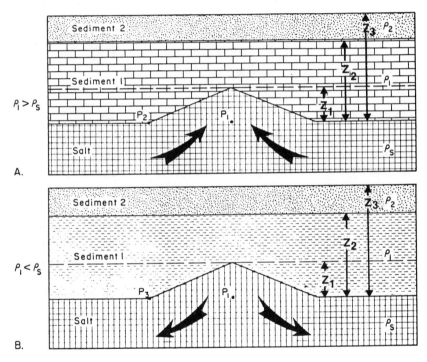

Figure 7.8
Horizontal strata overlying salt bed with positive surface irregularity. Flow is away from irregularity if enveloping sediment density is less than that of salt (b), and into the irregularity if the sediment density exceeds that of the salt (a). (Jackson and Galloway 1984).

tance. Both the gravity and pressure head are constant along Z_1 because it is a line of equal elevation and because strata of equal density distribution and thickness are present above the line. Because of this uniformity, the contribution to the hydraulic head from the material above Z_1 is uniform and thus can be neglected in the analysis. This is true regardless of whether the density of sediment layer 1 or layer 2 is greater than that of salt.

The configuration of strata below Z_1 in figure 7.8 is identical to that in figure 7.6 except that only one sediment layer onlaps the inclined salt-sediment interface in figure 7.8. Thus the results obtained in section 2 above apply without modification. Consequently, we can state that if the density in the Z_1 interval is less than that of salt, flow is downhill away from the irregularity (figure 7.8b) and the size of the irregularity diminishes. If the density in the interval is greater than that of salt, flow is uphill into the irregularity, the irregularity is magnified, and salt-dome growth has begun (figure 7.8a).

The Role of Overburden

The previous analysis demonstrates that a uniformly thick overburden without lateral density variations exerts no direct influence on the direction or rate of salt flow. In fact, because a uniformly distributed overburden does not contribute to differential stress, it cannot "cause" salt flow. Furthermore, additional increases in overburden thickness have the same result. However, an increase in overburden thickness does cause compaction in buried sedimentary layers, increasing their density. Once the density of the sediment immediately overlying the salt exceeds the density of salt, the direction of salt flow changes and flow occurs *into*, not away from, positive irregularities. At such time, salt-dome growth by buoyancy begins.

The bulk density of Gulf Coast sediments increases with burial depth and reaches the density of salt after about 914 m (3,000 ft) of burial (Dickinson 1953). This means that at burial depths of less than 914 m (3,000 ft) in the Gulf Coast, salt ought to flow away from positive irregularities, and the height of these features ought to diminish with time. Flow into these anomalies ought not to begin until the density of the overburden exceeds that of the salt. At about 1,219 m (4,000 ft), the density of Gulf Coast sediment is about 2.25 gm/cm^3 (Dickinson 1953), and flow into highs may begin, provided that the irregularities still exist. As burial continues, sediment density increases at lesser rates, reaching 2.4 gm/cm^3 at 3,353 m (11,000 ft) and approaching 2.5 gm/cm^3 at 6,096 m (20,000 ft). Bouyant forces that cause flow into highs, resulting in dome growth, increase along with this increase in sediment density. Consequently, the analysis shows that salt domes can form and grow because of increasing burial depths. The driving forces are very small, however, and the response of the salt is inadequate. Jackson and Talbot (1986) show that an irregularity must have relief in excess of 163 m to grow into a pillow of typical dimensions and within the normal time frame for the Gulf Coast. Thus other mechanisms may be more important in actual geologic settings.

The assumption in the preceeding examples that the overburden is composed of perfectly tabular layers whose density varies only with depth of burial is not representative of many natural settings. Starved basins may approximate this condition. Elsewhere, thickness variation is the rule, not the exception. For instance, if a basin is filled by a prograding delta, a substantial thickness anomaly is created in the overburden. The resulting uneven sediment load sets up hydraulic gradients in all underlying strata, with the result that salt (or shale) flows away from sediment thickness anomalies. Many other sedimentary load anomalies are common. For example, reefs prograding over sulfate-filled lagoons, outbuilding turbidite cones, subsea lava flows, oolite shoals, and even strike-fed beach sands and desert sand dunes are common to many sedimentary basins. Each will induce fluid gradients in an underlying salt layer.

Overburden Thickness Variation Resulting in High Fluid Head Gradients

The effectiveness of gravity instability as compared to other mechanisms for initiating dome growth is indicated by the gradient of the total fluid head. As an example, consider figure 7.8a, with $\rho_s = 2.2$ gm/cm^3, $\rho_z = 2.25$ gm/cm^3 (corresponding to a depth of burial of 1,219 m or 4,000 ft), $Z_1 = 30.5$ m (100 ft), and the breadth of the irregularity equal to 8 km (5 mi). Using equations 12a and 12b, the total fluid head gradient is calculated to be $0.0002\,g$, with the direction of flow into the irregularity.

Contrast this with the hydraulic gradient initiated by an elongate delta about 8 km (5 mi) wide prograding onto a shelf (figure 7.9). If the mass of sediment associated with the delta lobe is 30.5 m (100 ft) thick, this delta creates an irregularity on the surface of the shelf the same size and shape as the irregularity on the salt surface considered in the preceding paragraph. If a layer of salt exists at depth beneath the shelf, a hydraulic gradient is established in the salt beneath the prograding delta. Using equations 10a and 10b, the fluid gradient resulting from the delta is calculated to be $0.004\,g$, if the newly deposited deltaic sediment density is only $2.0\,\text{gm/cm}^3$. The gradient is twenty times as great as that induced by buoyancy on the positive irregularity considered in the preceding paragraph. It is important that the dimension of the salt-surface irregularity and the "birdsfoot" delta were made the same to remove any dimensional effects. Clearly, uneven sedimentary loading induces much larger hydraulic gradients within natural salt layers than do density inversions. Thus uneven sedimentary loading is more important in initiating salt structures and in governing salt flow. In fact, it can be demonstrated in many basins that salt flow begins with the first episode of uneven sediment loading.

The effect of the differences in the anticipated hydraulic gradient calculated for buoyancy vs sediment load is much amplified in the development of salt structures because of the mechanical characteristics of halite. Constant strain-rate experiments (Heard 1976) show that most permanent deformation is accomplished through steady-state creep (figure 7.10). The rate of creep is very sensitive to the magnitude of the differential stress (the hydraulic gradient). Figure 7.10 illustrates that an order-of-magnitude change in strain rate re-

---

Figure 7.9
Comparison of the effectiveness of buoyancy (a) vs differential loading (b) as a halokinetic agent. For the same size of structure, differential loading causes a higher flow gradient and is therefore more effective than is buoyancy (Jackson and Galloway 1984).

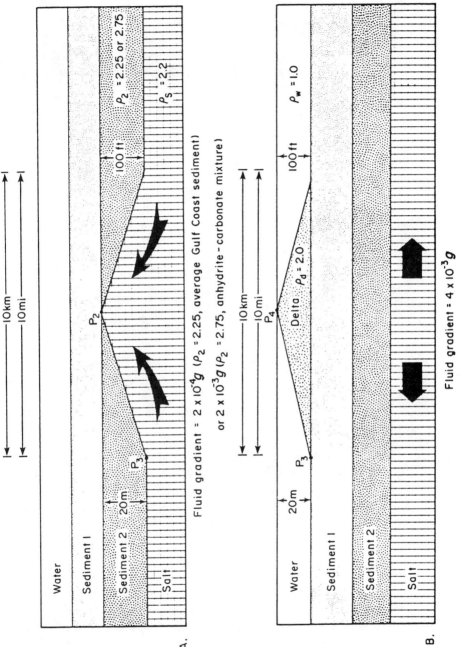

A.

Water

Sediment I

Sediment 2 $\rho_2 = 2.25$ or 2.75

$\rho_S = 2.2$

100 ft

20 m

P_2

P_3

10 km

10 mi

Salt

Fluid gradient = $2 \times 10^{-4}g$ ($\rho_2 = 2.25$, average Gulf Coast sediment)

or $2 \times 10^{-3}g$ ($\rho_2 = 2.75$, anhydrite-carbonate mixture)

B.

Water

Sediment I

Sediment 2

$\rho_w = 1.0$

Delta; $\rho_d = 2.0$

100 ft

20 m

P_4

P_3

10 km

10 mi

Salt

Fluid gradient = $4 \times 10^{-3}g$

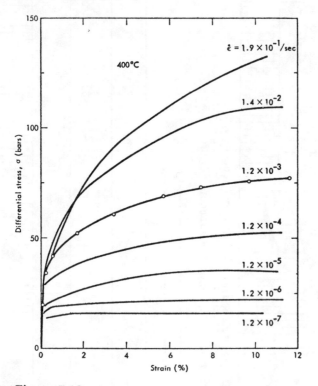

Figure 7.10
Differential stress-strain curves for polycrystalline halite extended at 2kb; 1.9×10^{-1} to 1.2×10^{-7} second^{-1} (Heard 1972).

sults in only a 15% difference in stress. The minimum difference in hydraulic gradient between buoyancy and sediment loading, 80%, translates into a difference in strain rate of more than four orders of magnitude. Under these circumstances the contribution of bouyancy to total deformation is negligible. Jackson and Talbot (1986) calculate that buoyancy induces growth 10^5 times slower than does sediment loading in low-relief structures.

The above analytic studies are supported by geologic evidence. Fisher (1973) found that dome growth in various Gulf of Mexico regions correlates with episodes of major deltaic progradation into the area. Growth of domes into the northern interior basins can also be correlated with delta progradation. Growth of these domes began during the influx of Late Jurassic Cotton Valley and Early Cretaceous Travis Peak and Tuscaloosa deltas (Loocke 1978; McGowen and Harris 1984).

When a salt layer is loaded unevenly, salt within the layer flows away from the loading anomaly. This lowers the anomalously thick column of sediment into areas of salt evacuation, providing space for additional sedimentation. This apparent subsidence tends to further concentrate deposition at that locality. The result is an increasingly unbalanced load which accelerates the flow of salt away from the site of sedimentation. Ultimately, if the sedimentation rate is limited, a closed topographic depression forms. As such, the facies tracts are fixed geographically either until sedimentation accelerates or until all salt is extruded from beneath the sediment anomaly.

The conclusion one draws is that the strutural style first developed in a near-horizontal salt-sediment sequence is one of closed depressions whose shape, size, and orientation reflect the sedimentary process causing the sediment loading anomaly. At the base of the continental slope of the Gulf of Mexico south of Louisiana and Texas, the style is one of vaguely elongated, strike-oriented, randomly-located closed depressions (Lehner 1969). Presumably these represent localized sites for the accumulation of turbidites and debris flows originating at various positions in the shelf at times of major deltaic progradation.

Inclined Salt-Sediment Sequences

We have already considered in section 2 an inclined salt bed with uniform parallel overburden layers. In that example, flow was found to be always downhill, eliminating the possibility of the formation of true salt domes and raising the possibility of a gravity slide involving the overlying strata. If the same sequence is overlain by a prograding delta system, salt flow will also occur from beneath the sediment wedge, principally downdip but also along strike, toward areas not yet covered by the new mass of sediment. It is not possible, however, for the delta system to create a depositional pattern that will cause uphill salt flow. Similarly, if a carbonate shoal progrades basinward, it will overload updip areas, accelerating downdip flow of underlying salt. Consequently, with or without uneven sediment loading, it is clear that the resulting fluid gradients within a buried salt layer dictate downhill, not converging, flow. Without converging flow, ordinary salt domes cannot form.

4. PIERCEMENT SALT DOMES, NORTHERN GULF OF MEXICO PROVINCE

"Piercement" salt domes that occur in the central, "flat" parts of the northern interior salt basins of the Gulf of Mexico do not form as a result of a mass of salt rising bouyantly upwards through an overlying sequence of sedimentary strata. Rather, such domes are actually salt pillars extruded through erosional holes in the sedimentary cover of predecessor concordant salt structures. The most common predecessors to piercement domes are salt "pillows," typically initiated by an episode of uneven sediment loading. The extrusion is caused by active sediment loading of the predecessor salt structure, as is the initiation of growth of the original structure. Consequently, piercement domes follow a predictable evolutionary sequence that terminates in a mature, stable configuration. A stable configuration is characterized by complete evacuation of salt from the original salt structure; steep or overhanging dome flanks; and the existence of a continuous sedimentary stratum, preferably Pliocene or older, across the crest of the feature. Bishop's (1978) analysis of the characteristics of capping layers is particularly illuminating. Once a dome reaches this stable configuration, growth ceases, regardless of whether or not the dome is buried beneath an increasing cover of sediments. Even if the dome is unroofed by erosion, no further salt extrusion occurs. Solution collapse is more likely in such circumstances. Destruction of an existing dome occurs only through erosion and solution by groundwater flow.

Growth History of Piercement Salt Domes

All interior domes that have been studied exhibit the same stages in their development. These are: 1) initiation, 2) salt pillow formation, 3) truncation; 4) extrusion and collapse; and 5) burial. This evolutionary history was first recognized by Trusheim (1960) and Sannemann (1968). These writers failed to recognize, however, that the salt core of a pillow (stage 2) was exposed subaerially as a result of erosional truncation (stage 3) before the extrusion of the salt pillar (stage 4). This fact was first recognized by Kehle (1971) and later confirmed by Loocke (1978). Trusheim (1960) schematically summarized this evolutionary growth sequence. A similar reconstruction for the Hainesville Dome, Wood County, Texas, by Kehle (1971) is shown in figures 7.11a and 7.11b. The extrusion phase is especially significant. During this period, the salt leaves its proper stratigraphic position (previously the structure was concordant) and passes through an eroded hole in the capping sediments. If sedimentation occurs simultaneously, the new deposits onlap the extruding salt pillar.

Not surprisingly, much salt is lost during this stage of the dome's development. Loocke (1978) calculated that of $184\,km^3$ (45 cubic miles) of salt in the initial salt pillow at Hainesville, only $37\,km^3$ ($9\,mi^3$) presently remain in the pillar. The rest was lost through erosion and dissolution. Thus, during the extrusion phase, salt does move upward past the newly deposited onlapping sediments. This process deforms the immediately adjacent sediments and destroys the original sedimentary contacts. The geometry of the contact is also changed during this phase. The sediments collapse into the void created through evacuation of the original salt pillow.

The final stage of the evolution of a dome occurs as the last volumes of salt are extruded from the underlying salt pillow. During this time, sediments commonly bury the dome, and subsequent growth is insufficient to raise the new sediments above the base level of erosion. Instead, these latest sediments are merely arched over the salt pillar. If further sedimentation occurs, these strata will extend undeformed across the structure, indicating that all growth has ceased.

Determining Growth History

The growth history of an interior basin salt dome is revealed by high-quality reflection seismic data, if interpretable data reach to the depth of the pre-salt surface. The seismic coverage must extend well beyond the primary withdrawal area—typically 19 to $24\,km$ (15–20 mi) from the center of the dome. Extensive coverage insures proper identifi-

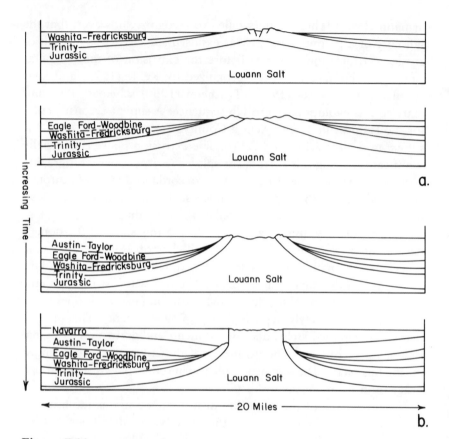

Figure 7.11
(a) Interpreted configuration of the Hainesville Dome at the end of Fredricksburg deposition (upper diagram) and Woodbine deposition (lower diagram). Reconstruction made using seismic time section (Kehle 1971). (b) Interpreted configuration of the Hainesville Dome at the end of Taylor deposition (upper diagram) and Navarro deposition (lower diagram). Reconstruction made using seismic time section (Kehle 1971).

cation of the principal withdrawal sink. The growth history is reconstructed by preparing isochron profiles for a sequence of seismic stratigraphic units and superimposing them sequentially. Migrated depth sections are best suited for the reconstruction process, although results derived from time sections yield similar results. Normal move-out velocities are inadequate for preparation of migration and depth sections. Unless more accurate interval velocities can be obtained, time sections should be used in reconstructions.

A Common Depth Point (CDP) profile of adequate quality for reconstructing the growth history of an interior basin piercement dome is given in figure 7.12. The profile clearly shows the pre-salt surface, and lateral coverage extends well beyond the primary peripheral salt withdrawal area. The record possesses good reflection quality throughout most of the post-salt sedimentary sequence. A profile must have all of these characteristics to allow reconstruction of the growth history. The profile crosses the Hainesville Salt Dome, Wood County, Texas, along a northeast-southwest line.

The pre-salt surface reflection is identified on figure 7.12. Experience in the northern interior basins shows that this surface is almost always planar and easily identified. It is undeformed except in a few locales where it is offset along normal faults of early to middle Mesozoic age. The apparent disruption of the pre-salt surface beneath the dome in figure 7.12 results from the method of presentation—horizontal distance vs reflection time. Because seismic energy travels faster through the salt dome than through the surrounding sediments, the pre-salt-surface reflection arrives in a shorter time interval below the dome than to either side (Tucker and Yorsten 1973).

The validity of this explanation is confirmed by a computer simulation of the seismic section shown in figure 7.12. A model interpretation of the depth section (figure 7.13) is introduced into an AIMS program, which computes the seismic profile that would be obtained across such a geologic feature (figure 7.14). Notice that the model has a continuous planar pre-salt surface beneath the dome, whereas the seismic section shows an elevated pre-salt surface beneath the dome similar to that recorded on the actual seismic profile.

The primary salt withdrawal sink is identified as the position of maximum isochron values for the deeper reflection packages. Typically, the position of maximum isochron value is the same for several of the lowermost stratigraphic units. The position of the maximum isochron value is shown in figure 7.12. Proper identification of these positions requires that the seismic profile extend well beyond the center of the primary sink. However, a profile need not extend as far beyond the sink as does figure 7.12.

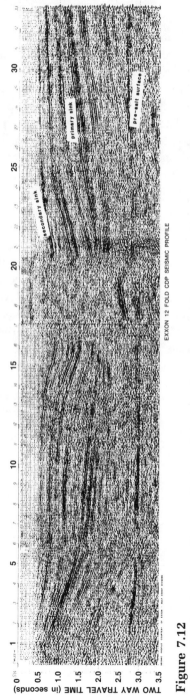

Figure 7.12

12-fold CDP seismic profile, trending northeast-southwest across the Hainesville Salt Dome, Wood County, Texas. Courtesy of Exxon Corporation (Loocke 1978).

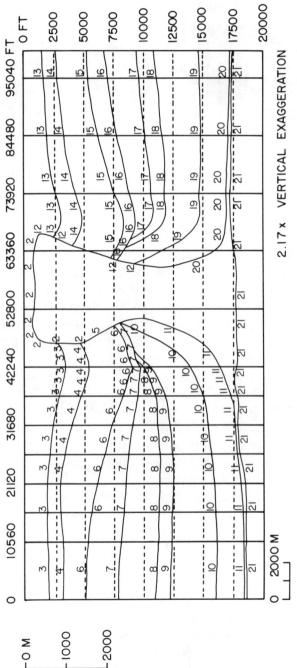

Figure 7.13
Model of a migrated seismic profile (figure 7.12) in depth-output by AIMS (Loocke 1978).

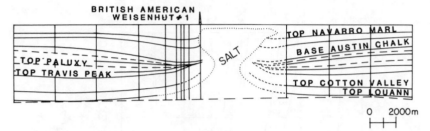

0   2000m
├─┼─┤

Figure 7.14
Depth model used in reconstructing growth history of the Hainesville
Salt Dome, Wood County, Texas (Loocke 1978).

Preparing a Reconstruction

A reconstruction of the growth history of an interior basin piercement salt dome is accomplished by adding, one at a time, each seismic unit on top of the last. In each step the top surface of the unit
is presumed to be horizontal. The isopach variation within the unit
is accommodated by conformable deformation of the underlying sediments and salt. It is assumed that all deformation in the post-salt
sediments is the result of deformation of the underlying salt. This
assumption is quite reasonable, because the pre-salt surface is planar
and undeformed, whereas the sediments overlying the salt are highly
deformed and exhibit significant variation in thickness across the area
of investigation. Because of the horizontal and vertical dimensions
involved and the limitations in precisely determining isopach thickness, the assumption of a flat surface is valid if sea-floor relief is less
than 61 m (200 ft) over the study area.

The most accurate reconstruction of a dome's history is accomplished using a migrated depth section. Preparation of such a section
requires accurate interval velocity data for the various seismic units.
Experience has shown that normal move-out velocities are of insufficient accuracy for this purpose. Therefore, either velocity surveys
or sonic logs at various sites within the area are needed for the required velocity information. To compound the problem, the interval
velocities reflect both the lithology and depth of burial of a unit.
From figure 7.14 it is clear that the thickness and burial depths of
most of the units vary widely across the area. Consequently, multiple
surveys are required to permit an adequate representation. Where

such data are unavailable, experience shows that isochron values may be used in the reconstruction with acceptable results. Migrated depth sections prepared with poor velocity data are commonly of poor quality and may make interpretation more difficult.

An Example of Reconstruction

To illustrate the method, reconstruction of the Hainesville Dome, Wood County, Texas, is presented below. A line drawing of the migrated depth section used in this reconstruction is shown in figure 7.14. The initial salt surface is presumed to be a horizontal plane extending continuously across the present site of the dome. Upon this surface is placed the first seismic stratigraphic unit, the Smackover to the Cotton Valley Lime sequence (Late Jurassic). This unit does not exhibit isopach variation, suggesting that no deformation accompanied its deposition (figure 7.15, section A). The surface of the unit is interpolated across the present site of the dome. This interpreted value is maintained throughout successive stages in the reconstruction, except where erosion may have truncated it.

The second seismic unit exhibits systematic isopach variation, suggesting contemporaneous deformation of the underlying salt and first sedimentary unit (figure 7.15, section B). The unit corresponds to the Cotton Valley Group (Late Jurassic) and the Travis Peak Formation (Early Cretaceous). Thus, initial salt movement corresponds to an episode of major deltaic sedimentation in the area. The resulting structure is a broad, low-relief salt pillow that is 32 km (20 mi) across with 550 m (1,800 ft) of structural relief. Again, the surface of the unit is interpolated across the no-data area. More importantly, the shape of the deformed unit below is also interpolated. This interpolation is done in a conservative way; the curvature of the surface is always increased toward the dome crest, thus minimizing its elevation. The isopach thickness obtained in this manner is maintained in successive stages of the reconstruction.

Each succeeding unit is then added sequentially in the same manner as before. This procedure is illustrated in figure 7.15, sections C, D, and E. The configuration of the structure in Eagle Ford time is illustrated in section F. At this time, the feature was 20.8 km (13 mi) wide, exhibited 3,660 m (12,000 ft) of structural relief, and poten-

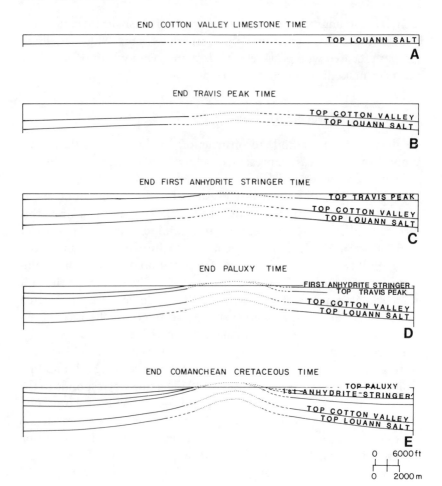

Figure 7.15
Sequential reconstruction of Louann Salt Dome growth (Loocke 1978).

tially as much as 762 m (2,500 ft) of topographic relief (assuming no erosion). If erosion of the crest kept pace with structural growth, the salt core of the structure would lie only a few feet below the surface at this time. This stage is critical in the growth of the structure. Shortly thereafter, the salt breaks to the surface and begins extruding from the crest of the feature, probably beginning during Austin Chalk time (figure 2.15, section G). Initial escape of the salt is probably accom-

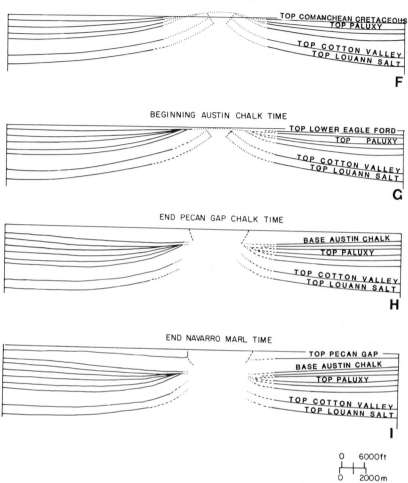

END LOWER EAGLE FORD

TOP COMANCHEAN CRETACEOUS
TOP PALUXY
TOP COTTON VALLEY
TOP LOUANN SALT

F

BEGINNING AUSTIN CHALK TIME

TOP LOWER EAGLE FORD
TOP PALUXY
TOP COTTON VALLEY
TOP LOUANN SALT

G

END PECAN GAP CHALK TIME

BASE AUSTIN CHALK
TOP PALUXY
TOP COTTON VALLEY
TOP LOUANN SALT

H

END NAVARRO MARL TIME

TOP PECAN GAP
BASE AUSTIN CHALK
TOP PALUXY
TOP COTTON VALLEY
TOP LOUANN SALT

I

0   6000ft

0   2000m

plished through exposure along the footwall of a crestal normal fault, and facilitated by stoping of smaller fault blocks from the extended crest of the pillow. Extrusion of salt at the surface is accomplished by collapse of the original salt pillow. Incidently, there are salt structures in the interior basin that presently exhibit this stage of evolution.

The development of piercement salt pillars is illustrated in figure

7.15, sections H and I. The salt extrudes from the eroded hole in the crest of the salt pillow, and the pillow collapses. The upper Cretaceous sediments deposited at the time of extrusion abruptly onlap the extruding spine. Outward flow into these low-density, mobile shales is predicted by the theory presented earlier and is the most likely origin of the large overhang on this dome (as in figure 7.14). The collapse of the original structure is reflected in the isopach character of these Upper Cretaceous units, which are thickest adjacent to the salt pillar. Thus, they demonstrate that salt was evacuated from the pillow to form the piercement pillar. As mentioned earlier, volumetric calculations by Loocke (1978) show that 45 mi^3 of salt existed in the original salt pillow, but only 9 mi^3 are now present in the salt pillar. The other 36 mi^3 were dissolved and eroded away during the formation of the piercement pillar during the Late Cretaceous.

The quality of the seismic data is insufficient to accurately define the time at which dome growth ceased. However, abundant shallow well-control in the area permits the mapping of the latest Cretaceous strata across the dome, establishing that activity ceased nearly 60 million years ago. A more detailed account of the reconstruction methods and the growth history of the Hainesville Dome is presented by Loocke (1978).

From this history it can be concluded that once the salt has been completely evacuated from the original pillow, growth of the piercement pillar ceases. In the interior basins of the northern Gulf of Mexico, most domes reach this evolutionary stage in the Late Cretaceous or early Tertiary. Because sedimentation also ceased at that time, undeformed capping sediments are rarely present across domes in this region. But sediments of considerable age (30 to 50 million years) remain intact across most of these structures, indicating that growth ceased during the early Tertiary at the latest.

Growth Rate of Piercement Salt Domes

Several investigators have estimated salt-dome uplift rates using structural relief data obtained from sediments enveloping a dome. This method, discussed below, underestimates the maximum uplift rates because it does not account for the extrusion of salt from the core of the structure. Extrusion rate can also be estimated, provided adequate subsurface and seismic data are available.

Dome growth at Hainesville began slowly, accelerated to a period of very rapid growth, and then decelerated to a negligibly slow rate, as is demonstrated by the graph of structural relief vs time (figure 7.16).

The maximum rate of structural uplift equaled nearly 800 ft per million years during the mid-Cretaceous, just prior to exposure of the salt core at the surface. Minimum growth rates occurred during the past 40 million years, during which time growth probably ceased. Cessation of growth cannot be proved because no strata younger than 40 million years cap the dome. Instead, the total uplift of the youngest strata on the dome is assumed to have occurred at the beginning of the time period, and the rate of uplift has probably decelerated since then. Because no salt remains in the original salt pillow to feed future growth, no acceleration or rejuvenation of dome growth is anticipated.

The method described above adequately measures uplift rate, provided that the salt core of the dome is covered by other rocks. During extrusion no rocks cover the crest of a dome, and salt rapidly escapes

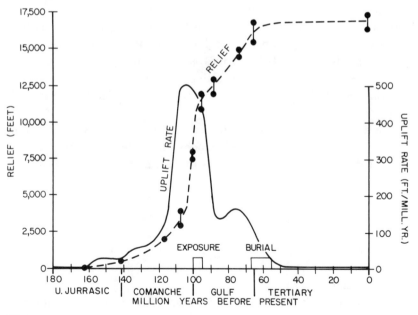

Figure 7.16
Structural relief—Hainesville Salt Dome.

through the hole in the cover of the enveloping sedimentary rocks. Rock masses located within the salt core would experience this rapid uplift, not the much slower uplift of the surrounding sediments.

The upward velocity of the salt escaping from the original salt pillow may be estimated reliably using a method that is outlined below, provided adequate subsurface and seismic data exist. The escape velocity of the salt cannot be measured directly. Rather, it is calculated from a knowledge of the volume of salt loss during a known time interval, divided by the cross-sectional area of the hole through which the salt escaped.

The volume of salt loss during a geologic time interval is estimated from the amount of collapse of the underlying salt pillow during that interval. The collapse volume is reflected directly in the isopach of the sedimentary unit deposited during the time interval in question. The minimum thickness of the unit, which occurs at the outer edge of the rim syncline, is assumed to reflect local subsidence of basement during the time interval. This assumption is reasonable provided that the locality has not suffered uplift because of salt flow. It is presumed that this thickness would have been deposited over the entire area if salt had not been withdrawn. The volume of salt withdrawal is assumed to equal the sedimentary volume in the rim syncline area that is in excess of what has been calculated from the minimum thickness observed. The volume of salt loss is taken to be equal to the salt withdrawn minus the salt remaining in the salt pillar. This information is plotted vs time before present and is labeled "loss rate" on figure 7.17.

The area of salt exposed at the surface during any geologic time interval is also estimated from isopachs of stratigraphic intervals deposited during extrusion. It is presumed that the zero depositional edge marks the position of the salt-sediment contact. The area within this zero edge is assumed to be the area through which salt is lost. The exposed area, plotted vs time before present, is shown in figure 7.17. The vertical salt velocity at the ground surface equals the quotient of total loss rate divided by the exposed area. This quotient is shown on figure 7.17 as "vertical salt velocity" and is plotted vs time before present.

In the Hainesville Dome, the maximum vertical rise rate of the salt during extrusion equaled 2,000 ft per million years and occurred dur-

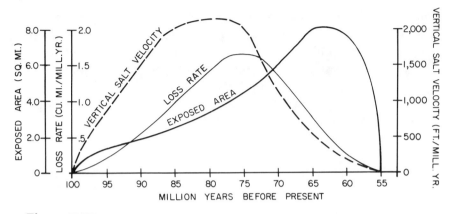

Figure 7.17
Salt loss—Hainesville Salt Dome.

ing a 10-million-year period that ended 75 million years before the present, as illustrated in figure 7.17. Extrusive uplift ceased 55 million years before the present, by which time most of the salt had been evacuated from the original salt pillow. The residual salt pillar was then buried beneath prodelta muds of the advancing Wilcox delta system. Although minor extrusion from the collapsed pillow has occurred during the past 55 million years, no measurable salt volume has been lost. The extruded salt simply added slightly to the height of the salt pillar.

Maximum rise rates at Minden Dome, Louisiana, have been estimated at 235 m (722 ft) per million years. The extrusion rates exhibited by these two domes may well be the highest for any domes in East Texas and northern Louisiana, because the initial salt pillows on these domes contained the largest volume of salt of any dome in their respective basins. Uplift rates following postextrusion burial are less than 6 m (20 ft) per million years.

5. EVOLUTION OF SALT STRUCTURES IN THE GULF OF MEXICO TERTIARY BASIN

According to current plate tectonic theory, the Gulf of Mexico is a small ocean basin formed during early Mesozoic continental rifting (Freeland and Dietz 1971; Beall 1973; Wood and Walper 1974). Initially, small fault-bounded topographic basins typical of such Atlantic-type, or trailing continental margins were formed. Then followed a period of basin filling, first with sediment derived from continental processes of erosion and deposition, and later with a sequence of thick evaporite deposits of probable Jurassic age. Later, the central part of the region subsided to oceanic depths. The continental margin has since been extended basinward by the deposition of successive offlapping wedges of clastic sediment during Tertiary time.

The original distribution and thickness of these evaporites has introduced an important element of structural mobility that has profoundly influenced the construction of the northern continental margin in the Gulf. Numerous salt-cored structures of varying size and shape have been identified beneath the continental shelf and slope (Lehner 1969, and figure 7.18). The varying morphology of these salt structures is a result of the deformational response of the original salt to variations in sedimentary loading and depositional history. The salt structures probably represent an evolutionary sequence of increasing morphological maturity from the foot of the continental margin to its northern rim. Salt structures beneath the northern continental margin have been grouped into a series of latitudinal belts on the basis of their shapes and sizes, which reflect their present stage of morphological development (Lehner 1969; Woodbury et al. 1973; Amery 1976).

On the lower continental slope, broad sediment troughs (figure 7.19, stage I) have subsided into a salt plateau, forming closed topographic depressions (Lehner 1969). Beneath the middle slope, the salt plateau is divided into a number of salt massifs separated by narrow, deepening, anastomosing troughs (figures 7.18 and 7.19, stage II). This contrasts with the common belief that it is the salt ridges that are responsible for the underlying structural grain of the slope region. Few ridges are evident. Instead, most positive salt features are broad, flat-topped massifs with a roughly polygonal outline.

Beneath the upper slope, the salt massifs have been deformed into a series of segmented ridges or semicontinuous salt uplifts (figure 7.20) that surround sediment-filled shallow basins (figure 7.19, stage III). This results in a regional honeycomb pattern (figure 7.18). On the continental shelf, large isolated salt spines or chimneys rise from the

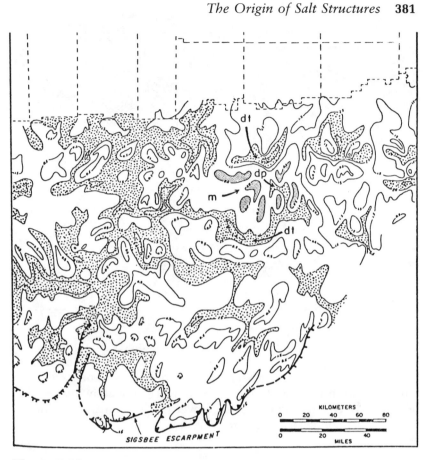

Figure 7.18
Isochron map of a portion of the continental slope, offshore Louisiana. Contours in seconds of two-way reflection time show sediment "thickness" above salt. M = salt massif with shaded salt structures; dp = depopod forming interior to massifs; dt = depotrough having an earlier depositional history on the slope (modified from Martin 1973).

old ridges or semicontinuous salt uplifts and seem to pierce thousands of feet of mostly shallow-water sediment (figure 7.19, stage IV). On the coastal plain, small isolated salt spines are representative of the final stages of salt-dome growth.

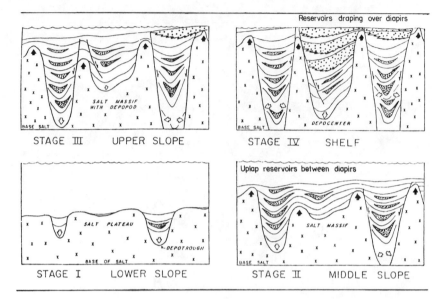

Figure 7.19
Effects of a diapiric evolution on reservoir geometry in a prograding
continental margin. Early evolution in depotroughs is characterized
by off-structure uplap reservoirs in a mobilizing sedimentary unit.
Later evolution is characterized by reservoirs of shallow-water sediments in an embanking unit draped over the structures and concentrated in depopods (Jackson and Galloway 1984).

Depotroughs and Depopods

Within the Gulf Coast geosyncline, two types of areas are envisioned,
here referred to as "depotroughs" and "depopods," where major,
albeit localized, accumulations of sediment may occur. Depotroughs,
initiated on the abyssal plain or lower continental slope, accumulate
significant thicknesses of sediment by the time the continental shelf
edge prograades to such an area. These are the topographically low
areas of thick sediments visible on figure 7.20. Because of continuing
salt flow from beneath the depotrough as deposition proceeds, these
sites may be devoid of underlying salt by the time the shelf edge progrades to them. If so, they will form a stable platform for shelf sedimentation (figure 7.19, stages II and III).

Depopods develop much later, atop salt massifs either on the upper slope or the shelf, whenever these areas are first inundated by deltaic or strand plain sediment. The massifs, on which the depopods form, are the "flat-topped" salt structures with thin sediment veneers between the depotroughs present in figures 7.20 and 7.21. The depopods form interior to the honeycombing salt ridges and are filled with shallow-water deltaic sediment. They may evolve into major depocenters as the center of shelf sedimentation progrades beyond these sites. The presence of abundant salt beneath a forming depopod permits major late-stage deformation (figure 7.19, stages II through IV), exemplified by the formation of salt spines or chimneys along the periphery of the depopod.

The Terrebonne Trough of southeastern Louisiana (figure 7.22) is a major deltaic depocenter thought to have been active in middle Miocene time (Limes and Stipe 1959). Late-stage depocenter growth, indicated by the thick accumulation of the massive sand lithofacies, suggests that this area is an example of a depopod or group of connecting depopods. Salt domes flanking the depocenter may mark the position of an older salt ridge (Atwater and Forman 1959), but are here interpreted to coincide with the edge of an old salt massif. The next domes occur far to the south, and the intervening area is stable and structurally high, presumably the site of a depotrough beneath which all salt had been evacuated prior to the time the area was inundated by mid-Miocene deltas. Faults paralleling the axis of maximum sand thickness are developed across both the northern and southern depocenter margins. They result from the stretching and elongation of these sediments as a result of their differential subsidence into the Terrebonne Trough depopod.

Another area illustrating the evolution of structures around a depopod is South Marsh Island, offshore Louisiana. The area is dominated by an elongate east-west trending basin with dimensions of 13 × 20 km that was delineated by structural and isochron mapping (Spindler 1977). The basin margin is structurally complex, with numerous salt structures and semiregional faults.

There are seven salt domes, lettered A through G in figure 7.23, somewhat uniformly spaced around the basin margin. Their presence indicates a significant thickness of salt underlying the area. The approximate locations of the salt domes, interpreted from the absence

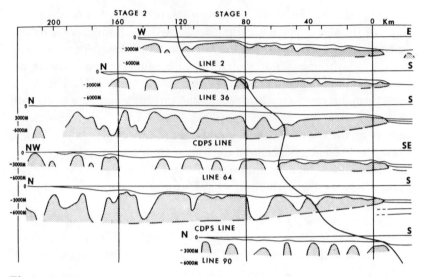

Figure 7.20
Panel of structural cross sections across the continental slope, showing interpreted salt features, developed from sparker and seismic profiles. Sections demonstrate stages in development of salt diapirs on
the slope. Stage 1 is the youngest; diapirs have not formed because
of insufficient sedimentation. Base of salt has been interpreted at the
southern end of two seismic profiles, which have much deeper penetration than sparker profiles. Sigsbee Escarpment is shown near the
right edge. Vertical exaggeration 2.6 (Humphries 1979).

of coherent reflections on seismic profiles, closely correspond to the
extent of salt at 3,050 m for the known domes in the area (Braunstein
et al. 1973, plate II, p. xi). Salt has been encountered on each dome
at depths between 180 and 3,050 m. Depth to the top of the salt layer
beneath the basin is probably in excess of 10,700 m, based on projections of the seismic data and from published regional cross sections (Woodbury et al. 1973:2437; Lehner 1969:2473). Three domes
(A, D, and F) are shallow, three are intermediate (B, C, and E), and
one (G) is deep-seated, according to a depth-of-burial classification
scheme (table 7.1) reviewed by Halbouty (1967:47). Dome A may
be a structure composed of several salt spines.

Most seismic profiles across or near domes in the study area show

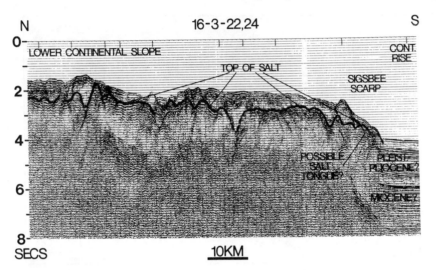

Figure 7.21
Seismic dip profile across the Sigsbee Scarp, the front of a large salt
nappe that has moved southward about 150 km (90 mi) into the Gulf
Abyssal Plain. Diapirically deformed upper surface of the salt nappe
is indicated as "top of salt" (Jackson and Galloway 1984).

no indication of migration of the axes of dome-related sediment thick
zones (rim synclines) toward the growing salt domes during subsequent depositional periods. Migration of a depositional axis indicates
that domal growth has progressed to a stage where most of the salt
has been evacuated from beneath the depositional thick. Subsequently, growth continues through a process of flank collapse. Stacking of the depositional axes (between 2.8 and 6.0 seconds, figure
7.24) strongly suggests the availability of a significant thickness of
salt beneath the thesis area that remained mobile into early Pleistocene time.

It is also a common feature of these domes for seismic reflectors
to continue dipping into a zone of incoherent reflections near the salt-
sediment interface, without turning up and onlapping the structures,
except possibly in the youngest stratigraphic intervals. This could indicate late domal growth or piercement of a thick stratigraphic section (Smith and Reeve 1970). However, it is more plausible that the
domes were always near the sediment-water interface (Halbouty and

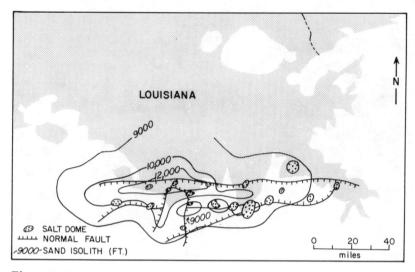

Figure 7.22
Structural elements and massive sand isolith, Terrebonne Trough, southeastern Louisiana (based on Limes and Stipe 1959).

Hardin 1950; Scholten 1959; O'Neill 1973; Johnson and Bredeson 1971). In such cases, the turn-up and trunction of reflectors would exist very close to the salt core, well within the zone of no coherent reflections. Spatially migrated seismic data would have helped to resolve questions about the truncation of the flanking sediments. This geometry is common to domes in the interior salt basins where salt is escaping through an eroded or structurally formed hole (Loocke 1978), but it is unlikely that domes on the outer continental shelf evolve in that manner.

Integrated Structural History

The growth histories of the three major structural elements (salt withdrawal basin, dome A, and the major basin-bounding fault) as recorded by the thickening or thinning indices show a remarkable parallelism with one another (figure 7.25), indicating possible syngenetic relationships among the structures. Interpretation of a sequence of structure and isopach maps and the growth history plots (figure 7.25) shows that the small, 13 × 29 km elongate basin labeled "depocen-

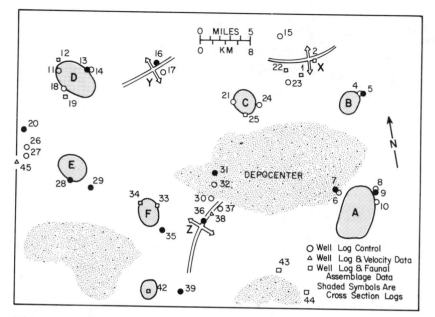

Figure 7.23
Map of positive and negative elements, South Marsh Island, Louisiana. Letters identify salt domes (A to G) and low-relief anticlines. Shaded areas are depocenters (Spindler 1977).

Table 7.1
Burial Depth of Salt Domes

Structure	Depth to Salt ft (m)	Classification
A	3,565 (1,090)	shallow
	11,670 (3,560)	deep
B	9,180 (2,800)	intermediate
C	8,043 (2,450)	intermediate
D	573 (180)	shallow
E	7,205 (2,200)*	intermediate
F	3,530 (1,080)	shallow
G	10,070 (3,070)	deep

Source of data: Base map provided by Pennzoil, Inc.

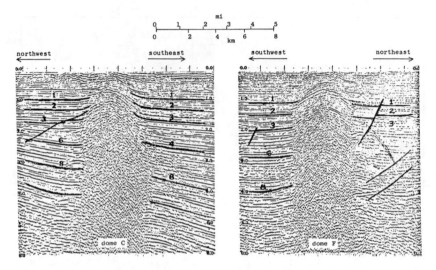

Figure 7.24
Typical seismic reflection profiles across salt domes. Note the lack of domeward convergence of reflectors and stacked depositional thickening along the axes (courtesy of Teledyne Exploration).

ter" in figure 7.23 originated as a small depopod perched atop an extensive salt massif on the upper continental slope during early Pliocene time. Major growth occurred from middle to late Pliocene time as the continental shelf edge prograded the salt massif.

The basin is a simple downwarp resulting from salt flow away from the site of depositional loading into surrounding positive salt features (seven salt domes and three low-relief anticlines), demonstrating that the area was one of significant original salt accumulation. Anticlinal structures are often separated from major sediment thicks by domes that align with one another parallel to the margins of the thickest sediment accumulations. Stacked axes of the depositional thicks adjacent to domes and a general lack of domal convergence of seismic reflections indicate that dome growth was initiated and maintained in relatively shallow water (less than 185 m) as progradation over an area of thick salt occurred. The largest salt dome in the area experienced major growth in early Pleistocene time that lags maximum depocenter subsidence, evidence that these domes are later-stage salt

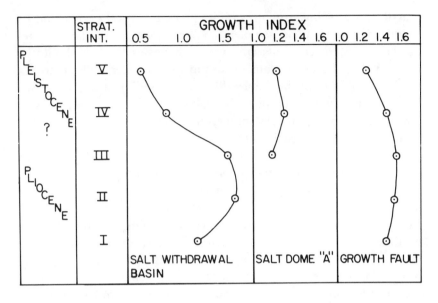

STRAT. INT.	GROWTH INDEX			
	0.5 1.0 1.5	1.0 1.2 1.4 1.6	1.0 1.2 1.4 1.6
PLEISTOCENE ?	Ⅴ			
	Ⅳ			
	Ⅲ			
PLIOCENE	Ⅱ			
	Ⅰ			
		SALT WITHDRAWAL BASIN	SALT DOME "A"	GROWTH FAULT

Figure 7.25
Comparative growth of each major structure vs stratigraphic interval.

spines rising from a preexisting salt ridge or rim surrounding the depopod.

Piercement Mechanism of Coastal Domes

The salt massifs on the continental slope are buried with only a thin veneer of abyssal plain sediment. Only the adjacent depotroughs receive sediment in the slope environment. The narrow salt spines or salt chimneys characteristic of domes along the coast originate along the periphery of these massifs during burial by deltaic sediments on the shelf. The mechanism by which an evolving salt spine penetrates these shelf sediments is of considerable interest, in that most of the hydrocarbon accumulations in salt domes occur within these "pierced" sediments.

When a salt ridge or massif is inundated by sediments of a prograding shelf, the sediment load forces salt out of the center of a massif into growing structures on its periphery. As the spine and its mantle of abyssal plain sediments moves upwards, the layer of recently deposited sediments is elevated above sea level. It should be recalled that the delta plain or strand line sediment fills in to sea level. Any growth of the dome will raise this sediment cap above sea level, where it will be eroded. Typically, the sediment cover is overstretched by the uplift, and numerous faults form that bound and segment the uplifted sediment.

Once uplifted above depositional base level, the newly deposited strata are eroded. Over the dome crest even the faults are eroded away. The resulting map view is that of a hole (later occupied by the salt core of the dome) from which radiate growth faults (figure 7.26). In this way, the salt core of the structure remains near base level at all times, and each successive sediment layer is removed by faulting and erosion shortly after deposition. During this step-by-step evolution, erosion need not entirely strip the sedimentary cover from the crest of the dome after each episode of sedimentation. However, the sediment cover must be kept sufficiently thin to permit uplift along the radial and trap-door faults; otherwise dome growth will cease. Once formed, it is likely that the same faults will be reactivated during each episode of uplift following each subsequent episode of sedimentation.

Compaction waters and hydrocarbon precursors draining into the uplifted sediment "lid" have direct access to the surface through the erosional breach. Large volumes of these fluids escape before subsequent deposition again seals the erosional wounds. Thus the depth of erosion (amount of uplift between deltaic progradations) plays an important role in determining which strata trap hydrocarbons.

6. SALT STRUCTURES ON BASIN FLANKS

The most generally accepted view of the origin and geometry of salt structures on the flanks of the interior basins of the northern Gulf of Mexico province was developed by Hughes (1968). He proposed four main types of structures progressing from the border to the center of an interior salt basin: peripheral salt ridges, low-relief salt pillows, intermediate salt pillows, and high-relief salt pillows (illus-

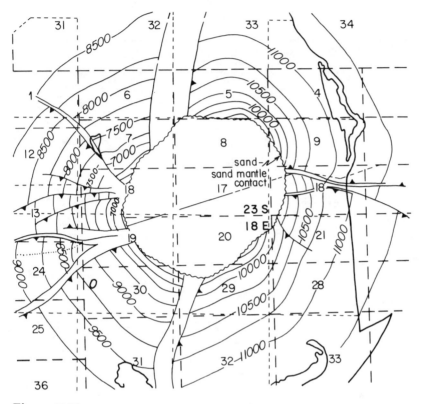

Figure 7.26
Structure map of Lake Pelto Field, Terrebonne Parish, Louisiana, showing large trap-door fault. Contoured on "25" sand. Contour interval is 250 ft.

trated from right to left in sections A, B, and C of figure 7.27). He also emphasized the following points:

• The base of salt is a fairly flat basinward dipping surface.
• The Louann Salt is thicker basinward.
• A porous zone at the top of the Smackover develops concomitant with structural growth.
• A fault zone (Pickens-Gilbertown-Polland) is associated with and occurs parallel to the peripheral salt ridges.

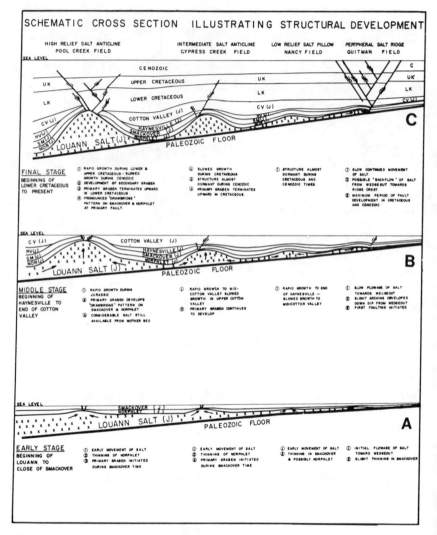

Figure 7.27
Schematic cross section illustrating the evolution of the salt structures discussed by Hughes (1968). Louann Salt was originally deposited as a basinward thickening wedge. Movement of salt began with deposition of Norphlet (Hughes 1968).

Hughes also noted that salt flow began only after a few hundred feet of Norphlet were deposited on top of the Louann Salt, but no mechanical explanation is given for this observation.

Modern seismic records across producing structures in the East Texas Basin reveal that structures in the position corresponding to Hughes' low-relief salt pillows and intermediate-relief salt anticlines commonly have large normal faults at or near the crest of the structures, with the downdropped basinward block far below regional dip. Similar structures are described as "salt rollers" by Bally (1981). Structures with this configuration cannot arise from simple arching over an accumulating salt swell. Rosenkrans and Marr (1967) recognized such structures and offered the explanation illustrated in figure 7.28. A graben forms along a regional hinge line along a basin margin (the "state-line" graben system along the Arkansas-Louisiana state line

Seismic Exploration of the Gulf Coast Smackover 201

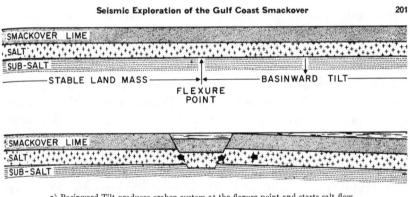

a) Basinward Tilt produces graben system at the flexure point and starts salt flow.

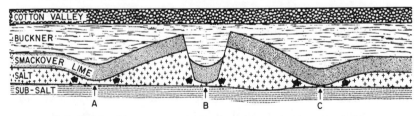

b) Buckner deposition accentuates salt flow. Salt flow stops when the supply of salt is exhausted at A, B, & C.

Figure 7.28
Postfaulting salt anticline tectonics (Rosenkrans and Marr 1967).

served as their model) because of flexure. After formation of the graben, subsequent deposition of Buckner sediments causes salt flow away from points A, B, and C for unexplained reasons. Haynesville Field, Arkansas, exhibits this structural style (figure 7.28). Using the techniques developed in sections 2 through 4 above, it can be shown that the minimum hydraulic head occurs midway along each fault and that salt ought to flow toward these points from either side, as suggested by Rosenkrans and Marr (1967).

Neither Hughes (1968) nor Rosenkrans and Marr (1967) suggest that basinal flow is an important factor in the development of these structures. This flow occurs in accordance with principles previously described. One of the results of this basinward subsidence is the development of a peripheral graben system that surrounds the interior salt basins of the northern Gulf of Mexico. This graben system has various names in different localities along the basin margin, including the Charlotte Fashing Person fault system (South Texas), the Mexia Talco fault system (East Texas), the South Arkansas fault system (Arkansas) and the Pickens Gilbertown fault system (Mississippi). Additional names probably are applied to this fault system in other localities.

Fault systems of this type originate because of downhill flow of salt, arising once basinward dip is established—though the opposite conclusion was reached by Jenyon (1985). Such systems are localized over areas where the salt thins rapidly or pinches out altogether, typically over contemporaneous normal faults or "basement" scarps against or across which the salt was deposited. The rate of downhill flow below the scarp (figure 7.29) is greater than above the scarp (where it may be zero), and consequently there is a net evacuation of salt from the vicinity of the old scarp. With time, this lowers the overlying section into the evacuated area, resulting in the formation of a graben over the scarp. The evolution of an identical fault system in abyssal plain sediments along the northern margin of the Campeche Escarpment is shown in figures 7.30 and 7.31. Additionally, these figures show some of the complementary structures that form downdip, in particular several anticlines, seen in figures 7.32, sections D to G. These structures are salt supported, but they are not salt structures. They result from a minor gravity slide that caused buckling in the sliding sediments. The distinction is that salt move-

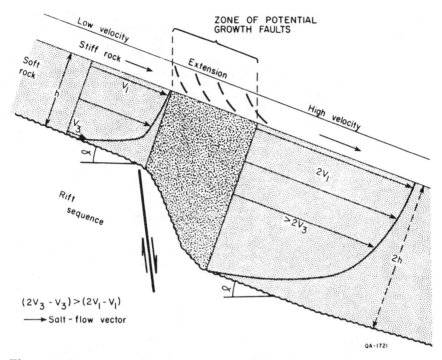

Figure 7.29
Formation of a zone of extension in a stiff glide sheet overlying a
soft glide zone (such as rock salt) of uneven thickness due to deposition across a fault scarp. Acceleration of the flowing glide zone causes
normal faulting in the stiff overburden. This model could apply either
to salt deposited on a rifted terrane and overlain by carbonate or to
an uneven thickness of shale overlain by sand (typical of Gulf Coast
Tertiary). (Jackson and Galloway 1984).

ment did not cause the structures. In fact, the growth of the structures
caused the salt to flow from the synclines to the anticlines.

The various structures occurring on the flanks of the interior salt
basins are interpreted as resulting from a gravity slide of the post-
salt sediments over the underlying salt. Sliding occurred early in the
depositional history of the area because no basal resistance to sliding
existed. This was because the salt itself was flowing basinward. Subsequently, these rapidly formed structures were buried by the first
major influx of paralic sediments.

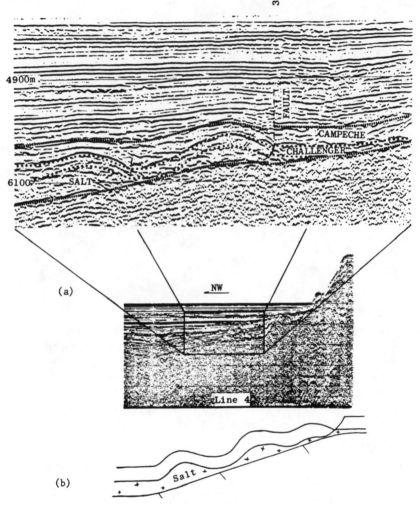

Figure 7.30
(a) Basinward movement of post-salt sediments along salt decolle-
ment on a Campeche salt horizon (southern Gulf); and (b) a diagram
of a similar salt feature on the northern Gulf Coast, as described by
Kehle (1970). Vertical exaggeration 5× (Long 1978).

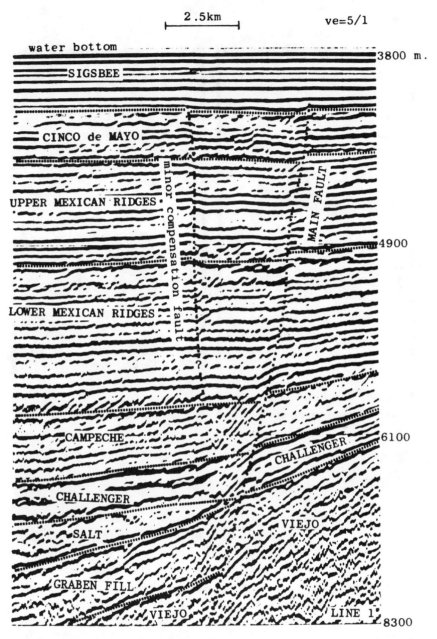

Figure 7.31
Peripheral faulting developed at the updip end of the Campeche salt creep system, resulting from downdip flow of fault away from buried fault-line scarp (Long 1978).

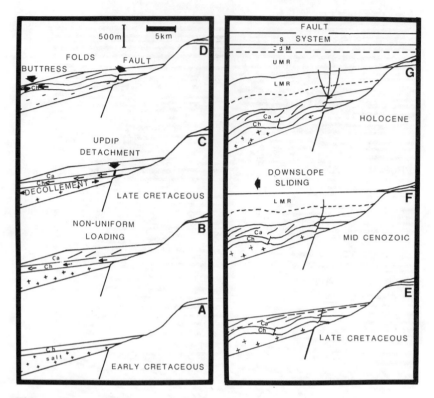

Figure 7.32
Development of fault system because of downslope creep of underlying salt, northern Campeche Abyssal Plain (adapted from Long 1978).

The updip grabens developed as pull-apart structures as the upper
plate slid downdip. Folds formed at the base of the continental slope
as a result of this movement. The flat basin centers, surrounded by
mobile, sliding flanks, were stationary, and acted as buttresses against
which the sliding plates crumpled. In this interpretation, the underlying Louann Salt does not cause the observed structures, although
its presence is necessary to permit sliding and to support the developing folds.

The gravity-slide interpretation explains the simultaneous development of grabens updip and elongate folds oriented parallel to depositional strike downdip. It also removes an enigma associated with

these salt-cored structures, which form prior to the accumulation of sufficient strata to cause bouyant salt movement (typically thought to require 1,220 m—4,000 ft—or more of overburden).

The structures arising from the gravity slide in turn influence the local depositional environments (figure 7.32). Horsts form in updip areas characterized by shoal-water deposition, principally skeletal or oolitic carbonates. Similar environments develop offshore over the crests of anticlines elevated to or above wave base. Dissolution and other secondary processes may alter the rock character significantly and may result in the formation of desired reservoir properties. The topographic expression of some updip structures may restrict circulation, resulting in the formation of hypersaline lagoons marginward of these structures. Percolation of these brines basinward (reflux) will encourage dolomitization of the more permeable carborate facies, while the lagoonal sediments themselves may serve as source rocks in certain circumstances. Should progradation occur, lagoonal sediments would serve as seals as well. In grabens and synclines further down-

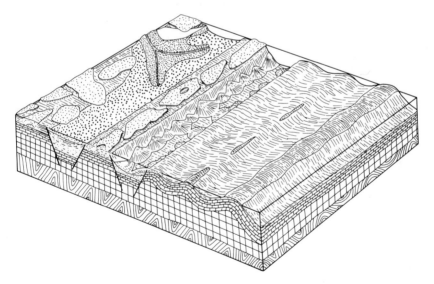

Figure 7.33
Schematic showing facies development during Haynesville time. Carbonate debris fans accumulate in offshore lows, while reefs and grain carbonates develop on highs. Deltas supply mud to grabens. Deltas also restrict circulation, causing evaporite deposition in interdelta areas.

dip, starved basin sediments, carbonate debris cones, and even euxinic shales may be deposited.

A schematic representation of this structural and sedimentologic complexity and the evolution of the salt-supported features in response to continued sediment loading is portrayed in figure 7.33. For example, in basins of this type, the era of carbonate sedimentation commonly ends abruptly when major delta systems prograde into the basin. The existence of an irregular bottom topography underlain by salt of variable thickness almost guarantees continual structural evolution. The deltas will differentially fill the topographic lows, thereby loading them more than adjacent highs. Updip, where little if any salt remains beneath the lows, minor adjustment of the structures occurs in response to new sediment load. Downdip, however, where thick salt remains even beneath the synclines, the differential loading causes major adjustment and growth of salt structures. In many areas, anticlines are uplifted above sea level and wave erosion cuts to the core of the fold. Subsequent growth results in the characteristic drawbridge structure (Hughes 1968) as illustrated in figure 7.27c.

REFERENCES

Amery, G. B. 1976. Structure of the continental slope, northern Gulf of Mexico. Paper presented at Beyond the Shelf Break. AAPG Annual Meeting and Gulf Coast Association Geological Societies. New Orleans, May 22, 1972.

Arrhenius, G., Sr. 1912. Zur Physik der Salzalagerstatten. *Meddelanded fran K. Vetenskabsakademians Nobelinstitute* 2(20):7–25.

Atwater, G. I. and M. J. Forman. 1959. Nature of growth of southern Louisiana salt domes and its effect on petroleum accumulation. *Bull. AAPG* 43:2592–2622.

Balk, R. 1947. Salt-dome structure. *Bull. AAPG* 31:1195–1299.

Balk, R. 1949. Structure of Grand Saline salt dome, Van Zandt County, Texas. *Bull. AAPG* 33:1791–1829.

Bally, A. W. 1981. Thoughts on the tectonics of folded belts. In K. R. McClay and N. J. Price, eds., *Thrust and Nappe Tectonics*, pp. 13–32. Oxford: Geological Society of London.

Beall, R. 1973. Plate tectonics and the origin of the Gulf of Mexico. *Gulf Coast Association Geological Societies Transactions* 23:9–114.

Biot, M. A. 1963a. Theory of stability of multi-layered continua in finite anisotropic elasticity. *Journal Franklin Institute* 276:128–153.

Biot, M. A. 1963b. Stability of multilayered continua including the effect of gravity and viscoelasticity. *Journal Franklin Institute* 276:231–252.

Biot, M. A. 1966. Three-dimensional gravity instability derived from two-dimensional solutions. *Geophysics* 31:153–166.

Biot, M. A. and Helmer Ode. 1965. Theory of gravity instability with variable overburden and compaction. *Geophysics* 30:213–227.

Bishop, R. S. 1978. Mechanism for emplacement of piercement diapirs. *Bull. AAPG* 62:1561–1583.

Braunstein, J. and G. D. O'Brien, eds. 1968. *Diapirism and Diapirs: A Symposium.* AAPG Memoir no. 8. Tulsa, Okla.

Braunstein, J., J. A. Hartman, B. L. Kane, and J. H. Amringe. *Offshore Louisiana Oil and Gas Fields.* Lafayette Geological Society/New Orleans Geological Society Special Publication no. 1. Lafayette/New Orleans.

Carter, N. L., and F. D. Hansen. 1983. Creep of rock salt. *Tectonophysics* 92:275–333.

Currie, J. B., H. W. Patnode, and R. P. Trump. 1962. Development of folds in sedimentary strata. *Bulletin Geological Society America* 73:5–674.

Danes, Z. F. 1964. Mathematical formulation of salt-dome dynamics. *Geophysics* 29:414–424.

DeGoyler, E. L. 1925. Origin of North American salt domes. *Bull. AAPG* 9:831–874.

Dickinson, G. 1953. Geological aspects of abnormal reservoir pressures in Gulf Coast region of Louisiana. *Bull. AAPG* 37:410–432.

Dobrin, M. B. 1941. Some quantitative experiments on a fluid salt-dome model and their geological implications. *American Geophysical Union Transactions,* vol. 22, part 2, pp. 528–542.

Fisher, W. L. 1973. Deltaic sedimentation, salt mobilization, and growth faulting in Gulf Coast Basin. *Bull. AAPG* 57:779.

Fletcher, R. C. 1967. A finite amplitude model for the emplacement of gneiss domes and salt domes. Ph.D. diss., Brown University, Providence, R.I.

Freeland, G. L. and R. S. Dietz. 1971. Plate tectonic evolution of Caribbean–Gulf of Mexico Region. *Nature* 6:55–62.

Gussow, W. C. 1968. Salt diapirism: Importance of temperature and energy source of emplacement. In *Diapirism and Diapirs,* pp. 16–52. See Braunstein and O'Brien 1968.

Halbouty, M. T. 1967. *Salt Domes: Gulf Region, United States and Mexico.* Houston: Gulf Publishing.

Halbouty, M. T. and G. C. Hardin. 1950. Types of hydrocarbon accumulation and geology of South Liberty salt dome, Liberty County, Texas. *Bull. AAPG* 35:1939–1977.

Hansen, F. D. and N. L. Carter. 1984. Creep of Avery Island rocksalt. In

H. R. Hardy and M. Langer, eds., *The Mechanical Behavior of Salt: Proceedings of the First Conference, Pennsylvania State University, November 1981*, pp. 53–69. Series on Rock and Soil Mechanics, vol. 9. Clausthal, Germany: Trans. Tech.

Hardy, H. R., Jr. and M. Langer, eds. 1984. *The Mechanical Behavior of Salt: Proceedings of the First Conference, Pennsylvania State University, November 1981*. Series on Rock and Soil Mechanics, vol. 9. Clausthal, Germany: Trans. Tech.

Heard, H. C. 1972. Steady-state flow in polycrystalline halite at pressures of 2 kilobars. In H. C. Heard, I. Y. Borg, N. L. Carter, and C. B. Raleigh, eds., *Flow and Fracture of Rocks*, pp. 191–210. Geophysical Monograph no. 16. Washington, D.C.: American Geophysical Union.

Heard, H. C. 1975. *Comparison of the flow properties of rocks at crustal conditions*. Report no. UCRL-76267, Lawrence Livermore Laboratory, June 25.

Heard, H. C. 1976. Comparison of the flow properties of rocks at crustal conditions. *Phil. Trans. Royal Soc. London* A283:173–186.

Hughes, D. J. 1968. Salt tectonics as related to several Smackover fields along the northeast rim of the Gulf of Mexico basin. *Gulf Coast Association Geological Societies Transactions* 18:320–330.

Humphries, C. C., Jr. 1979. Salt movement on continental slope, Northern Gulf of Mexico. In A. H. Bouma, G. T. Moore, and J. M. Coleman, eds., *Framework, Facies, and Oil-Trapping Characteristics of the Upper Continental Margin*, pp. 69–85. AAPG Studies in Geology no. 7. Tulsa, Okla.

Jackson, M. P. A. and W. E. Galloway. 1984. *Structural and Depositional Styles of Gulf Coast and Tertiary Continental Margins: Applications to Hydrocarbon Exploration*. AAPG Continuing Education Course Note Series no. 5. Tulsa, Okla.

Jackson, M. P. A. and C. J. Talbot. 1986. External shapes, strain rates, and dynamics of salt structures. *Geological Society America Bulletin* 97:305–323.

Jenyon, M. K. 1985. Basin-edge diapirism and up-dip salt flow in the Zechstein of the southern North Sea. *Bull. AAPG* 69:53–64.

Johnson, A. M. 1970. *Physical Processes in Geology*. San Francisco: Freeman, Cooper.

Johnson, H. A. and D. H. Bredeson. 1971. Shallow domes in Louisiana Miocene productive belt. *Bull. AAPG* 55:204–226.

Kehle, R. O. 1970. Analysis of gravity sliding and orogenic translation. *Bulletin Geological Society America* 81:1641–1664.

Kehle, R. O. 1971. Origin of the Gulf of Mexico. Geology Library (call no. Q557K260), University of Texas at Austin. Typescript.

Lachmann, R. 1911. Ker Salzauftrieb. In *Geophysikal Studien über den Bau der Salzmaassen Norddeutschland*. Halle: Wilhelm Knapp.

Lehner, P. 1969. Salt tectonics and Pleistocene stratigraphy on continental slope of northern Gulf of Mexico. *Bull. AAPG* 53:2431–2479.

Limes, L. L. and J. C. Stipe. 1959. Occurrence of Miocene oil in Louisiana. *Bull. AAPG* 43:2513.

Long, John. 1978. Seismic stratigraphy of part of the Campeche Escarpment, southern Gulf of Mexico. Master's thesis, University of Texas at Austin.

Loocke, J. E. 1978. Growth history of the Hainesville salt dome, Wood County, Texas. Master's thesis, University of Texas at Austin.

Lowell, J. D. and G. J. Genik. 1972. Sea-floor spreading and structural evolution of southern Red Sea. *Bull. AAPG* 56:247–259.

Martin, R. G. 1972. Salt structure and sediment thickness, Texas-Louisiana continental slope, northwestern Gulf of Mexico. U.S. Geological Survey, Prof. Paper no. 800B: B1–B8. Washington, D.C.

McGowen, M. K. and D. W. Harris. 1984. Cotton Valley (Upper Jurassic) and Hosston (Lower Cretaceous) depositional systems and their influence on salt tectonics in the East Texas Basin. In W. P. S. Ventress, D. G. Bebout, B. F. Perkins, and C. H. Moore, eds., *The Jurassic of the Gulf Rim*, pp. 213–254. Austin: Gulf Coast Section SEPM.

Nettleton, L. L. 1934. Fluid mechanics of salt domes. *Bull. AAPG* 18:1175–1204; also in *Gulf Coast Oil Fields* (1936), pp. 79–108.

Nettleton, L. L. 1943. Recent experiments and geophysical evidence of mechanics of salt dome formation. *Bull. AAPG* 27:51–63.

Nettleton, L. L. 1955. History of concepts of Gulf Coast salt dome formation. *Bull. AAPG* 39:2373–2383.

Ode, H. 1968. Review of mechanical properties of salt relating to salt-dome genesis. In *Diapirism and Diapirs*, pp. 53–78. See Braunstein and O'Brien 1968.

O'Neill, C. A., III. 1973. Evaluation of Belle Isle salt dome, Louisiana. *Gulf Coast Association Geological Societies Transactions* 23:115–135.

Parker, T. J. and A. N. McDowell. 1955. Model studies of salt-dome tectonics. *Bull. AAPG* 39:2384–2470.

Rogers, G. S. 1918–19. Intrusive origin of the Gulf Coast salt domes. *Economic Geology* 13:447–485; 14:178–180.

Rosenkrans, R. R. and J. D. Marr. 1967. Modern seismic exploration of the Gulf Coast Smackover trend. *Geophysics* 32:184–206.

Sannemann, D. 1968. Salt-stock families in northwestern Germany. In *Diapirism and Diapirs*, pp. 201–270. See Braunstein and O'Brien 1968.

Scholten, R. 1959. Synchronous highs: Preferential habitat of oil? *Bull. AAPG* 43:1793–1834.

Smith, D. A. and F. A. E. Reeve. 1970. Salt piercement in shallow Gulf Coast salt domes. *Bull. AAPG* 54:1271–1289.

Spindler, W. M., III. 1977. Structure and stratigraphy of a small Plio-Pleistocene depocenter, Louisiana Continental Shelf. Master's thesis, University of Texas at Austin.

Trusheim, F. 1960. Mechanism of salt migration in northern Germany. *Bull. AAPG* 44:1519–1540.

Tucker, P. M. and H. S. Yorsten. 1973. *Pitfalls in Seismic Interpretation.* Society of Exploration Geophysicists Monograph Series, no. 2. Tulsa, Okla.

Wolf, A. G. 1923. The origin of salt domes. *Engineering and Mining Journal Press* 115:412–414.

Wood, M. L. and J. L. Walper. 1974. The evolution of the interior Mesozoic basin and the Gulf of Mexico. *Gulf Coast Association Geological Societies Transactions* 24:31–41.

Woodbury, H. O., I. B. Murray, Jr., P. J. Pickford, and W. H. Akers. 1973. Pliocene and Pleistocene depocenters, outer continental shelf, Louisiana and Texas. *Bull. AAPG* 57:2428–2439.

8

Geologic Interpretation of Well Logs and Seismic Measurements in Reservoirs Associated with Evaporites

ROY D. NURMI

Numerous major hydrocarbon reservoirs are stratigraphically and genetically associated with evaporite lithofacies. They include sandstone reservoirs deposited both in arid terrigenous settings such as alluvial fans, fan deltas, wadis, and eolian sand seas, and also in marine depositional settings. However, carbonate reservoirs associated with evaporites are even more common and important, especially grainstone sequences capped by supratidal and subaqueous anhydrites, dolomite tidal mudflats, and reefs encased in basinal or shelf evaporites.

Although most geologists are comfortable using rock samples, either cores or "drilling cuttings," in their exploration efforts, the data routinely available for an explorationist are often indirect geophysical

Roy D. Nurmi is Chief Geologist, Middle East Division, for Schlumberger Technical Services, Dubai, U.A.E.

measurements. Surface, or reflection, seismic and well logs are the most common of these geophysical measurements. They are, in fact, complementary: seismic sections provide nearly continuous lateral views of the subsurface geology, whereas well logs yield a better vertical resolution of the geology penetrated by the wells.

Surface seismic data are very important in petroleum exploration because they can be acquired before the drilling of a well and often determine if and where a well should be drilled. However, it is rare that a seismic profile can be uniquely interpreted either in terms of structural or stratigraphic variables. Thus, it has become common practice to bring a part of a seismic survey as close as possible to the nearest well location, so that the logs from that well can be used to calibrate the seismic data. The well logs also improve the geologic understanding of seismic data by revealing the geologic origin of seismic reflections, such as changes in lithology or porosity.

Unfortunately, no single measurement provides anywhere near all the information needed for most subsurface geologic studies. However, examination of subsurface geology is improved by using the strengths of one measurement to overcome the limitations of another. Geologists thus need to know both the strengths and limitations of these measurements and also how gains can be made by integrating and using subsurface data together.

There are important differences in vertical resolution between subsurface data gathered by different methods. The general vertical resolution of some of the data used in exploration is shown in figure 8.1. Surface seismic profiles are roughly equivalent to examining rock outcrops from a considerable distance, such as a kilometer, whereas borehole seismic profiles provide a slightly better vertical resolution. Most well logs, although providing a better resolution than borehole seismic profiles, average the properties of rock surrounding the well bore, so that their vertical resolution is on the order of only one foot, which is roughly equivalent to examining the rock visually from a distance of more than ten meters. Only a few logging measurements, such as borehole electrical imagery, provide a vertical resolution comparable to examination of a core at a distance of one meter.

While seismic surveys have historically defined the gross geometry of exploration prospects, well logs provide the detailed information that ascertains whether a hydrocarbon reservoir has been found. During

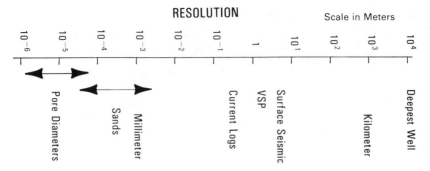

Figure 8.1
Logarithmic scale showing the vertical resolution of seismic, VSP, and current logs (resistivity, nuclear, acoustic).

the last decade, a number of well logging developments have significantly improved the evaluation of the complex rocks found in evaporite depositional sequences. These developments provide improvements for the evaluation of subsurface lithology, bedding, porosity, pore types, fractures, and fluids.

The reservoir search is commonly directed towards stratigraphic and subtle structural traps, as petroleum exploration reaches a mature stage in a basin. Considerable effort has been made by the seismic industry to detect such difficult traps. A number of seismic techniques have been developed to predict changes in facies, porosity, and/or hydrocarbons, including seismic amplitude measurements, such as "bright spots," "flat spots," and "hydrocarbon indicators"; inverted seismograms, such as "pseudovelocity logs"; and stratigraphic modeling. Unfortunately, most of the published studies involve terrigenous depositional systems associated with normal marine conditions; only a few papers deal with evaporite sequences.

The velocity-density relationships of common rock types encountered in evaporite sequences reveal a well-developed separation between terrigenous clastics and carbonates. Unfortunately, there is a substantial overlap of velocity and acoustic impedance for carbonates and anhydrite. The contact between a nonporous limestone and an anhydrite sequence will not yield a significant seismic reflection (figure 8.2). Porous dolomite and nonporous limestone may also produce the same acoustic impedance. These similarities of acoustic

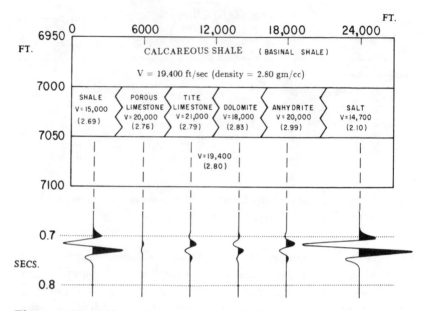

Figure 8.2
Seismic reflection character resulting from lithologic units with varying acoustic impedance contrast, which are overlain and underlain
by a high-velocity (19,400 ft/s) and dense (2.80 g/cc) calcareous shale
(Meckel and Nath 1977).

impedance frustrate the seismic interpretation of lithology, porosity,
and/or fluids in evaporite/carbonate sequences. Thus, in such sequences it is highly desirable to model the acoustic impedance in order to detect possible variations of lithofacies and porosity. Through
such efforts, stratigraphic traps may be discovered that would be otherwise left undrilled.

SEISMIC DETECTION OF REEFS

Reefs associated with evaporites form important and prolific hydrocarbon reservoirs in many areas of the world, including the Williston,
Paradox, Permian, and Michigan basins, South Texas, South Florida,
Mexico, western Canada, and the Gulf of Suez. Nearly all of these
reef reservoirs have been found using seismic reflection methods, and

this technique continues to be the primary exploration tool for finding subsurface reefs.

Although seismic reflection is the best exploration tool for reef reservoirs, a reef is not always obvious on a seismic profile, or seismogram. The recognition of a reef on a seismic profile may be enhanced by depositional topography and/or by contrasts in lithology, velocity, density, or stratification. A number of seismic profiles containing reefs from many different geologic sequences are presented in the hallmark AAPG Memoir no. 26 (see Payton 1977) on seismic stratigraphy. In this memoir, Bubb and Hatlelid (1977) provide an excellent list of direct and indirect criteria for recognition of reefs in seismic profiles (figure 8.3). Unfortunately, many other features may resemble a reef on a seismic profile, such as salt intrusions, igneous

(a) DIRECT CRITERIA   (b) INDIRECT CRITERIA

Geometry   Drape, Pull-Up, Diffractions

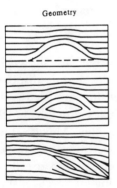 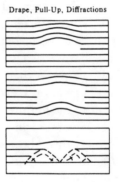

Reflection Disruption   Optimum Location

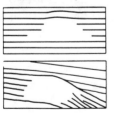 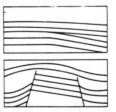

Figure 8.3
Criteria, direct (a) and indirect (b), suggesting the possible presence of a reef in seismic sections (Bubb and Hatlelid 1977).

intrusions, volcanic cones, salt dissolution edges, unconformities, large slumps, and faulted shelf margins. Thus, it is wise not only to be familiar with reef characteristics but also with the general geology of an exploration area. Furthermore, one must consider the size and characteristics of prospective reef targets, since they may be smaller than the minimum resolution of the seismic data.

Shelf-edge Reefs

In petroleum exploration the position of the shelf edge and especially any structural highs coincident with the shelf edge are usually good potential targets. Reefs along shelf margins are usually more easily identified in seismic profiles than isolated pinnacles or mounds because of the greater volume and extent of shelf-edge reefs. However, even when shelf topography and structural geology are clearly defined by seismic data, it is generally still difficult to determine whether a reef facies is actually present. The reef and associated lithofacies commonly do not have sufficient differences in acoustic impedance for the facies changes to be recognized.

Typical seismic characteristics of a shelf-edge reef are shown by Meckel and Nath (1977) in their modeling of the seismic response of a Permian Abo Formation reef. The Abo reef they modeled is the producing reservoir in the Empire oil field which lies along the northern edge of the Delaware Basin in southeastern New Mexico. Like many shelf-edge reefs this Abo reef is a narrow exploration target, 0.8 to 1.6 km (0.5 to 2 mi) wide. The Abo reef lens, up to 180 m (600 ft) thick, has a facies change landward passing into cyclical shelf carbonates and shales, and a basinward facies change into marine shales (figure 8.4a). Although both of these facies changes are somewhat abrupt, they are not very obvious in either the actual seismic data or the model. Fortunately the reef is overlain by basinal marine shales (Bone Spring Formation) and also "tight" shelf rocks, because the contrast in acoustic impedance of these overlying facies and the reef produces a strong reflection that reveals the geometry of this shelf-edge reef to a moderate degree.

This Abo reef example also demonstrates how subtle changes in seismic data can help indicate the probable presence of a reef facies. The slight difference between amplitudes of the seismic reflections from the back-reef facies and the overlying shale appears to be caused

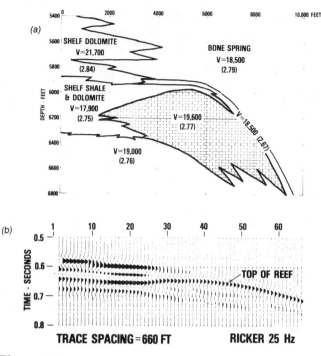

Figure 8.4
Seismic (inverse) modeling of a reef in the Permian Abo Formation in the Delaware Basin, New Mexico: (a) geologic cross section; (b) seismic model of the reef and adjacent facies (Meckel and Nath 1977).

by a facies change at the shelf edge. Also, the marked increase in amplitude of the seismic reflection immediately landward of the reeflike topography suggests either a lithofacies or porosity change (figure 8.4b). For the explorationist, such changes in the character of seismic reflections point to the possible presence of a reef reservoir. Similar seismic amplitude changes may reveal the presence of a reef facies even when no reef topography is present.

The modeling work by Meckel and Nath (1977) shows how the Abo reef and associated facies should appear on a seismic profile. Thus, the results of the modeling expand and quantify the geologic characteristics of Abo reefs that can be determined from the seismic data from this specific field, and these models may also be utilized in the general seismic exploration for other Abo reef reservoirs. Such seismic modeling is now a widely used tool for geologists exploring for reefs and other subtle structural and stratigraphic traps.

The physical character of seismic traces, such as amplitude, is a function of many factors that come into play during the acquisition phase in seismic exploration. These factors include the contrast in the acoustic impedance of the two layers at a boundary, the seismic source signal, and the nature of the rock sequence that the seismic energy must travel through. However, the character of the seismic traces in the final seismic profile is also determined by the manner in which the data are processed. In processing seismic data for structural traps such as folds and faults, the energy content is commonly sacrificed to some degree in order to enhance the continuity of the seismic reflections. The energy information in the seismic data must often be retained for reef and other stratigraphic exploration; otherwise little about the detailed geometry, lithology, or petrophysical characteristics can be determined.

The analysis of the energy content, or phase and amplitude, in seismic data can sometimes be invaluable to the detection of a reef reservoir. Rodrick and Poster (1987) used the energy content of both surface and borehole seismic data to successfully locate an Upper Cretaceous reef reservoir in the Middle East that had been missed. By analyzing the amplitude and phase characteristics of seismic reflections, they were able to determine that a porous fore-reef facies was located a short distance away from a well that had just been drilled. Based on their study, the well was sidetracked and drilled into the porous oil-filled fore-reef facies.

Pinnacle Reefs

Pinnacle reefs, mounds, and patch reefs are typically difficult to recognize in seismic profiles because of their small size and limited areal extent. A recommended reading is an excellent exploration case study of Devonian Nisku reefs in Alberta, Canada, by the Exploration Staff of Chevron Standard, Ltd. (1979). This paper documents the integrated geological and geophysical approach typically used in petroleum exploration for pinnacle reef reservoirs. Potential reef-bearing areas or trends identified by geologic studies were covered by seismic surveys to locate the elusive individual pinnacle reef targets.

In a seismic profile, the presence of a pinnacle reef is commonly suggested by a change in the seismic reflections from the strata surrounding the reef, rather than by some topographic expression of the reef itself. A break or disruption of the excellent seismic reflections from interbedded evaporites and carbonates that fill the areas between the reefs is sometimes the only indicator that a reef is present. This is often the case for both Canadian Devonian and Michigan Silurian pinnacles reefs (figure 8.5). The worst of exploration situations occurs where the lithology surrounding the reef has the same acoustic impedance as the reef itself and also lacks seismically detectable stratification. When reefs are small and/or when high-quality data acquisition or interpretation is difficult, explorationists have begun to use a relatively new technique, borehole seismic surveys, to locate reefs that have been missed by wildcat wells.

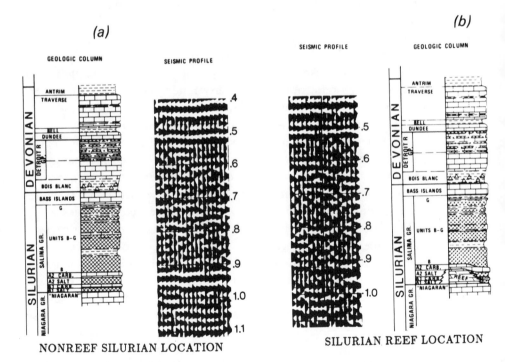

Figure 8.5
The geologic columns and seismic profiles for Michigan Silurian (a) nonreef and (b) reef locations (Caughlin et al. 1976).

Borehole seismic surveys were first used to remove some of the noise from surface seismic surveys. In borehole seismic surveys, the geophones are placed in the borehole to receive the seismic energy after it has reflected off of some subsurface horizon. Surface seismic techniques require sending energy down from the surface and then recording the energy reflected back to the surface. Borehole seismic surveys shorten the travel paths of the seismic energy, thus improving the quality of the seismic data. Furthermore, recording seismic data within a well provides a direct link between surface-recorded seismograms and a wide variety of well log data.

A significant shortcoming of surface seismic surveys is the uncertainty as to which geologic horizon a given seismic reflection actually represents, and whether the reflection has come directly to the surface or has further reflected off of one or more additional horizons, resulting in "multiple reflection." Multiple reflections are all too common in evaporite sequences, since such sequences generally contain interstratified lithologies of widely varying densities and velocities. Using borehole seismic data, however, it is relatively simple to determine which reflections are primary reflections because of the different time-depth relationships involved and also because of the characteristics of the actual horizon reflecting the seismic energy.

Borehole seismic techniques evolved from the velocity or "check shot" survey, which records the first arrivals of seismic energy directly from the surface source before they are reflected. These surveys are used to determine the velocity of the rock sequence that provides the relationship between time and depth for seismic analysis.

The first expression of borehole seismics was called a "vertical seismic profile," or VSP, because the number of geophones placed in the well bore provided some lateral coverage of the geology immediately around the well in the resulting seismogram. Later this approach was gradually modified for specific applications; for example, the surface energy sources were moved further away from the well location so that a larger interval of rock, further away from the well bore, was recorded in the seismogram. This borehole seismic approach, called an "offset borehole seismic profile," or offset VSP, provides a higher-resolution seismic picture of the geology in the vicinity of the well than does the standard surface seismic seismogram.

BOREHOLE SEISMIC REEF DETECTION

An experiment by Dupal and Miller (1985) to define small reefs, using multiple offset borehole seismic profiles in western Michigan (figure 8.6), demonstrated the potential value of an offset borehole seismic profile. The reefs in question have only a few hundred feet of relief. Although the seismic detection of these pinnacle reefs had been well established for other parts of the Michigan Basin (Caughlin et al. 1976), the western Michigan reefs are less than half the size of the reefs further to the north.

A reef detection experiment using borehole seismic methods was

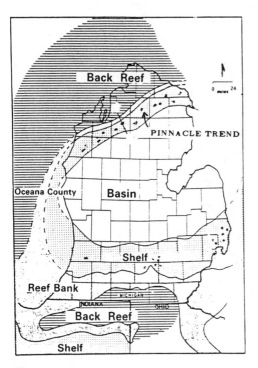

Figure 8.6
Niagaran (Silurian) paleogeography for the Michigan Basin (Lee and Budros 1982).

conducted using geophones placed in a well that missed a Silurian pinnacle reef. The control for the experiment was provided by a reef discovery well that was drilled 305 m (1,000 ft) north of the nonreef well. Numerous offset borehole seismic profiles were recorded with the surface sources laid out at first to the south, away from the reef, and then later to the north, toward the reef.

Previous exploration experience has shown that the presence of Michigan Silurian reefs is revealed by the termination and/or drape of the nonreef A-1 and A-2 Carbonate and associated evaporite units as they approach a reef (figure 8.7). The A-1 Salt, if present in the area, changes near the reef to anhydrite, which in turn pinches out on the lower flanks of the reef. The A-2 Salt also surrounds the reef and exhibits a progressive increase in anhydrite content approaching and overlying the reef. These facies changes and pinchouts cause extreme lateral changes in acoustic impedance around the reefs. The A-2 Carbonate unit also covers the reef, adding to a pull-up of the seismic reflections above the reef. Compaction of the carbonate muds underlying the reef may enhance the typical depression of seismic reflections from immediately below the reef. Although the presence

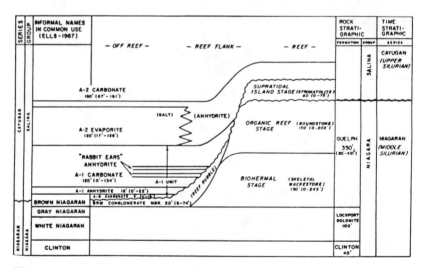

Figure 8.7
Idealized cross section of a Niagaran pinnacle reef and adjacent lithofacies in southeastern Michigan (modified from Lee and Budros 1982).

of the carbonate units is important, the local loss of A-1 and A-2 Salt reflectors is often the only criteria for detecting Michigan reefs on seismic profiles.

Reliable determination of the thicknesses of the A-1 and A-2 sequences and of the reef itself is made difficult by the usual pitfalls in defining geometry from two-way travel times and by the variation of velocity and density due to postdepositional chemical changes. A thick (183 m or 600 ft) surface cover of glacial deposits, valleys, and surface sand dunes also causes processing problems and "noisy" surface seismic records, which further complicate reef detection in the exploration of western Michigan.

Some of the results of Dupal and Miller's experiment with offset borehole seismics are displayed in figure 8.8. This data was generated from three offset surveys run northward from the nonreef well towards the reef itself. The borehole seismic profile in this figure is

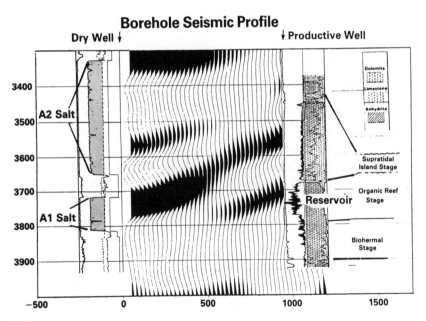

Figure 8.8
Borehole seismic profile across the flank of a Silurian pinnacle reef from West-Central Michigan, and well logs from wells at the approximate edges of the seismic profile (Dupal and Miller 1985).

plotted in depth at the same scale as the well logs. The well log data for the nonreef well (Well A) is presented at the left side of the seismic profile (with the salt units shaded), and the well log data for the reef well with no salt units present (Well B) are to the right. The well logs confirm that the A-1 Salt is absent over the reef. The reflection from the top of the A-1 Salt matches the depth and expected polarity of the well log data of the nonreef well. The approximate distance between each of the seismic traces is 7.6 m (25 ft).

The borehole seismic profile contains an appearance of a slope or draping in the vicinity of the reef well. In addition to the apparent drape at the reef flank, there is evidence of a seismic facies change and the appearance of a reflector within the reef itself, features that have not been previously detected by any surface seismic surveys. This new seismic reflection correlates with the top of the highly porous, hydrocarbon-bearing dolomite reef facies in the reef well (Well B). The change from the less porous overlying dolomitic limestone to the highly porous dolomite of the reef facies is the cause of this reflection, which is a black peak on the borehole seismic profile (figure 8.8).

Dupal and Miller (1985) demonstrate that the offset borehole seismic profile can detect and define a reef that was poorly defined in the surface seismic data (figures 8.9a and 8.9b). Their experiment shows that a much more detailed seismic profile is produced by an offset borehole seismic survey, because the borehole sensors receive higher-frequency data, premitting both a finer vertical and finer lateral resolution of the reef and surrounding lithofacies. The borehole seismic data is also less noisy because the acquisition in the borehole eliminates the travel path by which seismic energy returns to the surface through the zone of glacial weathering and erosion as well as through surface sand dunes. Similarly, Rodrick and Poster (1987) used the higher-resolution data of borehole seismics to define forereef facies of the Cretaceous Mishrif Formation in the United Arab Emirates, which were not well defined by the surface seismic profile.

WELL LOGS IN GRAINSTONE SEQUENCES

In most carbonate petroleum basins, major oil and gas reservoirs are present within carbonate grainstone shoaling-upward sequences. The

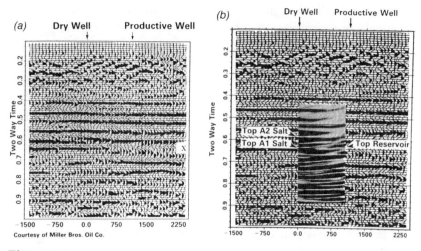

Figure 8.9

Silurian pinnacle reef. (a) Surface seismic section across the reef, including the region of the borehole seismic profile shown in figure 8.8. Disruption of the evaporite/carbonate reflectors suggests the presence of the reef (horizon marked by an "X"). (b) The same seismic section as in (a) with the borehole seismic profile superimposed and displayed in two-way vertical travel time. (Courtesy of D. E. Miller).

best known of these are the prolific Arab Formation of the Arabian Platform, the Smackover Formation surrounding the Gulf of Mexico, and parts of the San Andres Formation of West Texas and New Mexico. Recent well logging developments have made possible routine evaluation of lithology, bedding, porosity, secondary porosity, and fractures in such lithologies.

Lithology from Well Logs

Reliable well log interpretation of lithology improves routine correlation efforts, studies of depositional environment, and facies mapping. A general summary of basic lithologic interpretation of evaporite deposits is available in the notes of the 1978 SEPM Short Course on marine evaporites (see Nurmi 1978). Since the time of the SEPM short course, the addition of natural gamma-ray spectroscopy and the logging measurement of proton density have made it possible to produce reliable computer lithology analysis of well log data more routinely in complex evaporite sequences (see Serra 1986).

Some of the earliest efforts in analyzing complex lithologies using well logs were made in the Permian Basin. One of the pioneering efforts to determine the minerals present in the dolomitized grain-stones and tidal mud flats of the Permian Grayburg/San Andres formations was made by W. C. Savre (1963), a geologist in the Permian Basin. At shallow depths these formations commonly contain gypsum, which makes the rock appear to be porous in some log data because of the hydrogen within the molecular structure of gypsum (water of hydration). These early efforts in the Permian Basin led to a wider use of well logs in complex lithology sequences and later to a more reliable analysis of reservoir characteristics. Unfortunately, the limited number of well logging tools at the time did not permit a reliable computer analysis of lithology in such complex evaporite sequences. Local knowledge of the formations was required to compute reasonable evaluations of these complex rocks on a routine basis.

Historically, density, neutron, and gamma-ray logs were used to determine lithology in complex rock sequences. The gamma ray is generally used as a shale and clay indicator. The density and neutron tools, although designed for porosity determinations, are also used for basic lithology interpretation by integrating density and neutron porosity. Since such measurements are calibrated to compute an apparent limestone porosity, these logs provide true porosity values directly *only* if the formations are freshwater-filled limestone. This type of calibration is routinely taken into account when the rock is not a limestone. To interpret these logs quantitatively, generally density is plotted against neutron measurements on graph paper that has the limiting porosity values of the common rock types already marked on it.

Smackover Formation Lithology

One of the best-known grainstone sequences is the Jurassic Smackover Formation that rims the Gulf of Mexico. The Smackover Formation contains a number of large oil reservoirs that produce primarily from either calcareous or dolomitic grainstone lithofacies (figure

8.10). The Smackover Formation is underlain and sometimes also interfingers with the Norphlet Sandstone, which in turn is underlain by the Louann Salt. Salt structures in the Louann Salt produce many of the structural highs on which Smackover grainstone reservoirs are located. The regressive Smackover sequence is in turn overlain by the Buckner Anhydrite, a seaward prograding sabkha complex that serves as an excellent seal to the oil in underlying Smackover reservoirs. It has been revealed recently that porous reefs are also present in the lower portion of Smackover Formation (figure 8.11; Baria et al. 1982), although no oil has yet been discovered in these structures.

The well logs of a number of Smackover wells in Alabama and the Florida panhandle area have been studied to determine whether the interpretation of both the lithology and the pore system could be improved using some of the newer logging measurements (Nurmi and Frisinger 1983). Typical well log measurements of the Smackover Formation and Buckner Anhydrite are shown in figure 8.12. To provide insight into the quantitative interpretation of the lithology of such a sequence, a few of the data points are shown on a crossplot

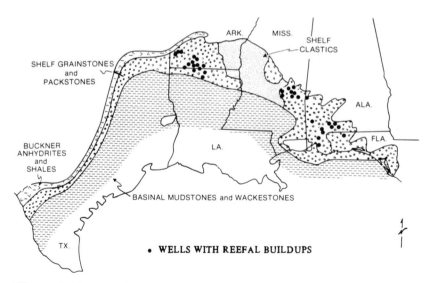

Figure 8.10
Jurassic Smackover lithofacies distribution in the Gulf of Mexico (Crevello and Harris 1984).

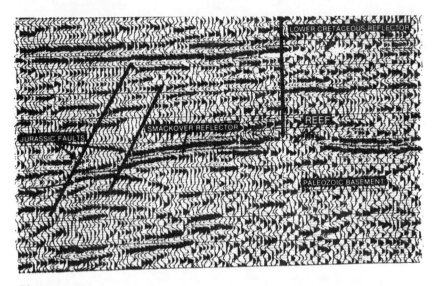

Figure 8.11
Seismic section through a porous Smackover reef, which is indicated by the amplitude anomaly. The Smackover interval thickens basinward (left) of the Jurassic faulting (Baria et al. 1982).

of density and neutron porosity in figure 8.13a. The position of point 1 on this crossplot indicates it to be limestone without porosity, whereas point 2 is that of anhydrite. Point 3 on the crossplot is an equal mixture of limestone and dolomite, assuming that no other minerals are present. Additional minerals would alter this lithologic interpretation. If the rock is, in fact, a mixture of dolomite and limestone, it is possible to further interpret the volume of each and the pore volume present.

A sonic measurement is often combined with density, neutron, and gamma-ray measurements during well logging. Although introduced prior to the nuclear porosity tools (neutron and density), the sonic tool fell into disuse because it is not very sensitive to changes in lithology. However, the use of the sonic tool later increased as its role in supporting seismic efforts developed. Recent research increasingly utilizes sonic measurements, tools, and interpretation, and is bound to further enhance the geologic data from sonic measurements.

A recent logging innovation for lithology determination is a new type of density tool that includes a measurement of the proton density in addition to the bulk density, or electron density, measured by standard density logging tools. This new measurement is, more exactly, the index of the photoelectric absorption cross section of the formation, or "Pe." Fortunately for the utility of log interpretation, the relationship between Pe and bulk density values differs for some of the common minerals within evaporite sequences.

These differences between Pe and bulk density make it possible to differentiate a clean limestone (basically calcite) from some mixtures of minerals that cannot be interpreted by density and neutron porosity alone, such as a mixture of dolomite and quartz. The Pe values of common Smackover minerals are: quartz, 1.8; calcite (limestone), 5.1; dolomite, 3.1; and anhydrite, 5.0. Although the Pe of anhydrite is close to that of limestone, anhydrite can be easily distinguished because of its high bulk density, which is also measured by this tool. Pe values significantly less than 3.1, the value for dolomite, are very

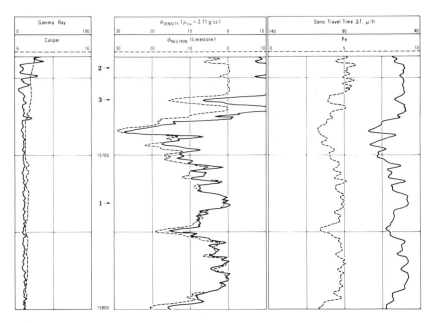

Figure 8.12
Well logs of a Smackover/Buckner interval from a Smackover reservoir in the Florida panhandle area (Nurmi and Frisinger 1983).

suggestive of the presence of quartz, which has a 1.8 Pe value. The Pe and other logging measurement values for the common minerals found in evaporite sequences are presented in table 8.1, near the end of this chapter.

In the Smackover example mentioned previously, the lithologic interpretation of the 15,664-ft depth interval, point 3, can be defined with greater confidence by referring to the Pe curve shown toward the right in figure 8.12, which indicates a 3.9 Pe value. This value is plotted in the crossplot of Pe vs bulk density (figure 8.13b), confirming the lithology at this depth to be a near-equal mixture of limestone and dolomite. The difference between the relative positions of the three common Smackover minerals in the two types of crossplots is very important for quantitative lithology interpretation.

A computer-generated lithology plot (figure 8.14), processed using all of the available logging data from the above Smackover example, accurately defines the complex mixture of minerals and also clearly reveals the correlation between an increase in porosity and an increase in the volume of dolomite. Furthermore, the computer correlation of dipmeter data from zones like this reveals that many of these porous dolomite zones in the Smckover are cross-bedded grainstone facies.

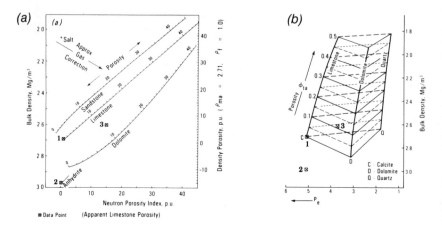

Figure 8.13
Plots of the log data from the Smackover Formation in figure 8.12: (a) a neutron porosity vs density crossplot; (b) a crossplot of the Pe vs density (from Nurmi and Frisinger 1983).

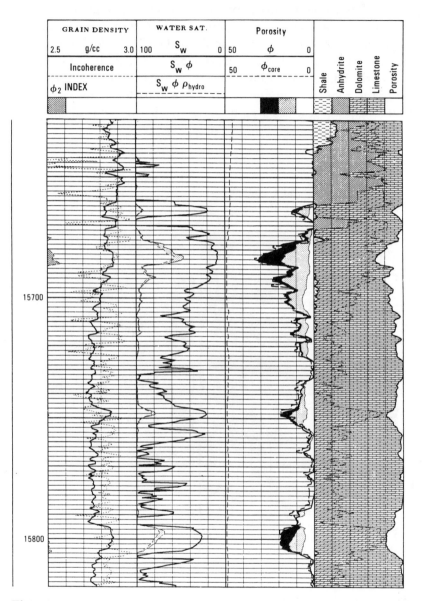

Figure 8.14
Log plot of the computer analysis of the fluid and lithology of the
Smackover/Buckner interval in figure 8.12 (Nurmi and Frisinger 1983).

An ambitious pioneering effort to define the lithology of the Cretaceous Ferry Lake evaporite sequence (northeastern Texas) using the data from a single spectroscopy tool, the G.S.T., is shown in figure 8.15 (Flaum and Pirie 1981). This tool provides data concerning hydrogen, silicon, calcium, iron, chlorine, and sulfur. An interesting aspect of this tool is that it makes it possible to perform lithology analysis through casing, which is particularly valuable for both old unlogged

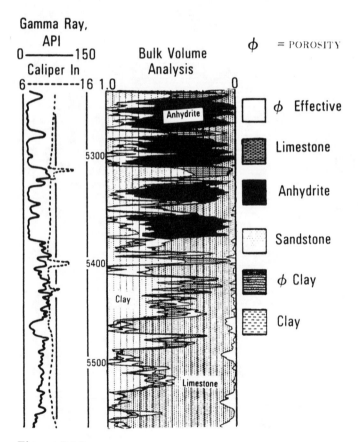

Figure 8.15
Lithology analysis of the Cretaceous Ferry Lake evaporite sequence (Panola County, northeastern Texas) using solely the elemental data from the induced gamma-ray spectroscopy tool (adapted from Flaum and Pirie 1981).

wells and problem wells that cannot be logged immediately after drilling. The feasibility of defining the lithology of the San Andres Formation of West Texas after casing has been placed in the well has been tested (see figure 8.16, and McGuire et al. 1985).

Smackover Pore System

In the upper portion of the previous Smackover example, the porosity determined by the sonic tool is less than the porosity defined by the nuclear density and neutron porosity (figure 8.12). A porosity relationship such as that seen in the upper 25% of figure 8.12, when present in a shale-free sequence, suggests the presence of what log analysts call "secondary porosity." Unfortunately, not all types of geologic secondary porosity cause the sonic-derived porosity reading to be less than the porosity defined by density-neutron logs. More-

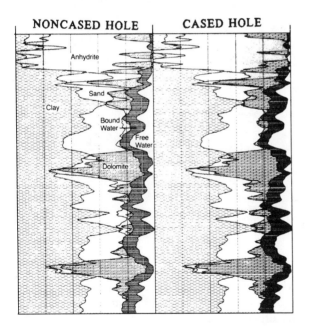

Figure 8.16
Comparison of a lithology interpretation of spectroscopy data from the Permian San Andres Formation measured in the same interval both before and after casing was present (McGuire et al. 1985).

over, it should be recognized that the sonic measurement is not only affected by the pore volume, or porosity, but is also affected by the geometric aspects of the pore system. Thus not all forms of secondary porosity, such as intercrystalline porosity in dolomite, cause the sonic tool to indicate a porosity less than the total porosity. Although use of the sonic tool to provide a qualitative indication of porosity has been known for almost twenty years, it is only recently that the quantitative use of the sonic tool has been refined (Brie et al. 1985).

The porosity/permeability relationship of nearly the entire Smack-over section is similar, even though the interval includes both limestone and dolostone (figure 8.17a). The low-porosity portion of this trend is almost entirely limestone, whereas the high-porosity portions of the trend are exclusively dolomite. The general trend of the porosity/permeability data is suggestive of the intergranular pore system of a grainstone (figure 8.17b). The association of higher porosity with dolostone intervals is often cited as evidence that the dolomi-

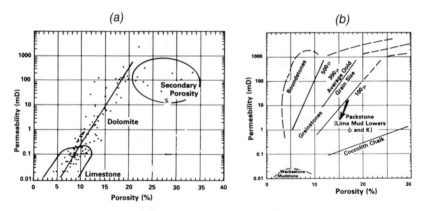

Figure 8.17
Smackover porosity and permeability data. (a) From the Smackover Formation (same interval as shown in figure 8.12). The trend of the data is essentially that of a grainstone (see 8.17b), although it is dolomitized. The low porosity points along the trend are predominantly limestone, whereas the higher porosity and permeability points are all dolomite. The points that have significantly higher porosity than the defined trend all contain secondary (oomoldic) porosity (see figure 8.18). (b) Porosity/permeability trends for various types of carbonate reservoir rock (modified from Nurmi and Frisinger 1983).

tization process directly produces porosity. However, in this example it appears that the dolomite replaced the calcite rock on a volume-for-volume basis, and that the preexisting intergranular pore system remained. Thus, the dolomitization preserved original porosity rather than having created secondary porosity.

The depths of these porous points correspond to the log interval in which the sonic measurement of porosity suggests the presence of secondary porosity when compared with the total porosity as determined by the density-neutron logs. Thin section and pore cast examination of the Smackover core samples from this interval confirm the presence of secondary porosity in the form of oomoldic pores (figure 8.18). The core samples, which contained oomoldic pores, all fall within the circle labeled "secondary porosity" in figure 8.17a.

Smackover Bedding

The dipmeter's ability to detect the bedding characteristics of large-scale cross-bedded grainstone facies has been confirmed in a number of cases. The bedding of the Smackover Formation is commonly well

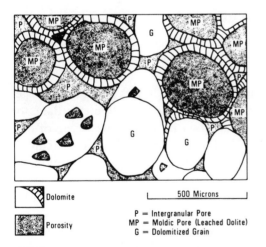

Dolomite

500 Microns

Porosity

P = Intergranular Pore
MP = Moldic Pore (Leached Oolite)
G = Dolomitized Grain

Figure 8.18
Drawing of a thin section of a Smackover core sample with secondary (oomoldic) porosity, marked with an "X" in figure 8.17a (Nurmi and Frisinger 1983).

defined by dipmeter data in the areas where it is predominantly a limestone sequence, as in southeastern Mississippi and the Texas-Arkansas-Louisiana region. However, even where the grainstone facies of the Smackover Formation is entirely dolomitized, it is still possible to recognize the large-scale cross-bedding characteristics of the grainstone facies by using the dipmeter, as in the Florida-Alabama area. This large-scale cross-bedding as defined in the dipmeter "arrow plot" (dip log) is identical to the bedding observed in the core from that same interval (figure 8.19a). The high-angle inclined bedding, with a consistent 15° to 20° northeast dip, is used in detecting and mapping Smackover oolite bars (figure 8.19b). These bars appeared to have formed a barrier that affected subsequent deposition and diagenesis in the Jay Field area. This interpretation of the depositional facies is significantly different from a previously published geologic study (Sigsby 1976) based primarily on the microscopic textures and fabrics of the rock observed in core chips and thin sections. The primary reason for the difference in interpretation is that the log data permitted the reservoir interval to be examined in both a different manner and scale.

Bedding in Paris Basin Grainstone

The study of Jurassic-age grainstone sequences from two wells in the Paris Basin demonstrates that a newer generation dipmeter can be used to recognize and interpret not only large-scale but also the more common small-scale cross-bedding (figure 8.20). Although few geologists are familiar with the interpretation of dipmeter data for stratigraphic or sedimentological applications, this new type of dipmeter tool provides images of bedding within a borehole in addition to arrow-plot directional data. The field tests of this new tool in the Paris Basin demonstrate that the bedding geometry of both large-scale cross-beds (figure 8.21a) and small-scale cross-beds (figure 8.21b) in the Jurassic Dogger grainstones are easily recognized using the electrical images. In addition, a vugular pore system in a deeper dolomitized Jurassic interval was also revealed by the electrical imagery, even though the vugs are on the order of a centimeter in diameter (figure 8.21c). More recently, with the commercial use of borehole imagery in the Middle East, similar features have been identified, in-

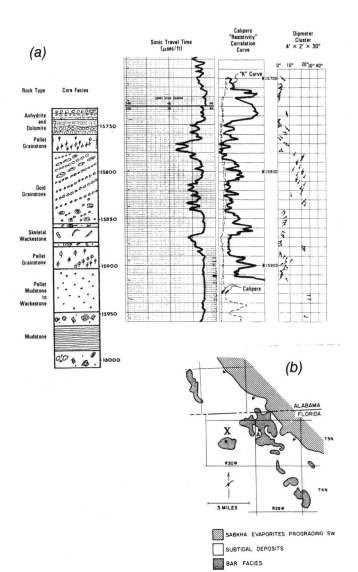

Figure 8.19
(a) Core and log comparison of the Smackover Formation showing the high-angle dip of the dolomitized grainstone facies in a well west of the Jay Field, Florida (Nurmi and Frisinger 1983; Lomando, 1981). (b) Jurassic lithofacies map showing the bar facies which contains the high-angle cross-bedding. The location for the well data shown in (a) is marked with an "X" (Nurmi 1978).

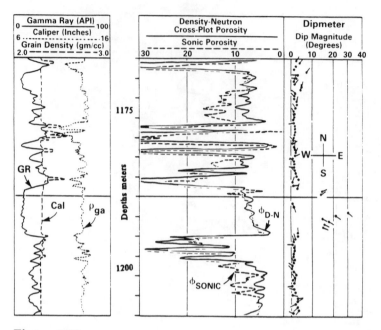

Figure 8.20
Well log measurements from a Jurassic Dogger grainstone sequence from the Paris Basin, Villejust, France (Nurmi 1984).

cluding a variety of bedding characteristics, vugs, fractures, slumping, mottling, and nodules (Nurmi 1986).

Well log measurements of the Jurassic Dogger sequence, from which the above-mentioned electrical images were taken, are shown in figure 8.20. The grain density, which is the bulk density adjusted for porosity, reveals the formation at the center of the figure to be a limestone. The close match of the density-neutron porosity and sonic porosity in this example indicates that there is little secondary porosity present. The dipmeter arrow plot and electrical images show the presence of large-scale cross-bedding that is dipping to the southeast. The fact that the formation contains large-scale cross-bedding allows us to infer that we are dealing with a grainstone. Since no secondary porosity is indicated, we may conclude that the pore system is dominantly intergranular. Thus, from routinely available log data alone we can determine quite a bit about the formation: lith-

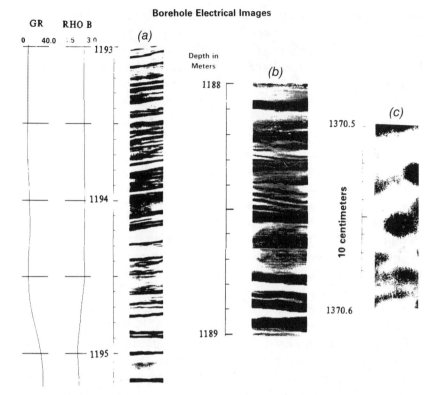

Figure 8.21
Borehole electrical images from three intervals from the same well as in figure 8.20: (a) the bottom portion of the large-scale crossbed set; (b) small-scale crossbeds and interbedded limestone and shale; and (c) vugs developed in a dolomitized zone deeper in the well.

ology, bedding, rock type, dip direction, and pore type. Drilling cuttings can and should be used to further enhance the understanding of the formations drilled.

SEISMIC MODELING OF GRAINSTONE RESERVOIRS

By and large, the exploration for reservoirs in these grainstone sequences has been primarily performed using seismic records for the detection of structural traps such as anticlinal structures, faulted structures, and salt-influenced structures (Rosenkrans and Marr 1967). The possibility that stratigraphic traps may be associated with regressive cycles in the Smackover has also been considered (Bubb and Hatlelid 1977), but few have yet been found. Unfortunately, stratigraphic traps in grainstones are difficult to detect using seismic analysis because they do not have an easily identified geometric form in seismic profiles. Efforts to define facies and porosity changes through seismic data modeling—including both inverse and forward modeling—of the Smackover and adjacent formations in Arkansas and East Texas has been somewhat successful (Ward 1984).

The Jurassic Arab Formation, the world's greatest oil-producing formation, is directly overlain by the Hith Anhydrite and is also interbedded with anhydrite. The Arab sequence is actually comprised of four grainstone reservoir units, each with an overlying sabkha anhydrite that represents the final depositional unit of a grainstone depositional cycle. The boundaries between the cycles can be identified on seismic profiles because of the high contrast between the lower acoustic impedance of the porous grainstones and the higher acoustic impedance of the anhydritic sabkha facies.

Because nearly all of the anticlines in prospective Arab Formation areas have been drilled, petroleum companies are now attempting to define possible stratigraphic traps. Although the high seismic contrast evident in the vertical section is also present laterally, it is not as well delineated because these lateral facies changes take place over distances of kilometers. There is evidence to suggest, however, that facies changes are sometimes more rapid on structures that were depositional highs during the deposition of the Jurassic Arab Formation.

Stratigraphic traps are commonly the result of subtle lateral lithology or porosity change. Unfortunately, it is generally impossible to provide unique interpretations of lateral changes in porosity or lithofacies from seismic profiles. Therefore, in the exploration for stratigraphic traps it has become quite common to model a sequence geologically and acoustically using seismic data.

One approach to overcoming the problems inherent in seismic data is to expand synthetic seismograms into two-dimensional models. Such models relate rock properties, in terms of acoustic properties, to the available seismic data in order to provide a better understanding of geologic and/or petrophysical characteristics. These acoustic models are generated from well log data that provides the direct bridge from the actual rock properties (lithology, velocity, porosity, fractures, and pore fluids) to attributes derived from seismic data (wave form, amplitude, and phase). By analyzing the synthetic seismograms created by using well log data, the feasibility of detecting a reservoir zone within the seismic data can be evaluated.

North Sea Zechstein Grainstone

Few modeling studies of seismic surveys of grainstone reservoirs have been published (Ward 1984; Gaynor and Stasko 1985; and Neidell et al. 1985). One of the earliest efforts in modeling the seismic response of grainstone facies in an evaporite/carbonate sequence is that of Maureau and van Wijhe (1979). Their report on their success with seismic exploration for porous grainstone facies of the Permian Z2 Carbonate, or Zechstein-2 Carbonate, in the eastern Netherlands (figure 8.22) is still very relevant today. Prior to their study it was possible to approximate only roughly the areal extent of the Z2 Carbonate, and it was also very difficult to predict the areas of best reservoir development because of sparse well control. Through the integrated use of both seismic and log data, Maureau and van Wijhe found that the presence of porous lenses of grainstone units in the Z2 Carbonate could be determined using acoustic impedance data.

Along the margins of the Zechstein Basin, just prior to the Z2 Carbonate, deposition was dominated by shallow-water gypsum (later converted to anhydrite), whereas the sequence in the basin center was dominated by salt (Taylor 1984). At the start of the deposition of the Z2 cycle (figure 8.23), a carbonate buildup developed along a preexisting sulfate (Z1 Anhydrite) platform. It appears that both the thickness and environment of deposition of Z2 Carbonate was controlled by the topography of this underlying sulfate platform. Core studies indicate that the Z2 Carbonate contains a typical range of shallow-water deposits on the platform and deeper marine deposits on the slope and basin with an overall regressive tendency. The basal Z2 Anhydrite overlying the Z2 Carbonate has a rather uniform thickness of 20 m on the platform that thins to 10 m or less in the basin center area.

The most favorable Z2 Carbonate reservoirs (average porosity 10%

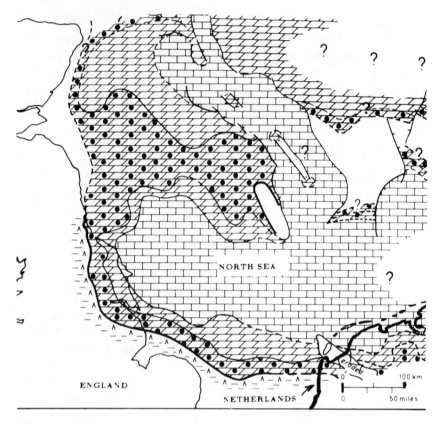

Figure 8.22
Lithofacies map of the Late Permian Zechstein 2 Carbonate (Taylor 1984).

to 15%; permeabilities ranging from 1 to 5 mD) are in shallow-marine, high-energy oolite deposits. As in the modern Bahamas, a variety of oolite accumulations occurred on the platform in the form of bar deposits, and resedimented ooids in the form of grainflow deposits were also present on the shelf. Although the principal reservoir occurrence is along the platform edge, there is a patchy distribution to these Z2 Carbonate porous grainstones, which is an exploration problem that lends itself to seismic exploration.

To detect and recognize these isolated porous Z2 Carbonate reservoirs on seismic sections was the ultimate goal of the study by

Maureau and van Wijhe (1979). They addressed the two basic questions relevant to all facies and porosity predictions using seismic data: how is a specific lithologic unit represented on a seismic profile? And can the lateral changes in this unit be identified and analyzed using seismic data?

Their initial effort was to construct sets of simple geologic models based on logs (sonic and density) and cores. For each of the three types of depositional locations in this Zechstein Basin (platform/upper slope, slope, and basin), model acoustic impedance logs were created using typical thickness, velocity, and density ranges (figure 8.24). The model log representative of each depositional environment was filtered with a broad range of zero-phase filters (figure 8.25), with filter ranges representative of the spectral range of the seismic data. Fortunately, the Z2 Carbonate on the platform/upper slope, the ma-

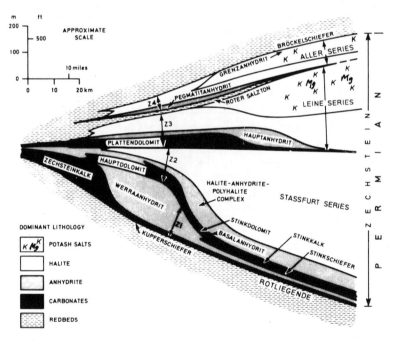

Figure 8.23
Generalized cross section of the Zechstein from the Norfolk coast of England towards the offshore Viking Field (modified from Taylor 1984).

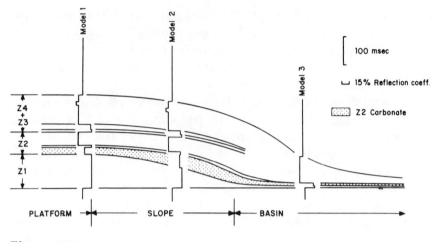

Figure 8.24
Model acoustic impedance logs representative of three basic Zechstein depositional environments (Maureau and van Wijhe 1979).

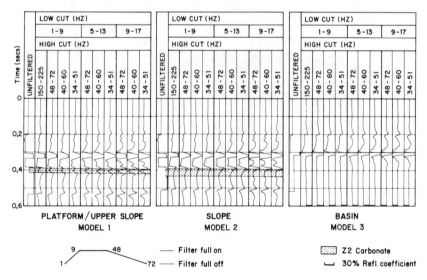

Figure 8.25
Selected zero-phase filtered versions of the model acoustic impedance logs of figure 8.24 (Maureau and van Wijhe 1979).

jor reservoir, showed sufficient acoustic contrast with respect to the overlying and underlying anhydrite to produce a prominent reflection. Model 1 in figure 8.25 represents the most favorable development of the Z2 Carbonate, with the relatively low-acoustic impedance of the Z2 Carbonate clearly visible on each of the filtered logs. In the slope and basin environment (models 2 and 3) the low-porosity Z2 Carbonate is virtually indistinguishable from the surrounding anhydrites. Thus, the Z2 Carbonate development in neither the slope nor basinal setting was likely to produce any significant seismic reflections. This situation is similar to the one described by Meckel and Nath (1977) in their modeling of a San Andres tidal-flat reservoir in New Mexico where the low-porosity or thin-porosity zones interbedded with sabkha anhydrites are not resolvable using seismic data.

From the modeling experiments, Maureau and van Wijhe concluded that within the seismic bandwidth one could qualitatively distinguish the highly porous Z2 Carbonate from its nonporous counterparts in the slope and basin environments. Their next step was to study a suite of actual logs, again from each depositional setting, to determine whether any additional complications might seriously downgrade the resolution shown by the ideal models. These logs were carefully edited to remove any bad or spurious data caused by zones of adverse borehole conditions. Acoustic impedance logs and synthetic seismograms were derived and subsequently filtered in the same manner as the model logs.

It was confirmed that the Z2 Carbonate porosity is most favorably developed in the platform/upper slope well, and that the low-acoustic impedance related to Z2 Carbonate porosity is clearly recognizable for each of the filter settings. The resolution of acoustic impedance is enhanced by broad-band filtering, with the low frequencies (0–10 Hz) being especially important as nearly all geologic character was destroyed by the narrow-band filters. Thus, it appears that when absolute values of acoustic impedance are diagnostic, broad-band recording and processing should be considered. In the case of narrow-band data, a good match between seismic and well log data is necessary because one can easily mistake a salt unit for a highly porous carbonate, because these deflections could be comparable in magnitude.

A petrophysical model of the typical development of a patchy Z2

Carbonate reservoir is shown in figure 8.26a. At the same horizontal scale, the corresponding synthetic acoustic impedance section is also shown (figure 8.26b). In the petrophysical model, the hatched area indicates a low-porosity carbonate surrounding the higher-porosity carbonate at the center. From the synthetic section, it is clear that the high-porosity carbonate is distinguishable from the low-porosity carbonate. Furthermore, the thin highly porous beds, which extend into the surrounding low-porosity zone, clearly increase the amplitude of the Z2 Carbonate signal. However, where the porosity is low (3%), the carbonate is not distinguishable from the surrounding anhydrite, and together they produce a prominent, low-frequency reflection. Maureau and van Wijhe indicate that this model was particularly useful for learning to interpret the seismic sections of the Z2 Carbonate in the eastern Netherlands (figure 8.27).

The combination of geologic modeling, well log measurements, and seismic acoustic impedance data, together with a basic geologic un-

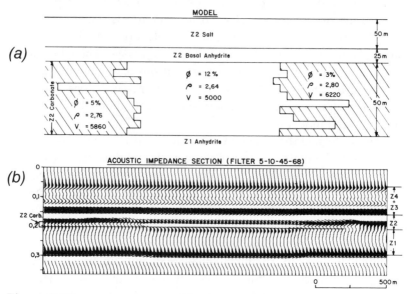

Figure 8.26
Inverse seismic modeling of a "patchy" Zechstein 2 Carbonate: (a) petrophysical model, showing parameters for the Z2 Carbonate Formation; and (b) the acoustic impedance section that corresponds to (a) (Marueau and van Wijhe 1979.)

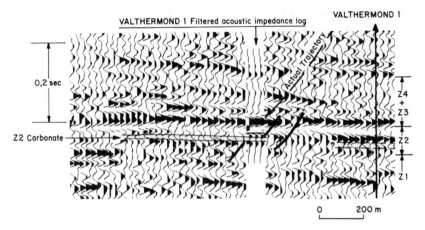

Figure 8.27
A seismic section of the Zechstein with a filtered acoustic impedance log superimposed. The Z2 Carbonate can be easily identified and traced on the seismic section (Maureau and van Wijhe 1979).

derstanding of the area, resulted in the successful prediction of Z2 Carbonate reservoir porosity. As documented by Maureau and van Wijhe, the feasibility experiments combining both model and actual downhole data are valuable in properly defining those factors important to the seismic detection of either a lithology or porosity facies change.

EOLIAN SANDSTONE FACIES

Ancient desert dune deposits make up some of the best-quality sandstone reservoirs associated with evaporite deposits. Not long ago, however, it was thought unlikely that oil or gas would be found in an eolian sandstone because it was assumed that the lack of organic-rich sediments in the desert setting precluded this possibility. It was also thought that if hydrocarbons did migrate into a sandstone of desert dune origin, the result would be a homogeneous and isotropic reservoir because of the absence of shale beds. Both assumptions have been found to be incorrect. There are, in fact, a large number of eolian sandstone reservoirs, and they are heterogeneous and anisotropic at all megascopic to microscopic scales of examination.

The largest number of oil-producing eolian formations are within the United States' Rocky Mountain region, which includes the Nugget, Tensleep, Minnelusa, Weber, Entrada, and Lyons formations. For nearly a hundred years it was questioned whether these large-scale cross-bedded sandstones were of eolian origin. Recently, however, the eolian sedimentological characteristics have been sufficiently defined that their dune origin could be confirmed (Glennie 1970; McKee 1979; Brookfield and Ahlbrandt 1983). Unfortunately, few of these eolian criteria are ever detected during either a core analysis or an examination of standard well logs.

Two other formations that contain major eolian sandstone reservoirs are the Norphlet and Rotliegendes. The Norphlet Formation underlies the Smackover/Buckner sequence and overlies the Louann Salt, and it produces from eolian sandstone facies in Mississippi, Alabama, the Florida panhandle, and more recently in nearby offshore areas. In contrast, the Rotliegendes Sandstone is the site of significant gas production from eolian dune facies in the southern North Sea. The Rotliegendes Sandstone underlies the evaporite/carbonate sequence of the Upper Permian Zechstein Group. The eolian sequence of the Rotliegendes Sandstone also contains evaporite minerals in the form of cements in the interdune deposits. There are also continental evaporites with interbedded shales at the center of the southern North Sea Basin that are time-equivalent to the Rotliegendes Sandstone.

It is a common occurrence for desert sandstone systems to be overlain by carbonate/evaporite deposits, for example as the Norphlet Sandstone is overlain by the Smackover Formation. Other such sequences are the Tensleep/Phosphoria, Minnelusa/Goose Egg, Glorieta/San Andres, Nugget/Gypsum Springs, and the Entrada/Todilto formations. This type of sequence is what would be expected if these desert sandstones were deposited in subsiding basins that eventually developed into dry, sub-sea-level topographic basins because of the arid climate. The common occurrence of subtidal carbonates immediately overlying many of the desert sandstone systems suggests that the ensuing connection with the sea resulted in rapid filling of these sub–sea-level basins. Thick salts associated with overlying carbonate sequences are also explained by aridity and, perhaps, by a degree of renewed isolation of the basin from the open sea.

Eolian Sandstone Bedding

The most distinctive feature of desert dune sandstones is very large-scale cross-bedding, with average cross-bed set thicknesses of 3 to 10 m and occasionally exceeding 30 m (100 ft). Other than the rare drilling of oriented cores of the commonly thick eolian sandstones, dipmeter "arrow plots" (dip logs) have been the only subsurface technique for obtaining continuous measurements of accurate bedding geometry and orientation. Borehole imagery, a new logging development, provides images which are often very similar in appearance to core photographs. These oriented images can be used to analyze bedding and fractures within eolian sandstones. Sometimes the electrical images resemble surface outcrops, where weathering has made the bedding characteristics more visible than those observed in slabbed cores.

The sedimentological variables that produce the bedding within sandstones are mineralogy, particle size, particle shape, orientation of particles, and packing (figure 8.28). For bedding to be defined by a dipmeter or to be visible in electrical images, there must be some variation in the electrical resistance. This may be caused by bedding characteristics such as alternating quartz sand laminae and shaly laminae. Alternating sand and shale laminae are distinctive in dipmeter data or electrical images because the sands are essentially resistive while the clays have the capacity to conduct electrical current.

Resistivity variations caused by bedding in clay-free sandstones are often much less distinctive than those in shaly sandstones. Core studies reveal that electrical definition of bedding is clearest in clay-free sandstones when the cementation has been controlled by textural variations of bedding laminae. Bedding-controlled cementation appears to be common in many porous sandstones. However, below 10% porosity, the cement distribution is generally less clearly controlled by the bedding.

Prior to the introduction of electrical imagery, it was possible to routinely define the bedding characteristics and geometry of eolian sandstones using dipmeter arrow plots. In one study of the Nugget Sandstone it was found that 15 individual bedding planes within a

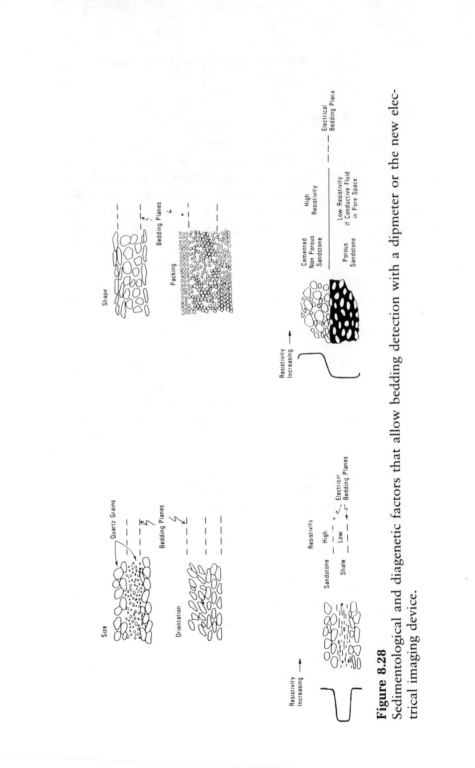

Figure 8.28
Sedimentological and diagenetic factors that allow bedding detection with a dipmeter or the new electrical imaging device.

1-ft interval were defined by the computer processing of a well in South-Central Wyoming (figure 8.29a). The interval contained thin laminae produced by wind-ripple migration along the bottom portion of a large cross-bed set. Even in this good example, however, not all of the individual laminae were detected.

The examination of the Nugget core of the above example revealed that adjacent sets of laminae, approximately a centimeter thick, were detected because of textural changes and differential cementation, which caused variations in both resistivity and permeability. The analysis of the dipmeter resistivity data indicates that the features detected and computer-correlated were on the order of a centimeter in thickness (figure 8.29b), which is essentially the detection resolution of this older-generation dipmeter tool. The vertical resolution

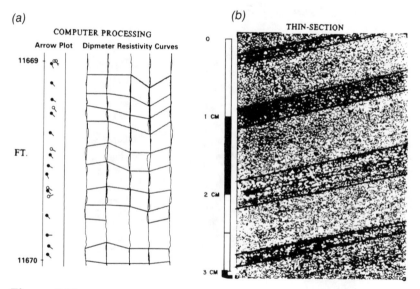

Figure 8.29
Bedding in an eolian sandstone: (a) a pattern-recognition computer processing for correlating the resistivity variations detected by the dipmeter; and (b) positive print from a thin-section of the Nugget Sandstone from the dune cross-bed set shown in (a). The shaded intervals are more poorly sorted, finer-grained, more highly cemented, less permeable, and more resistive.

of both newer-generation dipmeters and electrical imagery surveys is slightly better.

In electrical images, made in boreholes, the white laminae are the most resistive, whereas the black laminae are the most conductive. The various resistivity levels in between the two extremes are given the appropriate gray shades. The black or very dark grey laminae in a sandstone are either electrically conductive shale laminae or permeable sandstone laminae. The less porous, cemented (impermeable) sandstone laminae are very light in color.

During research field testing of electrical imagery, various bedding features in a Rotliegendes reservoir were identified and then later

Borehole Electrical Images

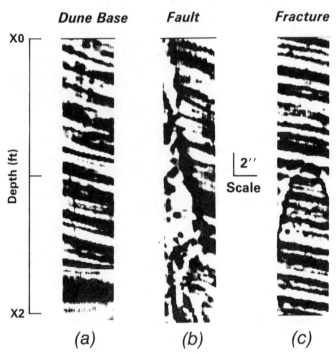

(a)   *(b)*   *(c)*

Figure 8.30
Borehole electrical images from the Rotliegendes Sandstone: (a) base of a transverse crossbed set; (b) fault zone through the Rotliegendes Sandstone; and (c) a high-angle fracture cutting the sandstone.

confirmed by comparisons with cores, including cross-bed sets, fractures, a fault zone (figure 8.30) with cemented fractures, and reactivation surfaces within these eolian dune sandstones.

Eolian Reservoirs

Although eolian sandstones are often thick, widespread blankets of sandstones that are relatively free of shale, they are far from being homogeneous and isotropic. Experience has shown that a sedimentological understanding of eolian facies is important for a realistic reservoir analysis. Sedimentological knowledge, especially bedding data, can guide the study of reservoir geometry and/or performance characteristics of sandstones of eolian origin.

Eolian reservoirs are usually comprised of multiple discrete porous zones whose characteristics and geometry are closely related to the depositional processes of sand dune migration. Only rarely are the shapes of individual desert dunes preserved in an eolian sandstone sequence. The actual dune forms that are preserved are found at the top of the sequence, as a result of a rapid marine transgression (figure 8.31).

The best reservoir units within a sandstone sequence of eolian origin are tabular sheetlike units that are the bottom parts of dunes left behind during migration (figure 8.32) across an ancient desert surface (Kocurek 1981). These tabular units are often distinct individual hydraulic zones within eolian sandstones. Some of the geometric characteristics of such reservoir units are a function of the depositional geometry and processes of the migrating dunes. Because there is a link between cross-bed geometry and dune shape, it is sometimes possible to interpret dune types by analyzing the relationships between dip magnitude and azimuthal variations of bedding within the vertical depositional sequences.

Although the internal reservoir framework is determined during deposition, much of the complexity of the eolian sandstones results from differential cementation that takes place after burial. Cementation along the almost horizontal contacts between the tabular units commonly causes these sandstones to be layered and multistoried (figure 8.33). The lower porosity of these reservoir barriers can be detected and recognized using density-neutron porosity logs. Fur-

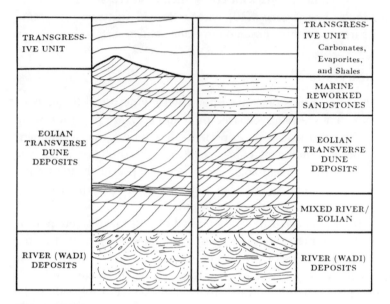

Figure 8.31
Idealized vertical sequences typical of ancient desert eolian systems (Nurmi 1985).

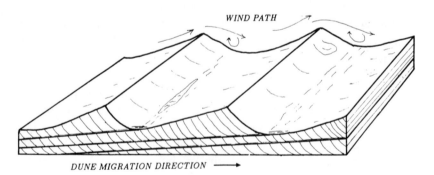

Figure 8.32
Multiple cross-bed sets (typical of most eolian reservoirs) caused by the migration and climbing of transverse dunes (Nurmi 1985).

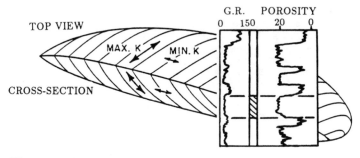

Figure 8.33
Idealized log through a typical eolian reservoir sequence and a reservoir unit with the probable orientation of the permeability anisotropy (Nurmi 1985).

thermore, as many of these interdune deposits have a high-density contrast when cemented by anhydrite and/or carbonate minerals, they are even more clearly defined by density logs. The detection of these low-porosity and low-permeability barriers between dune reservoir units is especially important, since they are generally of significant areal extent.

Cementation of many inclined laminae is a common characteristic of eolian sandstones and produces the pronounced directional permeability within some eolian sandstones. The cementation of foreset laminae may not be regularly spaced in a vertical sequence. Commonly, the bottom of a cross-bed contains a greater number of cemented, impermeable laminae; they decrease in both number and thickness towards the top of each cross-bed set. In a few examples, the permeability in the upper portion of a cross-bed set has been found to be relatively isotropic, whereas measurements of directional permeability in the bottom of the cross-bed set reveal a very high degree of anisotropy. The cemented cross-bed laminae within tabular reservoir units can sometimes be confused with horizontal barriers if the bedding geometry is not a part of the study.

The areal extent of the cross-bed set needs to be considered when studying an eolian sandstone reservoir, because each cross-bed set may form a discrete porous zone. In order to examine the areal extent of cross-bed reservoir units within the Roliegendes Sandstone, Weber (1979) examined the geometry of cross-bed sets in the De Chelly

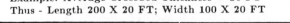

AREAL EXTENT OF CROSSBED SET RESERVOIR UNITS
Example: Average Crossbed Thickness = 20 FT
Thus - Length 200 X 20 FT; Width 100 X 20 FT

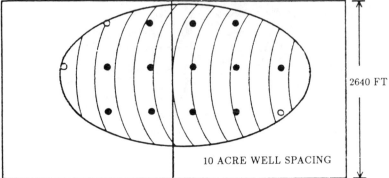

Figure 8.34
Model showing the possible areal extent of a cross-bed set produced during the migration of a transverse dune (Nurmi 1985).

Sandstone in Arizona. His study indicates that the length and width of the tabular cross-bed sets are related to the average cross-bed thickness. The length in the direction of dune migration and cross-bed dip direction is 200 times the average thickness, whereas the width is 100 times the average thickness (figure 8.34). This knowledge of the areal extent of cross-bed sets was later successfully applied to the study of the Rotliegendes in the North Sea Leman Field.

Nugget Sandstone Dune Types

Dipmeter examination of the Nugget Sandstone in southwestern Wyoming and eastern Utah suggests that much of the large-scale cross-bedding was produced during the migration of transverse dunes. However, the dipmeter data of the Nugget Sandstone in the Brady Field, South-Central Wyoming, suggests that the transverse dune ridges were not all linear. Some of the advancing faces of these dune ridges were probably more barcanoid, with crescentlike components along transversely migrating ridges. This is revealed by a statistical analysis of dip magnitudes and azimuth of the cross-beds. The dip magnitudes decrease as the dip azimuth radiates away from the effective wind direction. Such a relationship between the dip magnitude and the dip azimuth suggests that these ancient dune slip faces had crescentlike shapes. Smaller longitudinal ridges are also recognized, but analysis of the vertical sequences indicates that these longitudinal features were leeward projections of transverse ridges. The longitudinal ridges have bimodal dip directions of almost 180°. Significantly, the vector mean of this bimodal distribution is the same direction as the effective wind direction determined for the barcanoid ridges. The bedding study indicates that these longitudinal ridges were elongated in the downwind direction and projected perpendicularly out of the front of the barcanoid dune ridges. Similar longitudinal projections have also been observed in the Rotliegendes Sandstone in western Germany.

WELL LOG AND SEISMIC ANALYSIS OF EVAPORITES

Mineralogical evaluation of evaporites using the gamma-ray, density, neutron, and sonic logs was summarized by Tixier and Alger (1967). As mentioned previously, the Pe, or photoelectric effect measurement that is now part of most modern density logging tools, further improves the mineralogical evaluation of complex lithologies. In addition, spectral gamma-ray tools, which define the individual radioactive contribution of uranium, thorium, and potassium, are also commonly available. The quantitative determination of the potassium content is especially important where potassium-bearing evaporite minerals are present.

Log Analysis of Zechstein Mg-Salts

The Zechstein 3 Evaporite unit in the Netherlands (see figure 8.30) is being mined by solution for its magnesium salts. These relatively undisturbed salts are found in the uppermost Z3 Evaporite, where the sequence is either undeformed or only gentle pillows have formed. Where significant flow has occurred, as in doming or salt walls, there is little commercial interest in solution mining due to the early extrusion of the magnesium salts around the flanks of domes.

Well log analyses and core data compare very favorably in a recent mineralogical evaluation of the magnesium-bearing salts of the Zechstein 3 or Z3 Evaporite in the northern Netherlands (Haile and Blun-

den 1984; figure 8.35). In addition to the density and neutron porosity, the relatively new spectral gamma-ray and photoelectric effect
or Pe logging measurements were both included in this study.

The evaporite mineral assemblage in the Z3 Evaporite in the study
area consists of carnallite, bischofite, kieserite, halite, and sylvite (see
table 8.1 for the logging values of these minerals). A computer-generated model using these five principle evaporite minerals was developed to provide a precise quantitative mineralogical determination. Traces of anhydrite and langbeinite were also found in this
evaporite sequence but disregarded, because they contributed less than

CORE DATA | WELL LOG

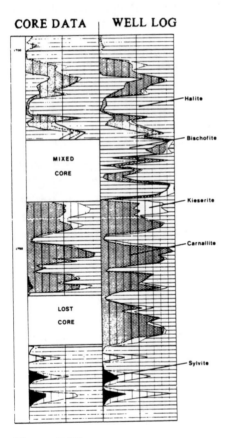

Figure 8.35
Core and log comparison of the Zechstein 3 Evaporite (Haile and
Blunden 1984).

Table 8.1
Petrophysical Properties of Evaporites and Associated Minerals

Name	Formula	M	ρ g/cc	ρ_{FDC}	P_e	Z	GR_k 10^{14} atoms/g	Σ c.u.	ϕ_{SNP} p.u.	ϕ_{CNL} p.u.
Silicates										
Quartz	SiO_2	60.09	2.65	2.64	1.81	11.78	—	4.26	−1.	−2.
Carbonates										
Calcite	$CaCO_3$	100.09	2.71	2.71	5.08	15.71	—	7.08	0.	−1.
Dolomite	$CaCO_3MgCO_3$	184.41	2.87	2.88	3.14	13.74	—	4.70	2.	1.
Ankerite	$Ca(Mg, Fe)(CO_3)_2$	240.25	2.9	2.86	9.32	18.59	—	22.18	0.	1.
Siderite	$FeCO_3$	115.86	3.94	3.89	14.69	21.09	—	52.31	5.	12.
Evaporites										
Halite	$NaCl$	58.44	2.17	2.04	4.65	15.33	—	754.2	−2.	−3.
Anhydrite	$CaSO_4$	136.14	2.96	2.98	5.05	15.68	—	12.45	−1.	−2.
Gypsum	$CaSO_4(H_2O)_2$	172.17	2.32	2.35	3.99	14.68	—	18.5	50+	60+
Trona	$Na_2CO_3NaHCO_3H_2O$	208.02	2.12	2.08	0.71	9.08	—	15.92	24.	35.
Tachydrite	$CaCl_2(MgCl_2)_2(H_2O)_{12}$	517.60	1.68	1.66	3.84	14.53	—	406.02	50+	60+
Sylvite	KCl	74.55	1.98	1.86	8.51	18.13	953.15	564.57	−2.	−3.
Carnallite	$KClMgCl_2(H_2O)_6$	277.86	1.61	1.57	4.09	14.79	255.51	368.99	41.	60+
Langbeinite	$K_2SO_4(MgSO_4)_2$	414.99	2.83	2.82	3.56	14.23	342.46	24.19	−1.	−2.
Polyhalite	$K_2SO_4MgSO_4(CaSO_4)_2(H_2O)_2$	602.94	2.78	2.79	4.32	15.01	235.72	23.70	14.	25.
Kainite	$MgSO_4KCl(H_2O)_3$	248.97	2.13	2.12	3.50	14.17	285.42	195.14	40.	60+
Kieserite	$MgSO_4H_2O$	138.38	2.57	2.59	1.83	11.82	—	13.96	38.	43.
Barite	$BaSO_4$	233.40	4.48	4.09	266.82	47.20	—	6.77	−1.	−2.
Celestite	$SrSO_4$	183.68	3.97	3.79	55.19	30.47	—	7.90	−1.	−1.

Source: Schlumberger, 1985 Log Interpretation Charts, Well Services.

5% of the total rock volume. The actual analysis was undertaken using a computer program to perform an iterative analysis. In such an analysis, a set of answers is first assumed and then used to progressively work through the logging response equations to produce a reconstructed data set. The incoherence in the original data and the reconstructed data is minimized by an iterative process until they converge. The final best-fit answer to the data from the Z3 Evaporite compares very well with that of a core analysis from the same interval of this well.

Seismic Analysis of Zechstein Salts

The structure of the Zechstein salts has been defined by petroleum explorationists who have run extensive seismic surveys throughout the North Sea and surrounding areas. Continued improvements in seismic acquisition and processing have gradually revealed more and more detail about the salt structures in the Zechstein (Taylor 1984; Jenyon 1985) and elsewhere. The halokinetic movement of these Zechstein salts of the southern North Sea is shown in figure 8.36.

Large salt structures are often a problem in petroleum exploration because they can mask potential underlying reservoirs such as the porous Rotliegendes and lower Zechstein zones underlying the salts of the upper Zechstein. The strong reflectors normally seen in the lower Zechstein sequence are often absent or are severely distorted below halokinetic structures. However, "undershooting" techniques for surface seismic acquisition have been developed to define the structure and stratigraphy below salt structures (Krey and Marschall 1975). Borehole seismic methods also can define reservoir horizons below salt structures, as has been well demonstrated in exploration below the Miocene salt walls in the Gulf of Suez, Egypt.

Large halokinetic structures and also smaller structures above or along their flanks can be prime targets in petroleum exploration (see also article 7, Kehle). The seismic recognition of these halokinetic structures is commonly based on a combination of features such as shape, virtual absence of reflections from the salt interior, lateral diffraction patterns, characteristic salt-base reflections, seismic-interval velocities, and the geometric relationship with the surrounding rocks. Although these criteria do not necessarily provide a clear distinction

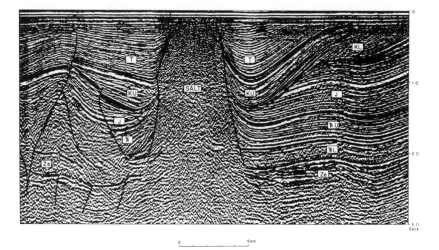

Figure 8.36
Seismic section from the southern North Sea showing a salt wall of Zechstein salt (Taylor 1984). This figure is remarkably similar to the seismic section of the Hainesville salt dome in Figure 7.12 of R. Kehle, in this volume.

between salt and shale diapirs, a knowledge of the stratigraphic sequence or evidence of dissolution of the salt usually permits the identification of salt structures on seismic profiles.

The accurate location of the interface between a high-angle salt structure and the surrounding rock has been a major problem in exploration around salt structures. The problem lies in the methodology of conventional data acquisition and processing, which normally attenuates signals arriving at high incidence angles, such as reflections originating from steeply dipping flanks of a salt dome. If the regional stacking velocities are used during the processing, the reflection from the salt flank can be so attenuated as to be unobservable. However, a recent study by May and Covey (1983) using forward and inverse modeling suggests that there is nothing in principle to prevent inversion of steeply dipping (70°–90°) structures from surface-recorded seismic data, including such extreme cases as overhanging salt domes. Care must be taken, however, during acquisition and processing that the high-angle data is not lost. Once a well is drilled adjacent to the salt structure, it is possible to use a borehole seismic survey to locate the flank of the salt more precisely (Paul 1984; Noponen 1985).

CONCLUSION

During the last decade, many new tools and measurements have been added to the already formidable arsenal available to petroleum ex-plorationists. These tools and measurements are revealing more details about subsurface geology. However, no single well log or seismic tool alone provides us with enough data about subsurface geology and reservoirs to answer all of the questions that need to be answered. The data should be integrated and used together so that the maximum information about the subsurface geology is extracted from them. Although this is difficult with the widening range of measurements that are and will be available, the results and rewards can be great.

The once-separate domains of well logging and surface geophysics are being increasingly merged, and as a result they are providing improved interpretations. Borehole seismic profiles are contributing to the merger of these two geophysical domains. Another contributing factor is the wider use of computers by seismic interpreters in modeling subsurface geology. The speed and great capacity of these computers gives the interpreters an expanded ability to interact with the various data available. The improvement gained by the merger of these once-separated domains is the enhancement of geology and reservoir characteristics.

REFERENCES

Baria, L. R., D. L. Stoudt, P. M. Harris, and P. D. Crevello. 1982. Upper Jurassic reefs of Smackover Formation, United States Gulf Coast. *AAPG Bulletin* 66:1449–1482.

Brie, A., D. L. Johnson, and R. D. Nurmi. 1985. Effect of spherical pores on sonic and resistivity measurements. *Transactions, 26th Annual Meeting of Society of Professional Well Log Analysts*, Paper N. Tulsa, Okla.: SPWLA.

Brookfield, M. E. and T. S. Ahlbrandt. 1983. *Eolian Sediments and Processes.* Vol. 38 of *Developments in Sedimentology.* Amsterdam: Elsevier.

Bubb, J. N. and W. G. Hatlelid. 1977. Seismic stratigraphy and global changes of sea level, part 10: Seismic recognition of carbonate buildups. In *Seismic Stratigraphy—Applications,* pp. 185–204. See Payton 1977.

Caughlin, W. G., F. J. Lucia, and N. L. McIver. 1976. The detection and development of Silurian reefs in northern Michigan. *Geophysics* 41:646–658.

Crevello, P. D. and P. M. Harris. 1984. Depositional models for Jurassic reefal buildups. In W. P. S. Ventress, D. G. Bebout, B. F. Perkins, and C. H. Moore, eds., *The Jurassic of the Gulf Rim,* pp. 57–102. Proceedings, 3rd Annual Research Conference of Gulf Coast Section of SEPM Foundation, Austin, Tex: Earth Enterprises.

Dupal, L. and D. E. Miller. 1985. Reef delineation by multiple offset borehole seismic profiles: A case study. *55th Annual Meeting of Society of Exploration Geophysicists, Transactions,* pp. 105–107. Tulsa, Okla.: SEG.

Exploration Staff, Chevron Standard, Ltd. 1979. The geology, geophysics, and significance of the Nisku reef discoveries, West Pembina area, Alberta, Canada. *Bulletin of Canadian Petroleum Geology* 27:326–359.

Flaum, C. and G. Pirie. 1981. Determination of lithology from induced gamma-ray spectroscopy. 21st Annual Meeting of Society of Professional Well Log Analysts, Paper O. Tulsa, Okla.: SPWLA.

Gaynor, J. A. and L. E. Stasko. 1985. Mission Canyon strat-seis on the east flank of the Williston basin. *55th Annual Meeting of Society of Exploration Geophysicists, Transactions,* pp. 350–352. Tulsa, Okla.: SEG.

Glennie, K. W. 1970. *Desert sedimentary environments.* Vol. 14 of *Developments in Sedimentology.* Amsterdam: Elsevier.

Haile, P. M., and H. A. Blunden. 1984. Zechstein magnesium-rich evaporite deposits of northern Netherlands and their volumetric analysis by GLOBAL. *Transactions, 9th International Formation Evaluation Symposium,* Communication no. 37.

Jenyon, M. K. 1985. Basin-edge diapirism and updip salt flow in Zechstein of southern North Sea. *AAPG Bulletin* 69:53–64.

Kocurek, G. 1981. Significance of interdune deposits and bounding surfaces in eolian dune sands. *Sedimentology* 28:753–780.

Krey, T. and R. Marschall. 1975. Undershooting salt-domes in the North Sea. In A. W. Woodland, ed., *Petroleum and the Continental Shelf of Northwest Europe,* vol. 1, *Geology,* pp. 265–273. New York: Wiley.

Lee, J. and R. Budros. 1982. Reef exploration in the Michigan Basin: Problems and solutions. In *Carbonate Seismology.* Society of Exploration Geophysicists Continuing Education Short Course. Tulsa, Okla.: SEG.

Lomando, A. 1981. Sedimentation and diagenesis of the Smackover Formation (Jurassic), Jay Field, West Florida. Master's thesis, Queen's College, New York.

McGuire, J. A., L. T. Rogers, and J. T. Watson. 1985. Improved lithology and hydrocarbon saturation determination using the gamma spectrometry log. *Proceedings of 59th Annual Fall Meeting, Society of Petroleum Engineers of AIME*, Paper SPE 14465.

McKee, E. D. 1979. *A Study of Global Sand Seas*. U.S. Geological Survey Prof. Paper no. 10522. Washington, D.C.: U.S. Government Printing Office.

Maureau, G. T., and D. H. van Wijhe. 1979. The prediction of porosity in the Permian (Zechstein 2) carbonate of eastern Netherlands using seismic data. *Geophysics* 44:1502–1517.

May, B. T. and J. D. Covey. 1983. Structural inversion of salt dome flanks. *Geophysics* 48:1039–1050.

Meckel, L. D. and A. K. Nath. 1977. Geologic considerations for stratigraphic modeling and interpretation. In *Seismic Stratigraphy—Applications*, pp. 417–438. See Payton 1977.

Neidell, N. S., J. H. Beard, and E. E. Cook. 1985. Use of seismic-derived velocities for stratigraphic exploration of land. In O. R. Berg and D. G. Woolverton, eds., *Seismic Stratigraphy: An Integrated Approach*, pp. 49–77. AAPG Memoir no. 39. Tulsa, Okla.

Noponen, I. 1985. Application of VSP observations of steeply dipping salt interface reflections to reservoir development. *55th Annual Meeting of Society of Exploration Geophysicists, Transactions*, pp. 105–107. Tulsa, Okla.: SEG.

Nurmi, R. D. 1978. Use of well logs in evaporite sequences. In W. E. Dean and B. C. Schreiber, eds., *Marine Evaporites*. SEPM Short Course no. 4. pp. 144–184. Tulsa, Okla.

Nurmi, R. D. 1984. Geological evaluation of high resolution dipmeter data. *Transactions, 25th Annual Meeting of Society of Professional Well Log Analysts*, Paper N. Tulsa, Okla.: SPWLA.

Nurmi, R. D. 1985. Eolian sandstone reservoirs: Bedding facies and production geology modeling. *Proceedings, 60th Annual Fall Meeting, Society of Petroleum Engineers of AIME*, Paper SPE 14172.

Nurmi, R. D. 1986. Carbonate close-up. *Middle East Well Evaluation Review* 1:22–35.

Nurmi, R. D. and M. R. Frisinger. 1983. Synergy of core petrophysical measurements, log data, and rock examination in carbonate reservoir studies. *Proceedings, 58th Annual Fall Meeting, Society of Petroleum Engineers of AIME*, Paper SPE 11969.

Paul, A. 1984. Use of offset vertical seismic profile for delineation of complex structures—Case of a salt dome. *Transactions, 9th International Formation Evaluation Symposium*, Communication no. 11.

Payton, C. E., ed. 1977. *Seismic Stratigraphy—Applications to Hydrocarbon Exploration*. AAPG Memoir no. 26. Tulsa, Okla.

Rodrick, D. L. and C. K. Poster. 1987. Reservoir delineation with VSP and well log data. *Proceedings, 5th Middle East Oil Show of Society of Petroleum Engineers of AIME*, Paper SPE 15743.

Rosenkrans, R., and J. Marr. 1967. Modern seismic exploration of the Gulf Coast Smackover trend. *Geophysics* 32:184–206.

Savre, W. C. 1963. Determination of a more accurate porosity and mineral composition in complex lithologies with the use of the sonic, neutron, and density surveys. *Journal of Petroleum Technology* 15:945–959.

Serra, O. 1986. Evaporites et diagraphies differées. In *Les Séries a Evaporites en Exploration Pétroliere*, chap. 2, pp. 13–71. Paris: Editions Technip.

Sigsby, R. J. 1976. Paleoenvironmental analysis of the Big Escambia Creek-Jay-Blackjack Creek Field areas. *Gulf Coast Assoc. Geol. Soc. Trans.* 26:258–278.

Taylor, J. C. M. 1984. Late Permian-Zechstein. In K. W. Glennie, ed., *Introduction to the Petroleum Geology of the North Sea*, pp. 61–83. London: Blackwell.

Tixier, M. P. and R. P. Alger. 1967. Log evaluation of non-metallic mineral deposits. *Transactions, 8th Annual Meeting of Society of Professional Well Log Analysts*, Paper R. Tulsa, Okla.: SPWLA.

Ward, J. A. 1984. Application of quantitative seismic stratigraphic exploration for regressive Smackover oolite bars. In M. W. Presley, ed., *Jurassic of East Texas*, pp. 43–51. Tyler, Tex.: East Texas Geological Society.

Weber, K. J. 1979. Computation of initial well productivities in eolian sandstone on basis of geologic model, Leman Gas Field, United Kingdom. *AAPG Bulletin* 63:549.

Contributors

Robert Evans
Mobil Research and Development
  Corporation
Dallas Research Division
P.O. Box 819047
Dallas, Texas 75381

C. Robertson Handford
Arco Research and Technical
  Services Center
Reservoir Analysis
2300 West Plano
Plano, TX 75075

Ralph O. Kehle
Hershey Oil Company
101 West Walnut
Pasadena, California 91103

Alan C. Kendall
Department of Geology
University of Toronto
Toronto, Ontario M5S 1A7
Canada

Christopher G. St. C. Kendall
Department of Geology
University of South Carolina
Columbia, South Carolina 29208

Douglas W. Kirkland
Mobil Research and Development
  Corporation
Dallas Research Division
P.O. Box 819047
Dallas, Texas 75381

Roy D. Nurmi
Schlumberger Technical Services
P.O. Box 926
Dubai, U. A. E.

Catherine Pierre
Département de Géologie
  Dynamique
Université Pierre et Marie Curie
4 Place Jussieu
75252 Paris, Cedex 5
France

B. Charlotte Schreiber
Department of Geology
Queens College, CUNY
65-30 Kissena Boulevard
Flushing, New York 11367

and
Lamont-Doherty Geological
  Observatory
Columbia University
Palisades, New York 10964

John K. Warren
Department of Geological Sciences
University of Texas at Austin
Austin, Texas 78712

Index

Capitan, 115, 118, 120, 123
Lamar, 118
Todilto, 282-84
Tonoloway, 29, 33
Limestone
basinal, 118
boxwork, 109
Lithofacies, 40
maps, 168
Loading, 25
Logs, *see* Well log
Lotzberg Salt, 38

McKittrick Canyon, 122
Macrophytes, 268
Magnesium
compounds, 270
solutions, 309
salts, 270
Mangrove, 72
Mats
algal, 202, 217
cyanobacterial, 217, 266
microbial, 258, 262, 267, 272
Mesosaline water, 274, 275
Messinian, 191, 235-37, 327, 331, 332
Metamorphism (progressive), 224
Metazoans, 275
Meteoric water, *see* Groundwater
Methanogenesis, 273, 276, 316, 317
Michigan Basin, *see* Basin
Miocene, *see* Messinian
Mixing, 42
Molecular
differences, 275
hydrogen, 276
Mud flats, coastal, 141, 167
Muskeg Evaporite, 57

Niagaran Carbonate, *see* Reef
Nitrogen compounds, 266, 271

Nodules, anhydrite, 37, 66, 82, 100, 102-3, 177
Nonevaporite sulfates, 5
Nonmarine water, *see* Continental water; Groundwater
Nucleation surfaces, 196
Nutrients, 266, 271, 284

Onlap, 21, 27, 28
Oolite
bars, 173, 430
gypsum, *see* Gypsum
halite, *see* Halite
reservoirs, 173, 436
shoals, 71
Organic matter, 256-59, 275-77, 282, 287, 290
decay, 312
destruction, 273, 318
preservation, 257, 273
production, 257
Organisms, aquatic, 268
Osmotic
desiccation, 256, 269
pressure, 268
Outflow, 34
Overburden
distribution, 36
uneven, 361
uniform, 360
Overprinting, diagenetic, 247
Oversteepened margin, 121
Oxygen, solubility, 270

Padre Island, 156, 263
Paralic lakes, 267
PDB, 303
Pekelmeer, 266
Pelecypod, 262, 263
Permian
Basin, *see* Basin
halite, 328